CONTROL and AUTOMATION of ELECTRICAL POWER DISTRIBUTION SYSTEMS

POWER ENGINEERING

Series Editor
H. Lee Willis
KEMA T&D Consulting
Raleigh, North Carolina

Advisory Editor
Muhammad H. Rashid
University of West Florida
Pensacola, Florida

1. Power Distribution Planning Reference Book, *H. Lee Willis*
2. Transmission Network Protection: Theory and Practice, *Y. G. Paithankar*
3. Electrical Insulation in Power Systems, *N. H. Malik, A. A. Al-Arainy, and M. I. Qureshi*
4. Electrical Power Equipment Maintenance and Testing, *Paul Gill*
5. Protective Relaying: Principles and Applications, Second Edition, *J. Lewis Blackburn*
6. Understanding Electric Utilities and De-Regulation, *Lorrin Philipson and H. Lee Willis*
7. Electrical Power Cable Engineering, *William A. Thue*
8. Electric Systems, Dynamics, and Stability with Artificial Intelligence Applications, *James A. Momoh and Mohamed E. El-Hawary*
9. Insulation Coordination for Power Systems, *Andrew R. Hileman*
10. Distributed Power Generation: Planning and Evaluation, *H. Lee Willis and Walter G. Scott*
11. Electric Power System Applications of Optimization, *James A. Momoh*
12. Aging Power Delivery Infrastructures, *H. Lee Willis, Gregory V. Welch, and Randall R. Schrieber*
13. Restructured Electrical Power Systems: Operation, Trading, and Volatility, *Mohammad Shahidehpour and Muwaffaq Alomoush*
14. Electric Power Distribution Reliability, *Richard E. Brown*

15. Computer-Aided Power System Analysis, *Ramasamy Natarajan*
16. Power System Analysis: Short-Circuit Load Flow and Harmonics, *J. C. Das*
17. Power Transformers: Principles and Applications, *John J. Winders, Jr.*
18. Spatial Electric Load Forecasting: Second Edition, Revised and Expanded, *H. Lee Willis*
19. Dielectrics in Electric Fields, *Gorur G. Raju*
20. Protection Devices and Systems for High-Voltage Applications, *Vladimir Gurevich*
21. Electrical Power Cable Engineering, Second Edition, *William Thue*
22. Vehicular Electric Power Systems: Land, Sea, Air, and Space Vehicles, *Ali Emadi, Mehrdad Ehsani, and John Miller*
23. Power Distribution Planning Reference Book, Second Edition, *H. Lee Willis*
24. Power System State Estimation: Theory and Implementation, *Ali Abur*
25. Transformer Engineering: Design and Practice, *S.V. Kulkarni and S. A. Khaparde*
26. Power System Capacitors, *Ramasamy Natarajan*
27. Understanding Electric Utilities and De-regulation: Second Edition, *Lorrin Philipson and H. Lee Willis*
28. Control and Automation of Electric Power Distribution Systems, *James Northcote-Green and Robert G. Wilson*
29. Protective Relaying for Power Generation Systems, *Donald Reimert*

CONTROL and AUTOMATION of ELECTRICAL POWER DISTRIBUTION SYSTEMS

James Northcote-Green
ABB Power Technologies AB
Vasteras, Sweden

Robert Wilson
Abasis Consulting Limited
Whitchurch, Shropshire, UK

Taylor & Francis
Taylor & Francis Group
Boca Raton London New York

CRC is an imprint of the Taylor & Francis Group,
an informa business

CRC Press
Taylor & Francis Group
6000 Broken Sound Parkway NW, Suite 300
Boca Raton, FL 33487-2742

© 2007 by Taylor & Francis Group, LLC
CRC Press is an imprint of Taylor & Francis Group, an Informa business

No claim to original U.S. Government works
Printed in the United States of America on acid-free paper
10 9 8 7 6 5 4 3 2 1

International Standard Book Number-10: 0-8247-2631-6 (Hardcover)
International Standard Book Number-13: 978-0-8247-2631-7 (Hardcover)

This book contains information obtained from authentic and highly regarded sources. Reprinted material is quoted with permission, and sources are indicated. A wide variety of references are listed. Reasonable efforts have been made to publish reliable data and information, but the author and the publisher cannot assume responsibility for the validity of all materials or for the consequences of their use.

No part of this book may be reprinted, reproduced, transmitted, or utilized in any form by any electronic, mechanical, or other means, now known or hereafter invented, including photocopying, microfilming, and recording, or in any information storage or retrieval system, without written permission from the publishers.

For permission to photocopy or use material electronically from this work, please access www.copyright.com (http://www.copyright.com/) or contact the Copyright Clearance Center, Inc. (CCC) 222 Rosewood Drive, Danvers, MA 01923, 978-750-8400. CCC is a not-for-profit organization that provides licenses and registration for a variety of users. For organizations that have been granted a photocopy license by the CCC, a separate system of payment has been arranged.

Trademark Notice: Product or corporate names may be trademarks or registered trademarks, and are used only for identification and explanation without intent to infringe.

Library of Congress Cataloging-in-Publication Data

Northcote-Green, James.
 Control and automation of electric power distribution systems / James Northcote-Green and Robert Wilson.
     p. cm. -- (Power engineering ; 28)
 Includes bibliographical references and index.
 ISBN-13: 978-0-8247-2631-7 (alk. paper)
 ISBN-10: 0-8247-2631-6 (alk. paper)
 1. Electric power distribution--Automation. 2. Electric power systems--Control. 3. Electric power systems--Automation. I. Wilson, Robert, 1951 Sept. 29- II. Title. III. Series.

TK3091.N65 2006
621.319--dc22   2006001531

Visit the Taylor & Francis Web site at
http://www.taylorandfrancis.com

and the CRC Press Web site at
http://www.crcpress.com

This book is dedicated to our wives for their loving support during its writing.

Series Introduction

Power engineering is the oldest and most traditional of the various areas within electrical engineering, yet no other facet of our modern industry is undergoing a more dramatic transformation in both technology and structure. This addition to Taylor & Francis's Power Engineering Series addresses a cornerstone of that modern revolution: the use of advanced monitoring, computation, and control to improve the reliability and the economy of power delivery to energy consumers.

As the editor of the Power Engineering Series, I am proud to include *Control and Automation of Electric Power Distribution Systems* among this important group of books, particularly because James Northcote-Green and Robert Wilson have been close friends for many years; co-workers in whose expertise and extensive knowledge of power distribution and automation technology I have come to respect greatly.

Traditionally, electric utilities operated power distribution systems on a type of "dead reckoning" basis, with little or no on-line monitoring or remote automation involved. Utility planners "bought" reliability for their customers by using large capacity margins and redundancy of circuits and equipment throughout the network. These long-established power distribution system designs were robust and reasonably reliable, but the utilities, and their customers paid a considerable price for the contingency margins needed to make them so. Modern automation technologies can reduce contingency margins, improve utilization and economy of operation, and even provide improved scheduling and effectiveness of maintenance and service. However, they must be applied well, with the technologies selected to be compatible with the system's needs, and targeted effectively for maximum impact, and integrated properly into the utilities operations and business enterprise, if the results are to live up to the promise.

This book provides the reader with a solid foundation to do exactly that. James Northcote-Green and Robert Wilson have put together well-organized, comprehensive, yet accessible discussion of distribution automation for the 21st-century electric utility. At both the introductory and advanced levels, it provides above-average insight into the capabilities and limitations of control and automation systems, and it helps the reader develop a rich understanding of how and why automation should be used, and of what is realistic in its performance. In particular, readers will find the extensive practical business cases reviewed by the authors useful in helping them evaluate their own needs and justification studies.

Like all the books planned for the Power Engineering Series, this book provides modern power technology in a context of proven, practical application; useful as a reference book as well as for self-study and advanced classroom use. Taylor & Francis's Power Engineering Series will eventually include books covering the entire field of power engineering, in all of its specialties and sub-genres, all aimed at providing practicing power engineers with the knowledge and techniques they need to meet the electric industry's challenges in the 21st century.

H. Lee Willis

Preface

This is a reference and tutorial guide covering the automation of electric power distribution networks. Automation of electric distribution networks in its broadest sense ranges from simple remote control to the application of automation logic and software-based decision tools. The possibilities are endless, and the cost of implementation is directly related to the possibilities.

A utility considering automation must be aware of and resolve a number of key issues. First, it must assess the cost and feasibility of adding automation to existing switchgear against replacing existing equipment with more "automation ready" gear. Second, the type of control infrastructure and level of automation it wishes to consider (central or distributed, system or local, or combinations of these) and its implications on the communications system weighted against its availability and practicality. Third, the ambition level it wishes to or is being forced (regulatory pressures) to achieve against the expenditure that is prudent. Ambition level is affected by the level of reliability and operating economies that must be achieved. It is necessary to search for the key functions that will deliver the performance cost-effectively without detracting from the bottom line of utility business performance. Finally, in order for any automation solution to be implemented, it must be rationally justified through a business case. Different business environments dictate very different evaluation, and such as utilities operating under the risk of performance-based penalties, will view energy not supplied with considerably more importance than those under traditional energy costing.

Control and Automation of Electric Power Distribution Systems addresses these four issues plus many related topics that should be considered in applying automation to power distribution networks. The fundamentals around which a control and automation solution should be based are introduced. These include such concepts as depth of control, boundaries of control responsibility, stages of automation, automation intensity level (AIL), distribution automation (DA), the distribution management system (DMS), substation automation (SA), feeder automation (FA), and automated device preparedness, all of which are introduced in Chapter 1. Many of these concepts are explored in considerable detail since FA or extended control, automation outside the primary substation, forms the predominant topic of the book.

Chapter 2 covers the role of central control in the DA solution by summarizing SCADA, control room operations management, advanced applications as operator decision support aids, and outage management (OM). A short section introduces the concepts of performance measures for real-time systems. The connectivity model of the distribution network is a foundation element in any DMS. Consequently, data and data modeling becomes the key in DMS implementations —

implications of which potential implementers should be aware. The importance of the data model and its implications of building interfaces with other business applications such as GIS are explained together, with the aims of the industry to standardize through the common information model (CIM) standard.

Chapter 3 introduces distribution design, planning, local control, comparison of network types, and network structure at an appropriate detail to assist in selecting the primary device and associated control. The latter leads into the concept of the network complexity factor, for which relationships are developed for use later in the book.

Chapter 4 covers the fundamentals of the distribution primary equipment, circuit breakers, reclosers, sectionalizers, and various types of sensors (CTs, VTs) that will become part of the DA scheme and from which the concept of feeder automation building blocks will be proposed later in the book.

Chapter 5 extends the groundwork of the previous chapter necessary for developing the FA building blocks. Basic protection requirements for distribution networks are explained and the considerations that must be accounted for due to different grounding (earthing) practices. Fault passage indicators (FPIs) and their application are explained in detail. Different types of intelligent electronic device (IED) that are suitable for automating primary devices are described and their possible roles. Finally, the need for automated switch power supplies, batteries, and their duty cycle are explained. The final section of this chapter selects and appropriately assembles the devices described in this and the preceding chapter to propose FA building blocks. Attention is given to all the interfaces between components that must be designed and tested to create an automation ready device.

Chapter 6 moves the discussion to distribution network performance calculations, and how different automation strategies and selection of different FA building blocks can deliver improved performance. The chapter summarizes the calculation of performance indices, the relationship between network complexity (NCF) and performance, together with different automation strategies.

The communication system is a key component for any DA implementation, and Chapter 7 introduces the subject in sufficient depth for the DA implementer to understand some of the intricacies of the topic. Having summarized different communication media, the topic of wireless communications is covered from antennas through configuration management to gain calculations. Wireless medium is followed by a thorough treatment of distribution line carrier (DLC). Types of communications that maybe suitable for DA are summarized, with advantages and disadvantages. The structure of protocols is explained, and finally, the requirements for dimensioning communications system are treated.

Chapter 8 develops techniques necessary to justify DA. It is started by introducing the concept of direct and indirect benefits both of which can be hard or soft. The ideas of generic benefits, the benefits opportunity matrix, and benefit flow charts are explained. The dependency of DA functions, not only implemented on the hardware but also the possibility of double-counting through shared benefits, is introduced. Methods for calculating benefits from capital deferral, energy not supplied, man-time savings, including a unique approach to crew travel time savings CTS (using Wilson's curve), are given. The final section draws the reader's attention to the importance of assigning the correct economic value when quantifying energy-related benefits. The chapter concludes by returning to the hard/soft classification of benefits as a way to present the quantitative results of the business case.

Chapter 9 concludes the book with two example case studies that draw on the ideas in the previous chapters to illustrate diverse situations in which the positive business case for distribution automation was successfully made.

As utilities continue to strive for better economies through improved management of their distribution network assets, DA is one of the tools at their disposal. All the topics in this book will give decision makers a useful guide to all the issues to be investigated and decided as they embark on the solution definition and justification.

This book covers a range of topics and would not have been completed without the tremendous input of some of our enthusiastic colleagues. The authors want to thank particularly the major contributions to Chapter 7 of Josef Lehmann, formerly of ABB and now Cipunet, of John Gardener, telecommunication expert within the U.K. railway industry, and Anders Grahn and Hans Ottosson of Radius Communications Sweden. The suggestions and contributions of Gunnar Bjorkmann and Carl-Gustav Lundqvist for Chapter 2 improved the SCADA, performance measurement, and data modeling sections significantly. We also want to acknowledge the input of Reinhard Kuessel and Dr. Ulrich Kaizer for the material in Chapter 2 on advanced applications. The book would never have been conceived if it were not for the strategic thinking of ABB senior managers, led by Andrew Eriksson, who identified the need to take a fresh look at feeder automation, which resulted in the funding of a project aimed at investigating DA. A further thank you is expressed to the late Ted Holmes, a senior member of the U.K. utility industry and author, for his worthwhile suggestions and review. The authors wish to thank members of the ABB team who were assigned to this project, namely, Dr. David Hart, Dr. Peter Dondi, Arnie Svenne, Matti Heinonen, Tapani Tiitola, Erkki Antila, Jane Soderblom, Duncan Botting, Graeme McClure, and Karl LaPlace, for their original contributions to many aspects of FA, which have been included in the book. The continued support of ABB Network Management in allowing significant reference and inclusion of technical topics has been invaluable. We also want to thank Jay Margolis and the other staff of Taylor and Francis for their involvement and efforts to make this book a quality effort. Last but not least, we thank our colleague and collaborator of many years, Lee Willis, who encouraged and cudgeled us to write down that which we had experienced and learned.

The Authors

James Northcote-Green, DFH, MSc E, MBA, IEE Fellow, C Eng, is a senior specialist for ABB Network Management, operating out of Vasteras in Sweden, specializing in distribution management systems, distribution automation, and network applications. He also has regional sales responsibility for a number of countries, primarily in the Far East.

An IEE Fellow with over 40 years in the power industry, he has held several responsibilities. In the late 1990s, he was part of the ABB Distribution Solutions Business Area Management Team as vice president of business development (Solutions) and technology manager. In the early 1990s in Europe, he was Product Manager, Distribution Management systems and real-time systems, and area sales manager for the British Isles. In the 1980s in the U.S. as Manager of Distribution Technologies, Advanced Systems Technology, he was responsible for the development team for electric network distribution planning software (CADPAD), pioneering advanced techniques that were used by over 200 power companies worldwide. He developed the concepts and led the team that resulted in the trouble call and control room management system, CADOPS. He was a member of the founding committee for the U.S. CIRED organizing committee and conference reporter and technical committee member for Distributech Europe.

He is the author or coauthor of more than 60 publications on electric power distribution systems.

Robert Wilson, BSc Eng, LLB, Ch Eng, IEE Fellow is the principal consultant for Abasis International Consulting based in the U.K., working in the U.K. railway industry, with responsibility for asset management policy.

An IEE Fellow for many years, he has more than 30 years of experience in the electricity industry.

For eight years he was the principal expert on distribution power systems for ABB based in Vasteras, Sweden, with responsibility for Asia and Europe. In the early 1990s, as principal engineer, he was responsible for specification, purchasing, and setting up all electrical plants in a major U.K. utility. In the 1980s, as senior engineer, he was responsible for the development of network reliability data from fault data and the development of system automation for the U.K.

He is the author or co-author of more than 40 technical papers on all aspects of distribution systems, presented at national and international conferences.

Contents

Chapter 1 Power Delivery System Control and Automation 1

1.1   Introduction 1
1.2   Why Distribution Automation? 1
       1.2.1   Incremental Implementation 4
       1.2.2   Acceptance of DA by the Utility Industry 5
1.3   Power Delivery Systems 7
1.4   Control Hierarchy 9
1.5   What Is Distribution Automation? 11
       1.5.1   DA Concept 11
1.6   Distribution Automation System 13
1.7   Basic Architectures and Implementation Strategies for DA 17
       1.7.1   Architecture 17
       1.7.2   Creating the DA Solution 19
       1.7.3   Distribution Network Structure 21
1.8   Definitions of Automated Device Preparedness 22
1.9   Summary 23
References 25

Chapter 2 Central Control and Management 27

2.1   Introduction 27
       2.1.1   Why Power System Control? 27
2.2   Power System Operation 28
2.3   Operations Environment of Distribution Networks 29
2.4   Evolution of Distribution Management Systems 31
2.5   Basic Distribution Management System Functions 35
2.6   Basis of a Real-Time Control System (SCADA) 39
       2.6.1   Data Acquisition 39
       2.6.2   Monitoring and Event Processing 41
       2.6.3   Control Functions 44
       2.6.4   Data Storage, Archiving, and Analysis 44
       2.6.5   Hardware System Configurations 45
       2.6.6   SCADA System Principles 47
       2.6.7   Polling Principles 48
2.7   Outage Management 50
       2.7.1   Trouble Call-Based Outage Management 52
       2.7.2   Advanced Application-Based Outage Management 57
       2.7.3   GIS-Centric versus SCADA-Centric 60

2.8	Decision Support Applications	60
	2.8.1 Operator Load Flow	61
	2.8.2 Fault Calculation	63
	2.8.3 Loss Minimization	66
	2.8.4 VAR Control	66
	2.8.5 Volt Control	67
	2.8.6 Data Dependency	68
2.9	Subsystems	69
	2.9.1 Substation Automation	69
	2.9.2 Substation Local Automation	72
2.10	Extended Control Feeder Automation	77
2.11	Performance Measures and Response Times	79
	2.11.1 Scenario Definitions	79
	2.11.2 Calculation of DA Response Times	81
	2.11.3 Response Times	85
2.12	Database Structures and Interfaces	86
	2.12.1 Network Data Model Representations	86
	2.12.2 SCADA Data Models	87
	2.12.3 DMS Data Needs, Sources, and Interfaces	89
	2.12.4 Data Model Standards (CIM)	93
	2.12.5 Data Interface Standards	100
2.13	Summary	100
Appendix 2A — Sample Comprehensive CIM Structure		103
References		104

Chapter 3 Design, Construction, and Operation of Distribution Systems, MV Networks..................105

3.1	Introduction	105
3.2	Design of Networks	107
	3.2.1 Selection of Voltage	109
	3.2.2 Overhead or Underground	110
	3.2.3 Sizing of Distribution Substations	110
	3.2.4 Connecting the MV (The Upstream Structure)	114
	3.2.5 The Required Performance of the Network	116
	3.2.6 The Network Complexity Factor	117
	3.2.7 Voltage Control	121
	3.2.8 Current Loading	128
	3.2.9 Load Growth	129
	3.2.10 Earthing (Grounding)	131
	3.2.11 Lost Energy	132
	3.2.12 Comparison of U.K. and U.S. Networks	137
	3.2.13 The Cost of Installation of the Selected Design	140
	3.2.14 The Cost of Owning the Network after Construction	141

3.3	LV Distribution Networks	142
	3.3.1 Underground LV Distribution Networks	142
	3.3.2 Overhead LV Distribution Networks	143
3.4	Switchgear for Distribution Substations and LV Networks	145
3.5	Extended Control of Distribution Substations and LV Networks	146
3.6	Summary	148
References		148

Chapter 4 Hardware for Distribution Systems 149

4.1	Introduction to Switchgear	149
	4.1.1 Arc Interruption Methods	150
4.2	Primary Switchgear	154
	4.2.1 Substation Circuit Breakers	154
	4.2.2 Substation Disconnectors	158
4.3	Ground-Mounted Network Substations	158
	4.3.1 Ring Main Unit	160
	4.3.2 Pad-Mount Switchgear	163
4.4	Larger Distribution/Compact Substations	164
4.5	Pole-Mounted Enclosed Switches	167
4.6	Pole-Mounted Reclosers	168
	4.6.1 Single-Tank Design	169
	4.6.2 Individual Pole Design	169
4.7	Pole-Mounted Switch Disconnectors and Disconnectors	170
4.8	Operating Mechanisms and Actuators	171
	4.8.1 Motorized Actuators	172
	4.8.2 Magnetic Actuators	173
4.9	Current and Voltage Measuring Devices	175
	4.9.1 Electromagnetic Current Transformers	177
	4.9.2 Voltage Transformers	180
4.10	Instrument Transformers in Extended Control	181
4.11	Current and Voltage Sensors	182
	4.11.1 Current Sensor	182
	4.11.2 Voltage Sensor	183
	4.11.3 Combi Sensor and Sensor Packaging	184
Reference		185

Chapter 5 Protection and Control 187

5.1	Introduction	187
5.2	Protection Using Relays	187
	5.2.1 Discrimination by Time	188
	5.2.2 Discrimination by Current	189
	5.2.3 Discrimination by Both Time and Current	189

5.3	Sensitive Earth Fault and Instantaneous Protection Schemes	190
5.4	Protection Using Fuses	192
5.5	Earth Fault and Overcurrent Protection for Solid/Resistance Earthed Networks	197
5.6	Earth Faults on Compensated Networks	198
5.7	Earth Faults on Unearthed Networks	203
5.8	An Earth Fault Relay for Compensated and Unearthed Networks	204
5.9	Fault Passage Indication	207
	5.9.1 The Need for FPI on Distribution Networks with Manual Control	207
	5.9.2 What Is the Fault Passage Indicator, Then?	209
	5.9.3 The Need for FPI on Distribution Networks with Extended Control or Automation	211
	5.9.4 Fault Passage Indicators for Use on Closed Loop Networks	212
	5.9.5 Other Applications of Directional Indicators	213
5.10	Connection of the FPI to the Distribution System Conductor	214
	5.10.1 Connection Using Current Transformers	214
	5.10.2 Connections Using CTs on Underground Systems	215
	5.10.3 Connections Using CTs on Overhead Systems	216
	5.10.4 Connection without CTs on Overhead Systems (Proximity)	216
5.11	Distribution System Earthing and Fault Passage Indication	218
	5.11.1 Detection of Steady-State Fault Conditions	220
	5.11.2 Detection of Transient Fault Conditions	221
	5.11.3 Indication of Sensitive Earth Faults	222
5.12	AutoReclosing and Fault Passage Indicators	222
5.13	The Choice of Indication between Phase Fault and Earth Fault	223
5.14	Resetting the Fault Passage Indicator	224
5.15	Grading of Fault Passage Indicators	224
5.16	Selecting a Fault Passage Indicator	225
5.17	Intelligent Electronic Devices	225
	5.17.1 Remote Terminal Unit	226
	5.17.2 Protection-Based IED	229
5.18	Power Supplies for Extended Control	229
5.19	Automation Ready Switchgear — FA Building Blocks	234
	5.19.1 Switch Options	237
	5.19.2 Drive (Actuator) Options	237
	5.19.3 RTU Options	237
	5.19.4 CT/VT Options	237
	5.19.5 Communications Options	238
	5.19.6 FPI Options	238
	5.19.7 Battery Options	238
	5.19.8 Interfaces within Building Blocks	238

5.20 Examples of Building Blocks 239
5.21 Typical Inputs and Outputs for Building Blocks..................................241
      5.21.1 Sectionalizing Switch (No Measurements)...............................241
      5.21.2 Sectionalizing Switch (with Measurements) 242
      5.21.3 Protection-Based Recloser for Overhead Systems 243
5.22 Control Building Blocks and Retrofit...244
5.23 Control Logic 244
      5.23.1 Option 1, Circuit A with 1.5 Switch Automation, FPI
              and Remote Control of Switches..245
      5.23.2 Option 2, Circuit B with 2.5 Switch Automation, FPI
              and Remote Control of Switches..246
      5.23.3 Options 3 and 4, No Fault Passage Indicators 247
      5.23.4 Options 5 and 7, Local Control Only 248
      5.23.5 Options 6 and 8, Local Control Only 249
      5.23.6 Special Case of Multishot Reclosing and Automatic
              Sectionalizing 249

Chapter 6   Performance of Distribution Systems 251

6.1 Faults on Distribution Networks 251
      6.1.1 Types of Faults 251
      6.1.2 The Effects of Faults 254
      6.1.3 Transient Faults, Reclosers, and Compensated Networks 254
6.2 Performance and Basic Reliability Calculations 259
      6.2.1 System Indices...259
      6.2.2 Calculating the Reliability Performance of Networks.................260
      6.2.3 Calculation of Sustained Interruptions (SAIDI).........................261
      6.2.4 Calculation of Sustained Interruption Frequency (SAIFI) 263
      6.2.5 Calculation of Momentary Interruption Frequency
            (MAIFI) 264
      6.2.6 Summary of Calculated Results..264
      6.2.7 Calculating the Effects of Extended Control 266
      6.2.8 Performance as a Function of Network Complexity
            Factor 267
      6.2.9 Improving Performance without Automation 268
6.3 Improving the Reliability of Underground Networks 272
      6.3.1 Design Method 1 — Addition of Manually Operated
            Sectionalizing Switches..272
      6.3.2 Design Method 2 — Addition of Manually Switched
            Alternative Supply..273
      6.3.3 Design Method 3 — Add Automatic in Line Protection............274
      6.3.4 Design Method 4 — Add Continuous Alternative Supply.........275
6.4 Improving the Reliability of Overhead Networks (Design
     Methods 5, 6, and 7)...278
6.5 Improving Performance with Automation 281

6.6 Improvements by Combining Design Methods 1, 2, 3, 4, and
    8 on Underground Circuits..282
References 287

Chapter 7 Communication Systems for Control and Automation...............289

7.1 Introduction...289
7.2 Communications and Distribution Automation 289
7.3 DA Communication Physical Link Options 292
7.4 Wireless Communication 293
    7.4.1 Unlicensed Spread Spectrum Radio..293
    7.4.2 VHF, UHF Narrow Bandwidth Packaged Data Radio
          (Licensed/Unlicensed) 293
    7.4.3 Radio Network Theory..293
    7.4.5 Trunked Systems (Public Packet-Switched Radio) 302
    7.4.6 Cellular 303
    7.4.7 Paging Technology 303
    7.4.8 Satellite Communications — Low Earth Orbit 303
7.5 Wire Communications..304
    7.5.1 Telephone Line 304
    7.5.2 Fiber Optics 304
    7.5.3 Distribution Line Carrier...304
    7.5.4 Summary of Communications Options....................................331
7.6 Distribution Automation Communications Protocols..........................333
    7.6.1 MODBUS 333
    7.6.2 DNP 3.0...336
    7.6.3 IEC 60870-5-101 342
    7.6.4 UCA 2.0, IEC 61850...345
7.7 Distribution Automation Communications Architecture 346
    7.7.1 Central DMS Communication...346
    7.7.2 Polling and Report by Exception..348
    7.7.3 Intelligent Node Controllers/Gateways.....................................349
    7.7.4 Interconnection of Heterogeneous Protocols............................349
7.8 DA Communications User Interface..350
7.9 Some Considerations for DA Communications Selection 350
7.10 Requirements for Dimensioning the Communication Channel.............351
    7.10.1 Confirmed and Nonconfirmed Communication.........................351
    7.10.2 Characterization of Communication Systems..........................351
    7.10.3 Communication Model...353
    7.10.4 Calculation of the Reaction or the Response Time....................353

Chapter 8 Creating the Business Case..357

8.1 Introduction...357

8.2	Potential Benefits Perceived by the Industry for Substation Automation	358
	8.2.1 Integration and Functional Benefits of Substation Control and Automation	358
	8.2.2 SCADA vs. SA	360
	8.2.3 Economic Benefits Claimed by the Industry	360
8.3	Potential Benefits Perceived by the Industry for Feeder Automation	363
8.4	Generic Benefits	364
8.5	Benefit Opportunity Matrix	367
8.6	Benefit Flowchart	367
8.7	Dependencies, and Shared and Unshared Benefits	367
	8.7.1 Dependencies	367
	8.7.2 Shared Benefits	371
	8.7.3 Unshared Benefits from Major DA Functions	372
	8.7.4 Benefit Summary	378
8.8	Capital Deferral, Release, or Displacement	379
	8.8.1 Deferral of Primary Substation Capital Investment	379
	8.8.2 Release of Distribution Network Capacity	383
	8.8.3 Release of Upstream Network and System Capacity	387
	8.8.4 Displacement of Conventional Equipment with Automation	388
8.9	Savings in Personnel	388
	8.9.1 Reduction in Substation/Control Center Operating Levels	389
	8.9.2 Reduction in Inspection Visits	389
	8.9.3 Reduction in Crew Time	390
	8.9.4 Calculation of Crew Times Savings Associated with Investment- and Operation-Related Savings	402
	8.9.5 Reduced Crew Time and Effort for Changing Relay Settings for CLPU	402
8.10	Savings Related to Energy	403
	8.10.1 Reduction in Energy Not Supplied Savings Due to Faster Restoration	403
	8.10.2 Reduced Energy Revenue Due to Controlled Load Reduction	404
	8.10.3 Energy Savings Due to Technical Loss Reduction	405
	8.10.3.1 Loss Reduction from Feeder Volt/VAR Control	405
8.11	Other Operating Benefits	407
	8.11.1 Repair and Maintenance Benefits	408
	8.11.2 Benefits from Better Information (DMOL)	408
	8.11.3 Improved Customer Relationship Management	410
8.12	Summary of DA Functions and Benefits	411
8.13	Economic Value — Cost	412
	8.13.1 Utility Cost	413
	8.13.2 Customer Cost	421

	8.13.3 Economic Value	422
8.14	Presentation of Results and Conclusions	426
References		428

Chapter 9 Case Studies 431

9.1	Introduction	431
9.2	Case Study 1, Long Rural Feeder	431
	9.2.1 Evaluation of Performance	431
	9.2.2 Crew Time Savings	433
	9.2.3 Network Performance and Penalties	434
9.3	Case Study 2, Large Urban Network	437
	9.3.1 Preparation Analysis — Crew Time Savings	437
	9.3.2 Preparation Analysis — Network Performance	439
	9.3.5 Summary of Cost Savings	446
	9.3.6 Cost of SCADA/DMS System	447
	9.3.7 Cost Benefits and Payback Period	448
	9.3.8 Conclusions	448

Glossary 451

Index 459

1 Power Delivery System Control and Automation

1.1 INTRODUCTION

Electric power utilities have strived to run their businesses as efficient enterprises providing energy at an acceptable level of quality. The emergence of deregulation has dramatically changed the business environment. This radical shift in business goals, now occupying power companies, in many countries as a result of deregulation, open access, and privatization, is causing a significant review of network design and operating practices. The resultant separation of production, supply, bulk transmission, delivery (distribution), and metering into different businesses has sharpened the focus of these organizations. In particular, the owners of the distribution networks are being required, whether directly through the regulator or indirectly through new rate structures or consumer awareness, to improve areas of the network with substandard reliability. The owners are also being required to maximize the use and life of their assets through improved monitoring and analysis. Power quality (PQ) is also an important issue for which they are accountable. Network control and automation will play a key role in enabling the network owners to adapt to the changing situation and opportunities to achieve their business goals while ensuring an adequate return for the shareholders. The objective of this book is to draw together all the components and systems that have been used in distribution network automation, to define many of the expressions used in the industry for automation, and to introduce new ideas and solutions now being proposed to facilitate control and automation implementation.

1.2 WHY DISTRIBUTION AUTOMATION?

Distribution companies implementing distribution automation (DA) are receiving benefits from many areas such as providing a fast method of improving reliability, making the whole operating function more efficient, or simply extending asset life. Acceptance of distribution automation across the distribution industry is varied and not universal, due to the limited benefit-to-cost ratios of the past. The legacy of past management perceptions that more efficient control of distribution

TABLE 1.1
Key Automation Benefit Classifications by Control Hierarchy Layer

Control Hierarchy Layer	Reduce O&M	Capacity Project Deferrals	Improved Reliability	New Customer Services	Power Quality	Better Info for Engr. & Planning
1. Utility	✔			✔		✔
2. Network	✔	✔	✔		✔	✔
3. Substation	✔	✔	✔		✔	✔
4. Distribution	✔	✔	✔		✔	✔
5. Customer	✔	✔	✔	✔	✔	✔

networks was neither required nor a worthwhile investment and is changing as a result of deregulation and the industry's experience with new, cost-effective control systems. Automation is first implemented at the top of the control hierarchy where integration of multifunctions gains efficiencies across the entire business. Implementation of downstream automation systems requires more difficult justification and it is usually site specific, being targeted to areas where improved performance produces measurable benefits. The benefits demonstrated through automating substations are now being extended outside the substation to devices along the feeders and even down to the meter. The utilities implementing DA have produced business cases* supported by a number of real benefits selected to be appropriate to their operating environment. The key areas of benefits down the control hierarchy† are summarized in Table 1.1.

Reduced Operation and Maintenance (O&M) Costs. Automation reduces operating costs across the entire utility, whether from improved management of information at the utility layer or from the automatic development of switching plans with a distribution management system (DMS) at the network layer. At the substation and distribution layers, fast fault location substantially reduces crew travel times, because crews can be dispatched directly to the faulted area of the network. Time-consuming traditional fault location practices using line patrols in combination with field operation of manual switches and the feeder circuit breaker in the primary substation are eliminated. Automation can be used to reduce losses, if the load characteristics justify the benefit, by regularly remotely changing the normally open points (NOPs) and dynamically controlling voltage.

Condition monitoring of network elements through real-time data access in combination with an asset management system allows advanced condition and reliability-based maintenance practices to be implemented. Outages for maintenance can be optimally planned to reduce their impact on customers.

---

* Chapter 8 covers the whole area of cost/benefit analysis and business case development.
† Described in Section 1.4.

Capacity Project Deferrals. Improved network operating information allows existing networks to be operated with reduced margins, thus releasing capacity that would otherwise be reserved for contingencies. Real-time loading analysis will allow component life to be optimized against operational needs. Automation of open points between primary substations will avoid, in many cases, the need for additional substation transformer capacity, because short-term load transfers to adjacent substations can be made remotely to maintain supply with little equipment loss of life.

Improved Reliability. Although reliability is a power quality issue, it is commonly treated separately because outage statistics are an important yardstick in distributions operations. Deployment of remote-controlled switching devices (reclosers and load break) and communicating fault passage indicators (FPIs), in combination with a control room management system, improve the whole area of outage management, substantially reducing both duration and frequency of outages. Customer demands and regulatory pressure, whether indirectly or through performance/penalty based rates (PBRs), for improved network reliability, are forcing utility management to review operating and design practices in areas of substandard performance. Automation provides the fastest way to reduce outage duration. Experience has shown that a 20–30% improvement in the average outage duration can be achieved for most well-maintained overhead feeder systems in one year through implementation of automation. It can even reduce the number of outages if an outage is recorded as an outage only if it is sustained beyond a certain interval.* This improvement is made on the basis that momentary interruptions due to autorecloser operation are acceptable. In contrast, for example, reconductoring to covered conductor would achieve the same improvement in duration and also in frequency but at the expense of significantly more cost and time, typically requiring a 3- to 4-year implementation period.

New Customer Services. Automation at the customer layer through remote meter reading allows the utility to offer more flexible tariffs and the customer more selectivity and control of consumption. This lowest control layer has to be coordinated with the customer information systems at the highest control layer to be an effective business system. Automation will be a prerequisite at the lowest level if distributed resources† are to be practical.

Power Quality. In addition to reliability, as measured by interruptions, power quality includes voltage regulation and unbalance, sags, swells, and harmonic content. These characteristics are receiving closer scrutiny with the increased penetration of electronic consumer loads. Automation of distribution networks increasingly includes osillographics in the intelligent devices, thus allowing true

---

* For statistical purposes used by national performance standards authorities or regulators for assessing utility performance, whether in the public or private environment. Outages are only counted if sustained over a certain time, typically between 1 and 5 minutes, depending on the country. Outages not corrected within 24 hours commonly incur a penalty per customer.
† Distributed resources refer to small generation systems (micro turbines, gas engines, windmills, photovoltaic arrays, etc.) usually connected directly to the medium-voltage (MV) or low-voltage (LV) network.

monitoring of quality. Automation also enables the dynamic control of voltage regulation through remote control of capacitor banks and voltage regulators.

Improved Information for Engineering and Planning. The increase in real-time data availability resulting from DA provides more visibility to planners and operators of the network. The optimization of the communications infrastructure is an important aspect of the automation implementation that will deliver the required data to the appropriate application. This data is fundamental to better planning and asset management under business objectives, forcing lower operating and capital investments.

1.2.1 INCREMENTAL IMPLEMENTATION

The benefits from computerized control and automation systems are obtained as each function is implemented. Implementation strategies are incremental, one function building on the previous stage; thus, the benefits accumulate over time. Figure 1.1 shows an example from a utility serving a predominantly rural area, with the improvement in interruption time and crew levels as each of the functions are implemented in stages over a 10-year period. These improvements carry an economic value assessed by the utility to provide a positive benefit-to-cost ratio for the business. The DA functions implemented were

- Supervisor control and data acquisition (SCADA)
- Communicating relays in substations
- Remote controlled disconnecting switches
- Distribution management system with integrated fault location function supported by the corporate network information system comprising asset data base and mapping system.

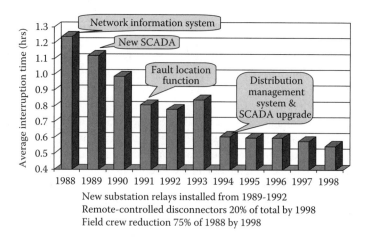

FIGURE 1.1 Summary of DA benefits accrued over an extended implementation time for a rural network. (Courtesy of ABB.)

Although the introduction of performance-based penalties (PBRs) is a very strong tangible economic driver for distribution automation, the case for distribution automation in the absence of PBRs has strengthened due to greatly improved equipment prices, emergence of standards in distribution automation, and the business pressure for improved utilization of assets. Considering that many of the hard and soft benefits within short-term business goals have made successful business cases for distribution automation, it should be noted that benefits in different categories could result from the implementation of only one automation function. Conversely, additional functions may be implemented within an established DA infrastructure at a small incremental cost. This interdependence should be maximized in developing a DA strategy, and particularly the weight of the soft benefits should not be overlooked in developing the overall business case. The correct assessment of these soft intangible benefits can make the difference in the ultimate contribution and value of distribution automation to the distribution power delivery business.

1.2.2 Acceptance of DA by the Utility Industry

The assessment of the degree of utilization of the DA concept in the industry is difficult not only due to the different interpretations of the concept but also due to the varied deployment strategies. Market and national power supply policies create different performance pressures on utility management, resulting in different business drivers. Some utilities are forced by regulatory pressure to take immediate action to improve performance of poor-performing parts of the network or supply to significant customers, whereas others are able to justify the gradual networkwide implementation of DA over a number of years. Further, component-based purchasing practices make identification of the volume of DA implemented difficult. DA though has been the subject of many market surveys, and a review (Figure 1.2) of this information confirms the increasing acceptance and implementation of DA.

In 1988, a survey (Survey Reference Number 1) of over 500 utilities in the United States revealed that only 14% had implemented DA and an additional 12% had in place a DA strategy. The context of DA in this period was the deployment of distribution SCADA down to the distribution substation using RTUs. The addition of extended control outside the substation was being considered by 70% of those implementing distribution SCADA. Surveys (Survey Reference Number 2) a decade later both in the United States and outside (predominately Canada, the U.K., and Australia) for substation automation showed a definite increase of automation implementation using a communicating bus within the substation rather than hardwiring. The utilities surveyed outside the United States showed a higher percentage of acceptance of substation automation. A survey conducted in 1999 (Survey Reference Number 3) focused on extended network control (feeder automation) and confirmed that over half of the 40 United States utilities questioned were actively deploying, and had planned to continue installing, remote-controlled switches in their primary distribution networks (medium voltage). This survey covered a total of 20,000 overhead and 3,500

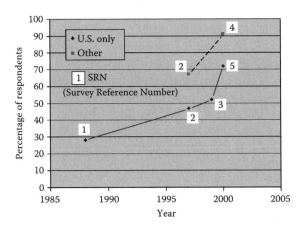

FIGURE 1.2 Survey of results giving percentage of distribution utilities in the United States and outside the United States, confirming implementation of a DA strategy. (Surveys conducted by Newton-Evans Research Company, Inc. of Maryland either commercially or on behalf of ABB; the summaries are made available courtesy of the two organizations.) (Courtesy of Newton Evans Research.)

distribution feeders operated by the utilities involved in the survey. This represented a 46.6 million customer base or a 20% sample of the entire United States. The DA survey conducted in 2000 (Survey Reference Number 5) covering DA practices in the United States was again focused on extended control and showed a significant adoption of remote-controlled switches or reclosers. The major reason that 50% of the respondents were not using DA was a lack of need on their systems, with 30% citing high cost as the deterrent. A survey of non-U.S. utilities (Survey Reference Number 4) conducted in the same year focused on both substation automation and extended control, where 75% of the respondents claimed that currently they had implemented automation on both substation and feeder switches and that this trend would continue for the period in the survey (2000–2002). The survey sample included representative utilities from South America (17), Europe (9), the Middle East (4), and the Far East (8).

The case for DA under a regulatory environment is clearly demonstrated by considering the U.K.'s experience. The regulator effectively introduced an indirect financial penalty by linking permitted income to the level of investment needed to improve the reliability of the worst-performing distribution feeders. The maximum income was restricted to below the level of general inflationary price rises by the factor "X," which was varied regularly according to recorded performance. The prime effect of this restriction was to ensure selected investment in reliability improvement, with the investment decision being compared against the income restriction. The penetration of automation outside the already SCADA-controlled substations is shown in Figure 1.3.

The number of switches automated (remote controlled) outside the substation increased rapidly once the goals had been set for each utility for the first regulatory

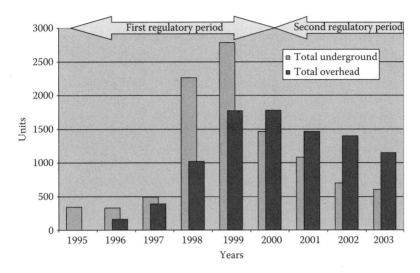

FIGURE 1.3 Number of switching devices outside the primary substations automated in U.K. distribution companies after privatization.

period. The automation strategy varied between utilities, some concentrating on high-density underground networks and others on the badly performing rural overhead circuits. Most companies proposed to continue improving performance during the second regulatory period, increasing penetration of automated switches to approximately 5% of the total switch population.

1.3 POWER DELIVERY SYSTEMS

The power delivery system is a continuous network linking the end user with the energy generator. This network is divided into bulk power for transmitting energy from the generators to the load centers, where it is distributed to the end user or customer. The larger the customer, the higher the level of voltage used to deliver power. The transmission system concentrates on the efficient and secure delivery of bulk power and the selection of the appropriate sources of generation. Until the recent introduction of the free market for energy supply, one company was responsible for all three areas. The assignment of a geographical service area to one company responsible for this vertical process created a monopoly for energy supply, which was regulated by an authorized energy authority. These authorities approved rates to the end users following submission of the financial and engineering plans of these utilities. Power generation economics balanced the operation costs of different energy forms (thermal, nuclear, natural gas and when available hydroelectric) with the cost of power delivery across the transmission network. Advantages of economies of scale were used to select the maximum sized generator that would retain the reliability policies in terms of loss of load probability. Transmission systems were designed through extensive deterministic single and multiple steady-state contingency analysis and dynamic stability analysis for investigating the impact of generation and bulk transmission loss on system security. The control of these systems was basic, where the grid operators using SCADA systems could remotely increase generation output by sending instructions to the power plant control systems in each generating plant and by remotely switching circuit breakers, tap changers, and capacitor banks. As time progressed, energy management systems were developed that provided automatic generation control (AGC) and continuously monitored the transmission network condition through state estimation and security analysis, to alert the operator of potential problems. Short-range forecasting of load increased the economic performance of the operation by allowing a minimum amount of generation to be scheduled on-line.

The control requirements have now become more complicated with the separation of generation from the network. Generation is now sized, located, and operated within a free market; thus, the transmission control system has to be able to respond to generation from different and varying locations depending upon the strategy of the generation owners. Control will have to be added to regulate energy flow across the bulk supply network to allow energy delivery at different price levels. This need has spawned merchant transmission lines solely for this purpose and the increased use of more controllable high-voltage direct current (HVDC) transmission interconnections. The separation of the various utility functions into independent businesses under deregulation is shown in Figure 1.4.

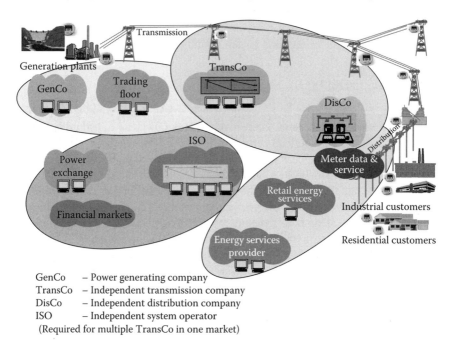

GenCo — Power generating company
TransCo — Independent transmission company
DisCo — Independent distribution company
ISO — Independent system operator
(Required for multiple TransCo in one market)

FIGURE 1.4 Utility business processes operating as separate legal entities in a deregulated environment. (Courtesy of ABB.)

Power Delivery System Control and Automation   9

This book will not dwell on the issues related to bulk generation and transmission control but concentrate on extended control and automation of distribution systems now operated in the deregulated environment by independent distribution companies (DisCo). Distribution systems have needed little real-time control because radial networks, which predominate, have been designed to operate within voltage limits and a range of anticipated loads. Networks are protected by feeder circuit breakers at the primary (distribution) substations and by an assortment of protection-operated devices down the feeder outside the substation boundary (reclosers, automatic sectionalizers, and fuses). Switching of the network outside the primary substation was performed manually by line crews sent out to locate, isolate, and repair the fault before restoring full service. The amount of lost energy compared to a bulk transmission line did not, of course, justify substantial investment in control systems. The first remote control introduced in distribution networks was simple SCADA implemented at large bulk substations where the economics were similar to a transmission substation. Small primary substations (<50 MVA) still, in general, remained under manual control similar to the distribution network.

Deregulation, though, has had its impact on the distribution network, emphasizing the need to reduce outage times. As a result, automation of the feeder system has emerged as one of the strategies to improve operational performance. This, coupled with enhancing the simple SCADA system with a distribution management system* (the distribution equivalent to a transmission system EMS), has added a much higher level of distribution network control. This added capability will become even more vital as distributed resources are deployed at the distribution level.

This whole discussion of power system control is best formalized as a control hierarchy.

1.4 CONTROL HIERARCHY

Network automation is applied within a structured control hierarchy that encompasses the need of the different delivery layers of the network. This requires the ability to control a network from one point, the control center, or a number of distributed control centers with delegated control. This process is called SCADA, or telecontrol, and relies on communication links from the control center to the primary device (generator, circuit breaker, tap changer, etc.) to be operated. Primary devices must be fitted with actuators or mechanisms to perform the mechanical opening and closing operation. These actuators must be interfaced with a secondary device — an intelligent electronic device (IED). The IED interfaces the actuator with the communication system. The relative size and sophistication of the IED depends on the control system configuration and its layer in the control hierarchy. The combination of control room system, communication, and IED comprises a SCADA system. SCADA systems are deployed

---

* See Chapter 2 for details.

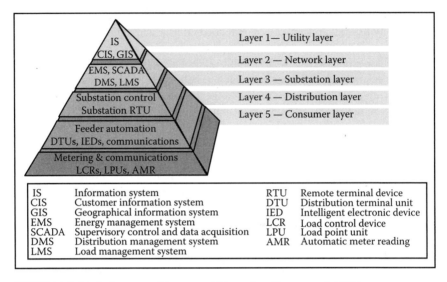

FIGURE 1.5 Typical power utility control hierarchy. (Courtesy of ABB.)

to control different layers of the network, either as one system integrated over a number of layers or as separate systems passing selected information to the control layer above. The actual selection of how central control is organized depends upon the ownership of the network layers. Owners of simple distribution networks with voltages below 33 kV tend to use one SCADA system to control the entire network. Even utilities with extensive networks covering a large geographical area are consolidating control from distributed control centers to one central operation.* Utilities with both medium-voltage and high-voltage (HV) subtransmission networks (230–66 kV) tend to operate the HV through a dedicated SCADA, integrating both voltage levels in one system.

A typical network control hierarchy is shown in Figure 1.5 and is comprised of five layers.

Layer 1. Utility: The upper level of the hierarchy covers all the enterprise-wide IT, asset management, and the energy trading systems.

Layer 2. Network: Historically, this layer has controlled the bulk power transmission networks, including the economic dispatch of the generators.

Layer 3. Substation: The integrated control of all circuit breakers inside the substation with the communication of all protection relay status.

Layer 4. Distribution: This layer of the control hierarchy covers the medium-voltage feeder systems and reflects the expansion of the real-time control capability, through remote control and local automation, of the feeder devices located below the primary substations.

---

* Improvement in computing technology and the introduction of DMSs has made this consolidation technically feasible.

Layer 5. Consumer: The lowest layer of control is where the delivery system directly interfaces with the consumer. It represents a growing activity where more flexible metering systems are required to allow convenient revision of tariffs and load control (demand side management — DSM). This functionality is being achieved by the implementation of automatic meter reading (AMR) systems integrated with new, easily configurable billing and accounting processes based on information technology (IT).

This division of the control process into control layers is made because in practice, the responsibilities for control within the utility are similarly organized. The power network, on the other hand, is a vertically integrated delivery system where each layer is a necessary part of the whole. The division of control layers and resulting architecture must assume an integrated enterprise perspective if the business needs of the network owner are to be met. Distribution automation as covered in this book comprises layers 3 and 4.

However, it would be incomplete if the other layers of the control hierarchy were not treated in sufficient detail to encompass their contributions and interactions with DA. Chapter 2 will review the technologies, applications, and contribution of these other control layers on the distribution network enterprise.

The discussion of hierarchical control cannot be concluded without introducing the concepts of depth of control and the boundaries of control responsibility. Depth of control refers to the control layer that any particular control system covers, for example, and transmission SCADA/energy management system depth of control may cover all devices down to the MV feeder circuit breakers. Distribution SCADA/DMS control may also start from the same circuit breakers and control all MV devices including measurements on the LV side of distribution transformers. It may even extend further up the HV network encompassing substransmision. The delineation of control responsibility, who controls what, must be defined through agreeing and setting appropriately the boundaries of control responsibility within the network organization.

1.5 WHAT IS DISTRIBUTION AUTOMATION?

The utility business worldwide has many perceptions of what is distribution automation, ranging from its use as an umbrella term covering the entire control process of the distribution enterprise to the deployment of simple remote control and communication facilities retrofitted to existing devices. Thus, for clarity, the umbrella term will be treated as the DA concept under which the other generally used terms of distribution management systems and distribution automation systems will be treated.

1.5.1 DA Concept

The DA concept simply applies the generic word of automation to the entire distribution system operation and covers the complete range of functions from protection to SCADA and associated information technology applications. This

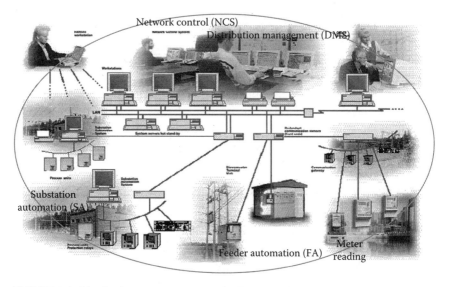

FIGURE 1.6 Distribution concept as an umbrella term. (Courtesy of ABB.)

concept melds together the ability to mix local automation, remote control of switching devices, and central decision making into a cohesive, flexible, and cost-effective operating architecture for power distribution systems. This is shown pictorially in Figure 1.6.

In practice, within the DA concept there are two specific terms that are commonly used in the industry.

Distribution Management System. The DMS has a control room focus, where it provides the operator with the best "as operated" view of the network. It coordinates all the downstream real-time functions within the distribution network with the nonreal-time (manually operated devices) information needed to properly control and manage the network on a regular basis. The key to a DMS is the organization of the distribution network model database, access to all supporting IT infrastructure, and applications necessary to populate the model and support the other daily operating tasks. A common HMI* and process optimized command structure is vital in providing operators with a facility that allows intuitive and efficient performance of their tasks.

Distribution Automation System. The DA system fits below the DMS and includes all the remote-controlled devices at the substation and feeder levels (e.g., circuit breakers, reclosers, autosectionalizers), the local automation distributed at these devices, and the communications infrastructure. It is a subsystem of the DMS essentially covering all real-time aspects of the downstream network control process. This book concentrates on this aspect of distribution control and automation; thus, a more detailed discussion of automation at this level is appropriate.

---

* HMI — human-machine interface.

1.6 DISTRIBUTION AUTOMATION SYSTEM

Distribution automation covers a wide range of implementations, from simple retrofitted remote control, or the application of highly integrated intelligent devices, to the installation of complete systems. The term *automation* itself suggests that the process is self-controlling. The electrical power industry has adopted the following definition:

> A set of technologies that enable an electric utility to remotely monitor, coordinate and operate distribution components in a real-time mode from remote locations.*

Interestingly, this definition does not mention an automatic function. This has to be inferred in the word *coordinate*. All protective devices must be coordinated to automatically perform the protection function satisfactorily by correct discriminatory isolation of the fault. Isolating the fault is only a portion of the possible functions of DA, because operation of the network would be improved if, having isolated the fault, as much of the healthy network as possible was re-energized. Further, the term *real-time* suggests that the automation system will operate in the 2-second response time frame typical in large SCADA control. This is overly ambitious for some parts of the distribution network where communication delays are significant. It is also not necessary or cost-effective for all DA functions where response times can be on a demand or demand interval basis. The terms of either *real-time* or *demand-time* provide flexibility to implement response times appropriate to achieving the operating goals for the network in a cost-effective manner. The one statement in the definition that differentiates DA from traditional protection-based operation (automatic) is that the relevant distribution components can be controlled from a remote location. This then necessitates integration of a communication infrastructure within the DA architecture. This is the key critical facility that offers increased information and control to the decision making required for smarter operation of the distribution network. Implementation and cost-effective integration of communications within the controlled distribution device and central control must be carefully planned.

DA, as stated earlier, also supports the central control room applications that facilitate the operations decision-making process for the entire distribution network of remotely controlled and manually operated devices — applications that are incumbent within the distribution management system. The number of the distribution assets not under remote control is in the majority for any distribution network. The proper management of these assets is vital to the business and requires the added facility offered within a DMS. These applications require support from corporate process systems such as the customer information system (CIS) and the geographical information system (GIS), which reside at the top layer of the control hierarchy.

Irrespective of which of the two control layers DA is applied to, there are three different ways to look at automation:

---

* Adapted from the IEEE PES Distribution Management Tutorial, Jan. 1998.

1. Local automation — switch operation by protection or local logic-based decision-making operation
2. SCADA (telecontrol) — manually initiated switch operation by remote control with remote monitoring of status, indications, alarms, and measurements
3. Centralized automation — automatic switch operation by remote control from central decision making for fault isolation, network reconfiguration, and service restoration

Any DA implementation will include at least two of these functions because communications must be a part of the implementation. There are, though, utilities that will claim to have operational distribution automation due to their early implementation of reclosers without or in combination with self-sectionalizing switches. The absence of communication to these devices does not fulfill the accepted definition of DA. Many utilities with such implementation do admit the need to have communication to these switching devices in order to know whether or not the device has operated.

Automation Decision Tree. The selection of the ways to automate a switching device can be illustrated through the decision tree in Figure 1.7. Once the primary device has been selected based on its required power delivery and protection duty in the distribution network, the degree of automation can be determined.

The implementation of automation to any manual switch can be described as a number of steps and alternative paths that lead to the degree and type of the control architecture. Some of the paths are optional but many are obligatory if automation is to be implemented.

Step 1: This is the basic step to provide a switch with a mechanical actuator, without which nonmanual operation would not be possible. Historically, switches have always been operated manually, but stored energy devices or powered actuators have been added to ensure that switch operation is independent of the level of manual effort and to provide consistency in operating speed. Safety is increased because the operator tends to be further away from the switch.

Step 2: Although the installation of an actuator will allow local manual operation, which is mandatory, simply by using pushbuttons, the main purpose is to facilitate the operation by local automation or by remote control.

Step 3: Once an electronic control unit has been installed for the actuator, one of the two main automation functions can now be selected. In the most simple choice at this step, the local automation can be interfaced to a communications system to allow control remotely. Alternatively, local intelligence can be implemented, allowing the device to operate automatically under some preset arrangement. A typical example of this alternative at step 3 would be a recloser without communications.

Step 4: This step builds on the two choices made at the previous step. Basically, remote control is added to local automation so that the operator will be informed of any operation of the device under local automation and can either suppress this local action or make the decision remotely. Local manual operation to override the intelligence is a mandatory feature. In the alternative path in step

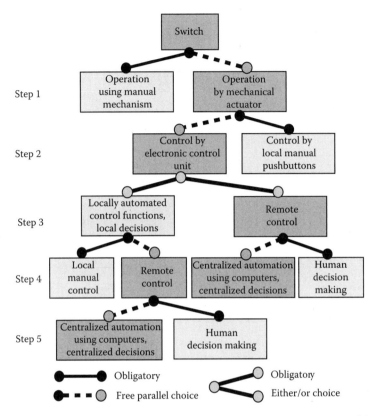

FIGURE 1.7 Decision tree showing the logical steps to the implementation of local or central automation of a primary switch.

3, where remote control was selected, two forms of decision making are possible, either a remotely located central process that incorporates a systemwide perspective or manually initiated remote control (human decision making).

Step 5: The final step applies the same options from step 4, remote control to the local automation. Although the ability to add central decision making to local automation offers the most advanced automation strategy, it is not commonly practiced because it is found sufficient and simpler to use remote control of intelligent devices.

The results of this decision tree in terms of meeting the basic definition of distribution automation are as follows:

- Switches must have remote-control operation capability.
- Decision making is implemented, either located locally in intelligent secondary devices (IEDs) centrally in a DA server, in combination with both local and central decision making or through human intervention remotely.
- Local operation must be possible either mechanically or by pushbutton.

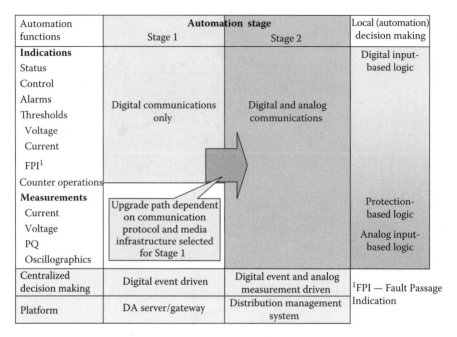

FIGURE 1.8 Stages of distribution automation for extended control.

Stages of Automation. The selection of automation level illustrated through the decision tree in the previous section can be viewed from a different perspective, taking into account the burden on the communication media. The more sophisticated the remote monitoring and automation requirements, the higher the burden and complexity of the messaging. This consideration has given rise to two different approaches to distribution automation (Figure 1.8), particularly for extended control down the feeder, where communications are predominantly radio based.

Stage 1: This is designated as meeting the basic requirements of distribution automation providing remote status and control functions. Remote status indication and control of switches has been the most justifiable stage of implementation of distribution automation outside the substation. This can be achieved by transmission of digital signals only. Other binary information such as alarms, FPI contact closing, and values above or below a threshold can be communicated digitally. Communication of digital values significantly lessens the complexity of the communication by reducing the data package length. Low-power radio systems have been developed and deployed to meet the needs of basic remote control.

Stage 2: This stage adds the transmission of analog measurements to status and control commands. This additional information moves the functionality of extended control close to that employed at the substation level; however, the burden on communications is increased and the capability of protocols used by

Power Delivery System Control and Automation 17

full SCADA systems is required. To reduce this burden, high-level protocols must have both unsolicited reporting by exception* and dial-up† capability.

Local automation can be applied under both stages and is only dependent on the sophistication of the power sensors and the IED. The restriction of only reporting status does not interfere with an analog/protection-based local decision process.

The degree of central decision making will depend on not only the amount and detail of the information passed to the server but the data transmission speed capabilities of the communication infrastructure.

There is not necessarily a natural upgrade path from one stage to another, because there could be a limitation in the protocol and communication infrastructure as a result of optimization for stage 1. The selection of the infrastructure for the first stage must consider whether an upgrade to stage 2 will be required within the payback period of the implementation.

Automation Intensity Level (AIL). AIL is a term employed to define the penetration of automation along the feeder system outside the substation. Two measures are commonly used: either the percentage of the number of manual switches placed under remote control, typically 5–10%, or the number of switches automated per feeder. Typically, the latter is designated as 1.0, 1.5, 2.0, 2.5, etc., where the half switch represents the normally open point shared by two feeders. One and a half switches per feeder denotes automating the open point and a midfeeder switch — an AIL that produces maximum improvement for the investment, because increasing AIL produces reducing marginal improvements to system performance. This is illustrated in Figure 1.9 for a set of actual feeders, the AIL being shown in both types of measure described above. A definite breakpoint occurs around an AIL of 1.5.

1.7 BASIC ARCHITECTURES AND IMPLEMENTATION STRATEGIES FOR DA

1.7.1 ARCHITECTURE

The basic architecture for distribution automation comprises three main components: the device to be operated (usually an intelligent switch), a communication system, and a gateway often referred to as the DA gateway — Figures 1.10a and 1.10b.

This configuration can be applied to both substation and feeder automation. In primary substation applications, the gateway is the substation computer capturing and managing all the data from protective devices and actuators in the switchgear bays. It replaces the RTU as the interface to the communication

---

* Unsolicited reporting by exception is the ability for a slave device to initiate a query to its master when any signal is out of the normal operating condition rather than waiting for the master to poll the slave.
† Dial-up capability is the function that allows the slave to initiate the communication link when necessary rather than having the communication channel operating continuously.

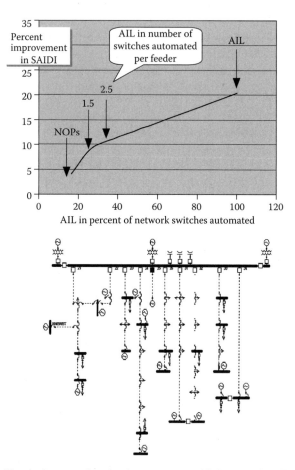

FIGURE 1.9 Marginal outage duration improvement with increase in AIL

system, which receives and sends information to the central control. Similarly, in feeder automation applications, the gateway manages the communication to multiple intelligent switches, acting as regards central control as a data concentrator. This, in effect, creates virtual locations for each switch and relieves central control of the need to establish every switch as a control point. The latter configuration is, of course, possible and used for automation where a few switches are remotely controlled, hybrid configurations where the substation computer or, in cases where there is no substation automation, the substation RTU acts as the gateway for all switches located on feeders emanating from the substation. The gateway can also be used to establish local areas of control where a more optimum communication infrastructure for extended control can be established separately from the SCADA system. The gateway becomes the conversion point from one infrastructure (protocol and communication system) to another. The gateway can also be extended from a simple data concentrator to one with limited graphical

Power Delivery System Control and Automation

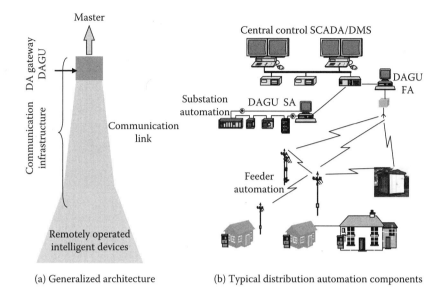

(a) Generalized architecture (b) Typical distribution automation components

FIGURE 1.10 (a) Generalized architecture and (b) major components of a DA system. (Courtesy of ABB.)

user interface to allow local control or even communicating selected information to multiple masters.

1.7.2 Creating the DA Solution

It is worth dwelling on a more detailed examination of the components of a typical distribution automation system at the hardware level because this exposes some of the challenges in implementing automation and the interplay of various components. The major components of a DA system, shown in Figure 1.11, are in the primary substation and the feeder devices outside the substation such as pole-mounted switches, ground-mounted units and secondary substations. They have to interface into the distribution control center (DCC). Across these three functions is the communication infrastructure that can use different media and protocols to the slaves in substations and on the feeders. The selection of communication method depends on the goals for each of the control layers, and the mixture has to be accommodated within the DA implementation through conversions at some point in the communication chain, usually at the gateway or SCADA front end.

Primary Substation. In a primary substation, the switchgear bays or cubicles are usually supplied as an integrated set consisting of a circuit breaker with actuator and a protection relay mounted and wired with terminal blocks ready for connecting to the station control bus. Remote control is achieved in two ways.

(1) By hardwiring the control, indication, and measurement circuits to a primary RTU. The RTU is supplied as part of the SCADA system with a communication structure uses a standard SCADA protocol over microwave radio or

20   Control and Automation of Electric Power Distribution Systems

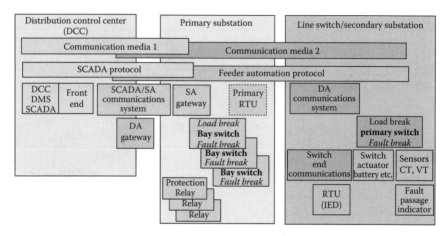

FIGURE 1.11 Components of DA that have to be integrated to make a working system.

dedicated land lines. This is the traditional method used to establish remote control of a substation via retrofitting the communication facility to existing primary gear.

(2) By implementing substation automation (SA), which establishes a local area network within the substation between communicating protection relays and a small PC-based SA gateway to manage the data within the station. This eliminates the need for a primary RTU and the hardwiring of station devices. The gateway provides the communication interface back to the DCC using the SCADA protocol, supports software-based internal substation interlocking and automation applications, and provides a GUI for local operation.

This examination can be treated separately for primary substations (substation automation) and line switches (feeder automation), even though the differences are marginal.

Line Switch/Secondary Substation. Remote control and automation have been used in substations for some time with the implementation of SCADA systems (certainly at the transmission and high end of the distribution control layer). Remote control switches outside the primary substation is now being implemented. There has been little standardization of the role and level of automation to be implemented at either a pole- or ground-mounted switch. This leaves room for many alternative configurations at the switch. What accuracy, quantity, and parameters of measurement are necessary? Will the implementation be an automation Level 1 or Level 2 application? These decisions configure the number of sensors, the type of intelligent device, and the communication burden. The type of local automation that will be required is also a major factor in first, the selection of primary switching device (recloser, load break switch, etc.), and second, the specification of the function of the intelligent device (full protection or simple communication interface). Thus, the selection of all the major components within the feeder device is key. Once defined, the communication media and the protocol have to be selected and integrated into the complete DA architecture. A device that has been fully specified and tested to plug and play in the defined environment is termed a "SCADA ready" or "automation ready" device (ARD).*

There are two basic approaches to implementing DA at this level.

(1) Retrofitting the remote-control facility to installed (legacy) switches, of which there are many. The majority of pole-mounted switches will be of the open air break type and are automated by supplying a DA control cabinet mounted at the bottom of the pole and containing the actuator and secondary equipment for control and communication. Retrofitting automation to existing ground-mounted equipment is purely dependent on the physical ability of the gear to accept a retrofitted actuator.

(2) Installation of new automation ready equipment specifically designed for remote control to replace an existing manual switch, the latter being returned to stores for deployment elsewhere in the network when required.

In any DA implementation, the probability of there being an existing SCADA system is high and will necessitate that any extension of the control system must interface with this legacy system. The further down the distribution network the control is extended, the higher the likelihood that integration of equipment from multiple suppliers, each with their own standards, will be necessary to achieve an integrated control system. It is these aspects with incumbent limitations that must be carefully understood as the DA implementation strategy is developed.

1.7.3 Distribution Network Structure

The power delivery process in the form of the network structure is, in effect, the ultimate user of any automation scheme. Implementation of automation must improve the performance of the network as an investment that is repaid by the improved performance and operating efficiencies of the business. Distribution networks are predominantly radially operated, and reconfiguration is one of the few ways of improving reliability. Once a fault occurs, the faulted section must be isolated, and any healthy circuit downstream of the isolated section can only be re-energized by closing the normally open point. Fault isolation can be achieved by local automation or by direct remote control; however, supply restoration by remote control using operator decision making, rather than automated logic, is the most accepted approach. Fully automated restoration schemes using local automation are possible but require the complete confidence of the operating staff before acceptance. Remote control is also applied to open loop schemes, which predominate in underground systems.

A closed loop underground distribution network with circuit breakers at secondary substations provides improved reliability at higher initial cost due to the need for more costly switches and directional, or unit protection. Fault isolation is achieved directly by protection (local automation), which in the case of cable systems isolates the fault without customer disturbance because customers

---

* Defined in Section 1.8.

are usually connected at the substation with switches on either side of the load point, there being only limited teed connections in typical networks.

The topology and type of the distribution network is an important consideration when deciding on the level and sophistication of the DA implementation. The impact of network design will be a recurring factor in later chapters covering fault location and economic justification of DA.

1.8 DEFINITIONS OF AUTOMATED DEVICE PREPAREDNESS

The implementation of DA is not only a function of the architecture and the level of automation required by the business case, but it is also influenced by the procurement practices at the distribution level. Distribution equipment has always been considered a volume product or component business rather than a system business. This has driven the procurement of DA to the component level in the majority of cases, particularly at the feeder level. Utilities tender for separate components such as IEDs to be installed in control cabinets, switches with actuators and a specified IED, communications (low-power radio, GSM, etc.). Each device must comply with an inferred system specification such as communication protocol and satisfactory operation over a particular communication medium. In addition, it may be required that the devices operate satisfactorily with the installed (legacy) SCADA system that will provide the control interface. Correctly configured and prepared devices allow the utility to implement independently the whole DA project. Application of this concept in an industry where standards are limited and where extension to the standards has been allowed resulted in many interoperability errors between components from different suppliers, which have to be resolved in the field. In system supply contracts such as a large SCADA system, a significant part of any system is configured in the factory and undergoes FAT* before shipping to the field for installation and SAT.† This procedure is not possible if a component procurement procedure is used, because the utility has taken the system responsibility. To avoid interoperability problems, pilot or proof-of-concept projects are used to iron out any incompatibilities. This allows the utility to select a number of devices that have been verified within the selected DA infrastructure for volume component procurement.

Interoperability errors can be reduced if a more precise definition of the preparedness of a device is made. The following levels of preparedness are proposed.

- *Automation infeasible device (AID)* — This describes a primary device for which it is either technically or economically infeasible to install an actuator for remote control. It applies to older switching devices

---

* Factory Acceptance Test (FAT), Site Acceptance Test (SAT).
† Also referred to as "SCADA Ready."

that mechanically could not support nonmanual operation. Most typical of these are old ring main units.
- *Automation prepared device (APD)* — This describes a primary switching device that has been designed specifically to be automated, thus it has provision for an actuator to be easily attached as part of the original design. It can be supplied without the actuator for later retrofitting. The device may also be designed to have an integral control facility (internal box or external cabinet). This control may or may not be populated with an intelligent electronic device, power supply, and provision for the communications transceiver. This control facility maybe be populated by a third party or supplied later as a retrofit.
- *Automation ready device (ARD)* — This refers to an automation prepared device that has been fully populated with all the necessary control equipment to allow it to operate in a DA scheme as specified by the customer (correct protocols for the communications media specified). It will also be protection prepared if required to operate independently within the network (i.e., recloser).
- *Automation applied device (AAD)* — This is an ARD that has the communications receiver installed and configured to work in the DA system of which it is part. Local automation logic is included where appropriate.
- *Automated distribution system (ADS)* — This describes the complete DA system, including all intelligent switching devices, communications infrastructure, gateways, integration with central control systems (SCADA), and the implementation of automation logic.

The different levels of inferred system responsibility (system integration testing) that must be undertaken at each definition is shown in Figure 1.12.

The concept of verification centers where different intelligent switch assemblies can be configured into a complete logical DA system are now being created by some suppliers to establish standard approaches and packages that, once verified, can be deployed in the field in volume with substantially reduced commissioning time and risk.

1.9 SUMMARY

The growing deployment of automation in distribution networks within the DA concept introduced in this chapter is irrefutable. However, the tendency for individual utilities to customize their solutions will not produce the economies of scale possible through acceptance of a more standardized approach. Further improvement of the device and implementation costs, together with applications that maximize network asset utilization, will lead to increased automation. This chapter has introduced concepts and terms such as control layers, depth of control, control responsibility boundaries, stages of automation, AIL, and the important concept of device preparedness. The remainder of this book builds on these

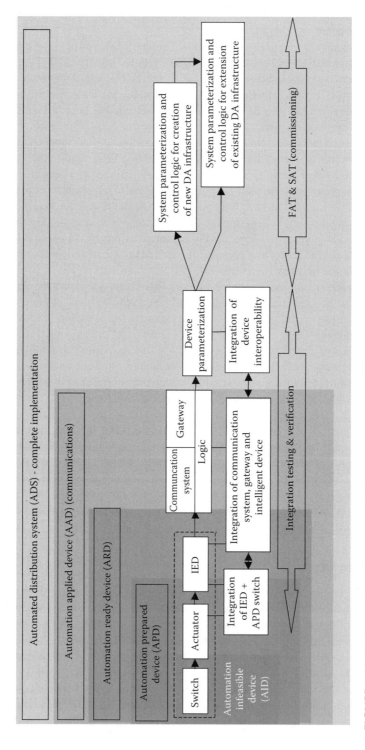

FIGURE 1.12 DA integration testing and verification requirements for different levels of automation preparedness.

concepts to describe the components and application logic required for distribution automation and explores a standardized approach through functional building blocks, all aimed at delivering benefit to the power delivery process through a more standardized approach.

REFERENCES

Bird, R., Business Case Development for Utility Automation, DA/DSM Europe 96, Volume III, Oct. 8–10, 1996, Austria Centre, Vienna.

Chartwell, Inc., *The Distribution Automation Industry Report,* 1996.

McCaully, J.H., Northcote-Green, J.E.D., Distribution Control Centers — Extending Systems Operational Capabilities, Third International Conference on Power System Monitoring and Control, IEE Savoy Place, 26–28 June 1991, Conf. Pub. #336, p. 92–97.

McDonald, J., Delson, M., and Uluski, R.W., Distribution Automation — Solutions for Success, Utility University 2001, UU 208 Course Notes, Distributech, San Diego, Feb. 4, 2001.

Philipson, L. and Willis, H.L., *Understanding Electric Utilities and De-Regulation,* Marcel Dekker, Inc., New York.

2 Central Control and Management

2.1 INTRODUCTION

The nerve center of any power network is the central control and management function, where the coordination of all operational strategies is carried out. Even if distributed control and operation is implemented, the results of such action must be communicated to the central coordination point. The control hierarchy introduced in Chapter 1 is a common way of defining strategies for power network control and presents a structure for the organization of the control, and management process. The upper two layers are covered by central control, with input from the substations and feeder subsystems. This chapter will cover the elements of the upper three control layers as they influence control and automation of distribution networks. It will introduce areas where system design is crucial and further study recommended.

2.1.1 WHY POWER SYSTEM CONTROL?

Electricity networks worldwide are entering a period of change that is necessitating improving methods of control and management. The business process being imposed on the companies as a result of the changed environment is compounding the complexity of the network.

The two major aspects of this change are as follows:

- The moves to privatize, deregulate, and unbundle, which provide consumers open access to independent power suppliers outside the network companies' service territory, together with the establishment of other legal entities for the trading and supply of energy.
- The increased awareness, either regulator or public led, of both business and residential consumers' perception of the utilities' operation, leading to a greater emphasis on quantifying the cost of providing services and improvements on quality, real or perceived.

Utilities will need to focus more on the needs and perceptions of their customers in achieving business goals. The regulators in the privatized environments are actively rewarding utilities with demonstrable improvements in customer satisfaction through power quality metrics. Although the core business functions are unlikely to change dramatically, they will need to be performed more efficiently and still meet these consumer-oriented goals. The inevitable outcome is for the utilities to employ a combination of real-time and information technology (IT) systems to support improved efficiencies and "right sizing" of their operations activity.

2.2 POWER SYSTEM OPERATION

Power system operation requires the balance to be maintained between security, economy, and quality while delivering electrical energy from the generating source to satisfy the demands of the end user. This balance from a purely technical viewpoint is dependent on the structure of the types and size of generating plant, the structure and condition of delivery network, and the demand characteristics of the end user. The new business environment now superimposes on previous technical constraints the need to balance the rules of regulation and a free market in combination. The free market operates at the supply and retail level, whereas regulation influences the operation of the monopolistic network companies that deliver the power.

Historically, control systems have been implemented on bulk power systems where it was economical to monitor all incoming and outgoing points of the network. Such real-time systems provide the facility for supervisory control and data acquisition and are known as SCADA systems. Advances in computation technology and power system modeling enabled fast applications to be fed with real-time data from SCADA to provide additional decision-making information for the operators. Natural evolution of these applications allowed increased levels of automation in the decision process. The system operation functions required at the control center can be split into three groups, each reflecting a time horizon. These have been summarized as follows:

Instantaneous operation: This involves the real-time monitoring of system demand and loads, power generation, network power flow, and voltage levels. The values of these parameters are continuously compared against defined technical and economic loading limits as well as contractual thresholds to ensure satisfactory operation. Any transgressions of these limits or thresholds either in the normal state or as a result of protection action have to be responded to in order to restore operation within the defined boundaries.

Operations planning: This for both the short (a few hours) and longer term (a few months). Short-term planning is crucial to economic dispatch of generating plant. Accurate short-term load forecasting techniques are key to this function. In the past, the heat rates of generating plants were the dependent parameter, but in today's deregulated environment, the bid price into the pool becomes the major parameter. Short-range load forecasting is crucial for distribution utilities operating in environments, usually generation-deficient markets, where exceeding a certain maximum contracted value will trigger penalty rates for the excess. Forecasting when and how much load to shed becomes crucial to the enterprise.

Central Control and Management

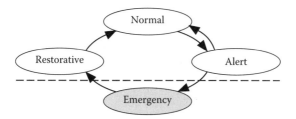

FIGURE 2.1 State modes of power system operation.

Operations reporting: This reflects the need to keep statistics on performance, disturbances, and loadings as input into planning and accounting functions. Postmortem analysis is key to determining disturbance causes. The reporting of quality is generally a legal regulatory requirement.

The four states of a power operation shown in Figure 2.1 are commonly used when describing bulk power systems, the emergency state reflecting the collapse of the power system usually from cascading protection intervention as a consequence of major generation or transmission line loss.

The alert state signifies that a disturbance has occurred and action should be taken directly (automatically or through operator intervention if time permits) to alleviate the situation. In bulk supply systems, the alert state can move very rapidly into the emergency state, making it impossible for an operator to prevent system collapse. The objective of power system operation is to keep the system within the normal state and return it to that state through the restoration process as soon as possible. The operator, using all the facilities of the control center, is the main decision maker in system restoration.

Distribution systems occupy the lower end of the control hierarchy, and the level of control possible is restricted by the specific structure of the distribution network and the penetration of real-time monitoring and control facilities. The level of SCADA implementation in distribution networks has historically controlled around 10% of switching devices and has been limited to circuit breakers at the larger primary substations. Adoption of the DA concept, where control is extended to small substations and primary feeders, will substantially increase the reach of real-time control.

2.3 OPERATIONS ENVIRONMENT OF DISTRIBUTION NETWORKS

Presently, the manner in which distribution networks are operated is influenced by the lack of remote control and real-time monitoring, requiring considerable manual intervention for decision making and restoration. The extent and multitude of elements that comprise a distribution network impose the need for handling considerable information to ensure satisfactory operation and crew safety.

This operations environment imposes the following three conditions on distribution system control personnel:

- *Normal conditions.* During normal system conditions, the operator is able to prepare switching plans for planned maintenance, monitor the system for out-of-tolerance operation, consider configurations to establish optimum operation, and initiate remedies to bring overloads or low voltages within limits. General maintenance of control room information such as the network diagram updating and management statistics are completed at this time.
- *Emergency conditions.* Failures on the network are unplanned and establish a stressed state in which the operator must perform. The primary objective is to organize restoration of the network as quickly as possible. This will involve preparing and executing switching plans to isolate the fault and restore supply by
  - Operating remote-controlled switching devices
  - Dispatching and controlling repair crews to operate manual devices and verify fault locations
  - Managing trouble call information and inform customers to preserve perceived customer satisfaction
- *Administration.* The everyday tasks of logging events, preparing standard management reports, and supplying performance statistics are time consuming. Privatization and changing external pressures will require improved auditing of system performance, audit trails of customer contacts, and increased attention to safety documentation. All these issues will require more effort and accuracy from the operating personnel. In addition, the generation of equipment statistics for improved asset management.

A distribution management system (DMS) must perform under the two major network-operating states, normal and emergency, by reflecting the workflows through which this is accomplished. The reporting function required for system audit, post-mortem analysis, and safety requirements must be continuous and be able to supply all the data legally specified by regulation.

Operating a system for which typically only 10% of the switching devices are remotely controlled necessitates manual operation by field crews dispatched to the switch to be operated. This process requires use of support systems or information outside traditional SCADA such as

- Operating diagrams and geographical maps showing the location of the network and devices
- Crew management methods to track and dispatch the correct resources and skills
- Repair truck inventory of network spares
- Trouble calls from customers to identify probable location of faults
- Mobile communications and data systems to allow command and data interactions between control center and field

TABLE 2.1
Four Key Functional Organizations within a Distribution Utility Summarizing Responsibilities

Function	Responsibility
Operations	This function is responsible for the daily running of the network with the primary object of maintaining continuity of supply. Traditional SCADA systems could be justified at the top of the distribution network hierarchy. For the remainder of the network, paper maps or large wallboards are used to manage operations.
Assets	All activities to do with the assets of the utility, essentially the electrical network, such as inventory control, construction, plant records, drawings, and mapping, are covered under this category. The major application to be introduced to facilitate this activity is the geographical information system (GIS), previously known as automated mapping facilities management (AM/FM) systems.
Engineering	Engineering department carries out all design and planning for network extensions. As one of the efficiency improvements, tools for network analysis and planning of systems are used to permit system operation audits for short-term solutions and optimal expansion planning systems to achieve system reinforcement at minimum cost.
Business	The business function covers all accounting and commercial activity within the utility. Of particular importance to distribution operators is customer information so they can respond to trouble calls. Such information is maintained in a customer information system (CIS) or customer relationship management (CRM) system.

All these functions must work in a coordinated manner and be synchronized between the control center and field operations.

2.4 EVOLUTION OF DISTRIBUTION MANAGEMENT SYSTEMS

Distribution companies had, before the advent of integrated DMSs, managed their networks through four key functions reflected by the organization of work within the company. These functions (Table 2.1) implemented independent applications to serve their own needs, thus creating the classic islands of control or work process.

Current distribution management systems are extensions of segments of these different applications specifically packaged for use in the control room and accommodating the unique characteristics of distribution networks. Although present DMSs are now converging on a common functionality, the evolution to this point has taken different paths. The starting point of the path is dependent on the dominant driver within the utility. Typical evolutionary paths seen in the industry are illustrated in Figure 2.2. The important element of the creation of the DMS structure is the ability to share data models and interface different data sources to form an integrated system that serves the needs of the operator. To

32 Control and Automation of Electric Power Distribution Systems

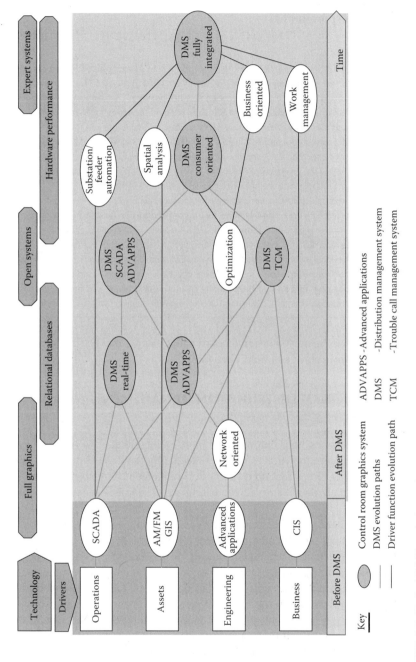

FIGURE 2.2 Typical DMS evolution paths.

achieve this, the system must be intuitive, imitating the traditional control room process with speed and simplicity of command structure.

The paths in Figure 2.2 merely demonstrate that DMS configurations and implementation strategies will vary depending on the "champion" within the organization. Trouble call management systems (TCMSs), for example, were driven by the need for the customer facing part of the utility to improve customer satisfaction. This was achieved by inferring, from a buildup of customer calls, where the fault was and then being able to respond to subsequent calls with information about the actions the utility was doing to restore power. TCMSs were not strictly real-time systems and could operate without SCADA. They were a pure IT solution to outage management (OM) and could be classified as customer-oriented DMSs. In contrast SCADA systems are real-time systems, but without the addition of a distribution graphical connectivity model of the whole network, they cannot be classified as DMSs due to the restrictive real-time cover of the network. The inclusion of a graphical model of the manually operated portion of the network allowed full operational management in terms of electrical load. This approach, lacking any customer representation, could be termed noncustomer oriented. Clearly, the bringing together of all the functions into an integrated system would achieve a system that was both customer oriented while allowing optimum utilization of the network assets. Common to all DMS configurations is the need for a detailed model of the network in terms of connectivity and operating diagrams. The latter, in the majority, take the form of geographic or geo-schematic continuous diagrams imitating existing wall maps or diagrams; however, there are still implementations that reflect existing operating practices using pages for each feeder loop and source substation. Thus, a fundamental requirement of all modern DMSs is that of a continuous world map with fast navigation and sizing. This full graphical system forms the vital part of the control room operations management (CROM) function used by operators to successfully perform their tasks. A sampling of key functions implemented (Figure 2.2) in early distribution management systems shows the importance of this function.

A full DMS is the focus of new management systems. It resides at the intersection of vertical integration (real power delivery process) and horizontal integration (corporate IT systems) of utility enterprise systems. Vertical integration is the domain of the operation's organization of the utility, and extended control of the network beyond traditional SCADA is within their responsibility. The horizontal integration element provides the source of corporate asset data (material and personnel) needed to support a full DMS implementation. A DMS requires interfaces with many different enterprise activities within the utility (Figure 2.4).

Implementation of a full DMS touches so many of the activities of the enterprise that justification can be lengthy. The more legacy systems within the enterprise to be interfaced or discarded, the more onerous the decision process. Justification is difficult, and a phased approach is usually adopted. It requires that a DMS must be modular, flexible, and open with a final solution as a seamlessly (to the user) integrated application to operate remote-controlled switching devices.

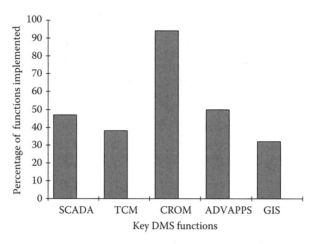

FIGURE 2.3 Implementation of key DMS functions in terms of percentage at responding utilities (SCADA — Supervisory Control and Data Acquisition; TCM — Trouble Call Management; CROM — Control Room Operations Management; ADVAP — Advanced Applications; GIS — Geographical Information System. CRM provides the operators view of the network used by TCM, ADVAP and SCADA).

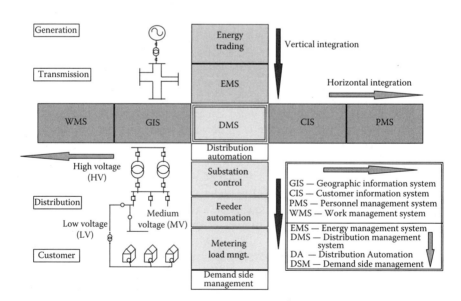

FIGURE 2.4 Horizontal and vertical integration in distribution management systems.

Central Control and Management

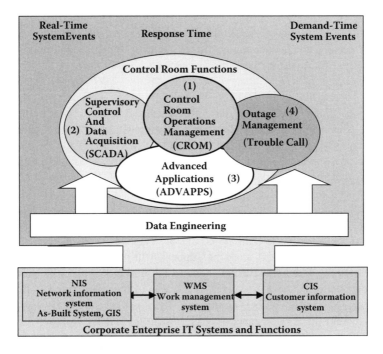

FIGURE 2.5 Key high-level distribution management system functions.

2.5 BASIC DISTRIBUTION MANAGEMENT SYSTEM FUNCTIONS

A modular DMS for network control and automation is described by four main functions, each with the ability to be fully integrated with the other yet to have the possibility to operate independently (Figure 2.5). The DMS is supported by other separate applications within the corporate information technology strategy.

(1) Control room operations management (CROM)* — CROM is the user environment vital to a DMS and is an umbrella function covering the facilities provided to the operator in the control room through the operator's console (HMI).† The following are typical CRM functions:

- Control room graphics system (CRGS) for network diagram display
- Interface to SCADA (in fully integrated systems, traditional SCADA is expanded to provide the CROM function)
- Switching job management

---

* Sometimes referred to as a distribution operations monitor (DOM).
† Human-Machine Interface.

- Access to advanced applications (ADVAPPS) including trouble call (TCMS) or outage management
- Interface to the data engineering application for DMS data modifications and input from enterprise IT systems (EIT)

The foundation of the DMS is the MV/LV network connectivity database, which is assumed to be part of the CROM because it has very little real-time element. This is a pictorial and data representation displayed through the control room graphics system forming the human-machine interface for the system in the form of a universal operator's console. Full graphics, windowing, and multi-function platforms support access to all functions under the control of operators with different authority levels within the DMS. It has the editing capability to allow maintenance of the control room diagram and MV network connectivity database. The current "as operated" state and also the facility to effect incremental update to the normal "as built" state are prerequisites. Full tagging, topology analysis, and safety checks must be supported through dynamic coloring and tracing. Having stated that CROM includes the distribution network model, the actual connectivity model may reside in any one of the other three key functions depending on how the DMS is configured. The types of graphic displays implemented within the CROM function vary to meet individual customer's requirements. Often, the physical displays used in the control room prior to implementing a DMS are repeated. For example, utilities using vast wall diagrams representing their entire network in geo-schematic form require world maps with excellent navigation features (pan, zoom, and locate), whereas those using multiple feeder maps reproduce them as a set of pages. The trend, however, is to use a combination of representations in a continuous world map form (pure geographic, geo-schematic, orthogonal schematic operating diagrams, or a combination of any) and individual pages for substations.

(2) Supervisory control and data acquisition — This provides the monitoring and control of the distribution system in real-time. Traditional SCADA extends down to the HV/MV distribution substation (primary substation) MV feeder circuit breaker with control room displays limited to substation single schematics. Under the concept of a DMS, traditional SCADA is being extended to include representation of the entire MV network in the form of a connectivity model and control of feeder devices outside substations (FA). The foundation of a SCADA system is the data acquisition system for gathering data from remote locations and the central real-time database that is the repository of this data to be processed and displayed for the operator's use.

(3) Advanced applications — Analytical applications that rely on the MV connectivity database provide the operator with a means to evaluate in real-time and study the loading and voltage conditions (load flow) of the system in advance of a switching sequence. The consequences of any network configuration on fault levels (short circuit) can also be determined with basic applications familiar to planning engineers. The potential for applying advanced applications to other problems is considerable, such as the use of expert systems to determine preferred

Central Control and Management

restoration sequences. Fast optimization and search techniques hold the key to developing the best system reconfiguration for minimum losses and supply restoration. As privatization emphasizes the business issues, applications that concentrate on meeting the contract constraints of the network business within the engineering limits will be required. The network model is again fundamental to these applications, and if the model is not held within ADVAPPS, it relies on a synchronized copy from either the SCADA or outage management (OM) functions. These applications are considered decision support tools.

(4) Outage management — Outage management spans a number of functions and can encompass the entire process from taking a customer's call, diagnosing the fault location, assigning and dispatching the crew to confirm and repair the fault (job management), preparing and executing switching operations to restore operations, and closing the outage by completing all required reports and statistics on the incident. During this process, additional trouble calls from customers should be coordinated with the declared fault if appropriate or another incident is initiated. When the call-taking function is included, the term *trouble call management* is often used. TCM systems have been implemented as stand-alone applications without interfaces to SCADA when the CROM has been implemented in advance, because it relies on the CRGS and the MV network model database. A DMS with full TCMS implementation implies the idea of a consumer-oriented DMS. Some of the OM functions can be regarded as included in the ADVAPPS area, because it relies on fast network topology and network analysis.

The box marked "Data engineering" in Figure 2.5 represents a vital component of any DMS. This activity populates the required data into the DMS. It must have either stand-alone functionality to populate data for the real-time and advanced applications together with supporting data requirements for the graphic displays or an interface to accept as-built data from a GIS. Data from the latter source have to be augmented with additional data required by the real-time system (SCADA).

This definition of DMS functionality into minimum stand-alone modules allows assembly of different DMS configurations that have been implemented by the industry, each having varying capability, each expandable in stages to a full implementation. The key to all functions is that a connectivity model must be resident in the first module to be implemented and that the model must be capable of supporting the performance requirements of successive functions.

Typical combinations of these key functions that can be found today in the industry are shown in Figure 2.6.

As an example, utilities with recently upgraded traditional SCADA systems are implementing new control room management systems to improve the efficiency of operations of the MV network. Those under pressure to improve their image to customers are adding the trouble call management function, often as a stand-alone system loosely interfaced to existing SCADA. The combinations are many and varied, being purely dependent on the function(s) for which the utility is able to develop an acceptable business case.

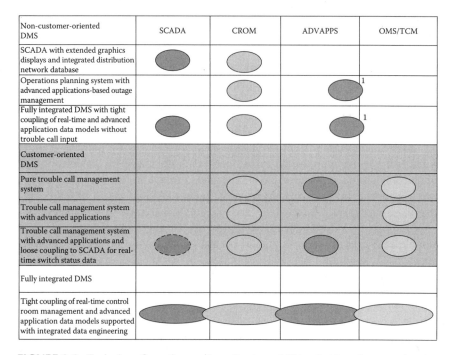

FIGURE 2.6 Typical configurations and resultant capabilities that have been used under the heading of distribution management systems (Note 1 indicates that the outage management function is performed within the advanced applications distribution network model without direct input of trouble calls).

It can be seen that the real-time requirements and those of manual operation must be carefully accommodated to allow seamless navigation between the functions by the operator. The ADVAPPS should be able to straddle both environments on demand and use both real-time and demand-time (trouble calls) data to maximize the quality of decisions.

It is evident that as the full DMS functionality is implemented, overlaps between applications within the functions will occur; thus, the architecture of the ultimate DMS must achieve a near seamless integration of the following:

- Operation of the real-time and manual controlled portions of the network considering
  - SCADA
  - Crew and job management
  - Switching scheduling and planning
  - Operating diagram maintenance and dressing (notes and tagging)
  - Economic deployment of network resources
  - Temporary and permanent changes to the network
  - Introduction of new asset and plant on the network

Central Control and Management

- Timely synchronization of as-operated and as-built network facility databases
- Data sources outside the control room such as GIS, personnel, work management, trouble call taking, and personnel (crew) management systems
- Asset management systems
- Common data engineering for all portions and components of the network within the DMS

The four main components of a DMS will now be treated in more detail in the remainder of this chapter.

2.6 BASIS OF A REAL-TIME CONTROL SYSTEM (SCADA)

The basis of any real-time control is the SCADA system, which acquires data from different sources, preprocesses, it and stores it in a database accessible to different users and applications. Modern SCADA systems are configured around the following standard base functions:

- Data acquisition
- Monitoring and event processing
- Control
- Data storage archiving and analysis
- Application-specific decision support
- Reporting

2.6.1 DATA ACQUISITION

Basic information describing the operating state of the power network is passed to the SCADA system. This is collected automatically by equipment in various substations and devices, manually input by the operator to reflect the state of any manual operation of nonautomated devices by field crews, or calculated. In all cases, the information is treated in the same way. This information is categorized as

- Status indications
- Measured values
- Energy values

The status of switching devices and alarm signals are represented by status indications. These indications are contact closings connected to digital input boards of the remote communication device* and are normally either single or

---

* Remote terminal device (RTU), communicating protective device, or substation automation system.

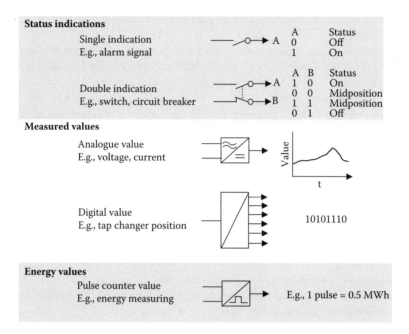

FIGURE 2.7 Examples of acquired data types. (Courtesy of ABB.)

double indications (Figure 2.7). Simple alarms are represented by single status indications, whereas all switches and two-state devices have double indication. One bit represents the close contact, and the other bit the open contact. This permits the detection of false and intermediate values (00 or 11 state), which would be reflected by a stuck or incomplete switch operation, resulting in a maloperation alarm. Also, errors in the monitoring circuits will be detected.

Measured values reflect different time varying quantities, such as voltage, current, temperature, and tap changer positions, which are collected from the power system. They fall into two basic types, analog and digital. All analog signals are transformed via an A/D converter to binary format; because they are treated as momentary values, they have to be normalized before storing in the SCADA database. The scanning (polling) of metered values is done cyclically or by only sending changed values respecting deadbands (report by exception) and recorded on a change-of-value philosophy. Digitally coded values are typical of different settings such as tap changer positions and health checks from IEDs.*

Energy values are usually obtained from pulse counters or IEDs. RTUs associated with pulse meters are instructed to send the pulse information at predefined demand intervals or, if required, intermediate points. At the prespecified time interval, the contents of the continuous counter for the time period is passed on and the process repeated for the next interval.

---

* IED — intelligent electronic device.

Central Control and Management 41

Different arrangements for data gathering systems are described in more detail in Section 2.6.5 and Section 2.6.6 as part of SCADA system configuration and polling design principles.

2.6.2 Monitoring and Event Processing

The collection and storage of data by itself yields little information; thus, an important function established within all SCADA systems is the ability to monitor all data presented against normal values and limits. The purpose of data monitoring varies for the different types of data collected and the requirements of individual data points in the system. Particularly if it is a status indication change or limit violation, it will require an event to be processed.

Status monitoring requires that each indication be compared with the previous value stored in the database. Any change generates an event that notifies the operator. To expand the information content, status indications are assigned a normal condition, thus triggering a different alarm with an out of normal condition message. Status indication changes can be delayed to allow for the operating times of primary devices to avoid unnecessary alarm messaging.

Limit value monitoring applies to each measured value. When the status changes, an event is generated, but for this to happen, the change must exceed some limit value. Different limit values with associated deadbands (Figure 2.8) can be set above and below the normal, each limit being used to signify different levels of severity producing a corresponding alarm category. Deadbands associated with the measuring device around each limit prevent small fluctuations activating an event. Also, they reduce the transmission traffic under report-by-exception because the RTU signal is blocked unless the parameter variation is greater than the deadband from the previous measurement. Deadbands can be specified at individual collection points (RTU) for each measured value. A delay function similar to that used in status monitoring is also implemented. A good practical example for the use of limit deadbands is to avoid extensive alarming caused by waves in a hydro reservoir when the water level is on the limit value.

In order for complex power system disturbances to be correctly analyzed, a very accurate time-stamping of events is necessary. Some RTUs have the availability to time-stamp events down to millisecond level and send information with this time-stamp to the SCADA master. In this case, it is a necessity that all RTU clocks are synchronized with the SCADA master, which in turn must be synchronized with a standard time clock. This type of data forms the sequence of events (SOE) list.

Trend monitoring is another monitoring method used in SCADA systems. It is used to trigger an alarm if some quantity is changing in magnitude either too quickly or in the wrong direction for satisfactory operation of the device or network (e.g., a rise in voltage by, say, 7% in a minute may indicate an out-of-control tap changer).

The need to continually provide the operator with information among a multitude of collected data has resulted in the idea of applying quality attributes

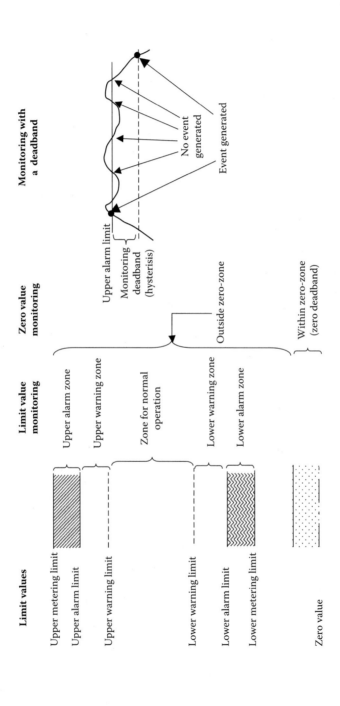

FIGURE 2.8 A typical limit diagram for a monitored quantity, like voltage, that should remain within certain bounds above zero, and the concept of deadband to limit event generation from small fluctuations of the quantity. (Courtesy of ABB.)

Central Control and Management

to data, which in turn invokes a method of flagging the data either in a particular color or symbol in the operator's display console. The following are typical attributes:

- Nonupdated/updated — data acquisition/manual/calculated
- Manual
- Calculated
- Blocked for updating
- Blocked for event processing
- Blocked for remote control
- Normal/non-normal state
- Out-of-limit, reasonable/alarm/warning/zero
- Alarm state
- Unacknowledged

Event processing is required for all events generated by the monitoring function or caused by operator actions. This processing classifies and groups events so that the appropriate information can be sent to the various HMI functions to represent the criticality of the alarm to the operator. Event processing is a crucial function within the control system and significantly influences the real-time performance, particularly during alarm bursts. The result of event processing is event and alarm lists in chronological order. In order to assist the operator, events are classified into a number of categories, the most significant being alarms, which generate an alarm list. The following categories are the most usual:

- Unacknowledged and persistent alarm categories determine a particular alert on the display such as flashing of the color presentation, and in some cases an audible signal is generated. The unacknowledged alarm remains until operator acknowledgment is made. The persistent alarm category remains until the state disappears (usually through operator action) or is inhibited.
- An event associated with a particular device type in which an attribute is assigned for each data point such as a bus voltage or relay protection operation.
- Reason for the event occurring by assignment to the monitoring function (e.g., spontaneous tripping of a circuit breaker or recloser, a manual or control command).
- A priority assigned for ranking all events into different priority groups often determined by combining the device type and the reason for the event.

The whole purpose of these classifications is to filter important events from less-important events, so in times of multiple activities, the operator is assisted in resolving the most important issues first.

2.6.3 Control Functions

Control functions are initiated by the operators or automatically from software applications and directly affect power system operation. They can be grouped into four subclasses.

Individual device control, which represents the direct open/close command to an individual device.

Control messages to regulating equipment that requires the operation, once initiated by the control room, to automatically be conducted by local logic at the device to ensure operation remains within predetermined limits. Raising or lowering tap changer taps is a typical example or sending of new set points to power generators.

Sequential control covers the automatic completion of a linked set of control actions once the sequence start command has been initiated. A set of sequential switching steps to restore power through a predefined backup configuration typifies sequential control.

Automatic control is triggered by an event or specific time that invokes the control action. Automatic control of voltage through on load tap changing responding automatically to the voltage set point violation is a common example. Time switched capacitor banks are another.

The first three control categories above are initiated manually except when sequential control is initiated automatically. Manually initiated control actions can be either always on a select-confirm-before-operate basis or immediate command.

2.6.4 Data Storage, Archiving, and Analysis

As stated earlier, data collected from the process are stored in the real-time database within the SCADA application server to create an up-to-date image of the supervised process. The data from RTUs are stored at the time received, and any data update overwrites old values with new ones.

Performance statistics captured by SCADA systems are extremely important in supplying customers and the regulator with actual figures on power quality of segments of the network as well as the network as a whole. The stored sequence of events (SOE) list provides the basis for developing these statistics.

This time tagged data (TTD) is stored in the historical database at cyclic intervals, e.g., scan rates, every 10 seconds or every hour. Normally, only changed data are stored to save disk space. Data can be extracted at a later date for many forms of analysis such as planning, numerical calculations, system loading and performance audits and report production.

Post-mortem review (PMR) is another important area and is usually performed soon after an interruption or at a later date using the historical database. To facilitate PMR, the data are collected by making cyclic recording of either selected sets of values within the PMR group or by recording all data. This segregation of data allows each PMR group to be assigned the appropriate collection cycle time and

Central Control and Management

associate it with the interruption cause event, making it possible to "freeze" the associated data before and after the interruption event for later analysis.

This requirement for data mining is driving more sophisticated data archiving functions with adaptable ways to select data and events to be stored. These historians, utility data warehouses, or information storage and retrieval (ISR) systems with full redundancy and flexible retrieval facilities are now an integral part of any DMS. They are normally based on commercial relational databases like Oracle.

2.6.5 Hardware System Configurations

SCADA systems are implemented on hardware comprised of a multichanneled communications front end that manages the data acquisition process from the RTUs. This traditionally has been achieved by repeated polling of RTUs at short intervals (typically, every 2 seconds). The data received are then passed to the SCADA server, over a local area network,* for storage and access by operators and other applications. Control is invoked through operator consoles supporting the HMI command structure and graphic displays. The mission criticality of SCADA systems demands that redundancy is incorporated, thus hot standby front ends and application servers based on dual LAN configurations are standard. The general configuration of a typical SCADA system is shown in Figure 2.9.

The front ends support efficient communications arrangements over a wide area network† to the RTUs, for the collection of process data and the transmission of control commands that can be optimized for both security and cost. Communication front ends support a variety of configurations. The most popular in use today are as follows:

- Multidrop is a radial configuration where RTUs are polled in sequence over one communications channel. This results in a cheaper solution at the expense of response time.
- Point-to-point dedicates one communication channel to one RTU. It is commonly used for either major substations or data concentrators having RTUs with large I/O requirements. This configuration gives high response levels with the added expense of many communication channels. In applications requiring very high reliability, an additional communication path is added to form a redundant line point-to-point scheme.
- Loop operates in an open loop configuration supplied from two communication front ends, each channel being of the multidrop type. The advantage is one of reliability, because the loss of any communication

---

* LAN.
† WAN.

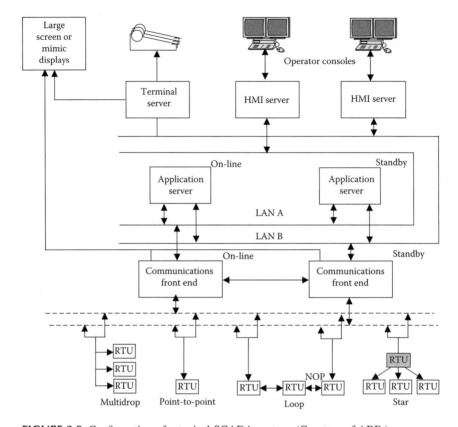

FIGURE 2.9 Configuration of a typical SCADA system. (Courtesy of ABB.)

segment path can be overcome by switching the normally open point (NOP).
- Star configuration is a combination of point-to-point to data concentrator RTU, which controls data access to slave RTUs configured as point-to-point or multidrop. Such configurations are used in distribution automation where a mixed response time can be economically engineered.

The central system comprising of everything above the communications front end is called the "master station." In the industry, there exist many communications protocols and their variants in use between the master station and the RTUs. Most protocols* are based on centralized polling of the RTUs.

---

* See Chapter 6, Communications.

Central Control and Management	47

2.6.6 SCADA SYSTEM PRINCIPLES

It must be understood that performance has been at the heart of all SCADA systems due to the historical limitation of communication speeds. The slow speed affects the data acquisition function, which has formed the foundation of system architecture of all traditional SCADA systems for distributed processes. Because of this traditionally very limited bandwidth of data transmission, the whole design of two asynchronous and independent cycles of data processing in SCADA has been formed. It would be impossible for the SCADA applications to receive the information from metered values and indications with reasonable response times directly in the substation when only 50 baud (bits per second) was available for communication. Therefore, the data acquisition cycle collects data as fast as the communication allows and mirrors the state of the process into a real-time database. Presentation of process state to the operators is made from this database as described in an earlier section. All applications work on this mirrored image of the process, thus being totally independent of the data acquisition function.

A particularity of the slow baud rates and high security demands is that RTU protocols were designed with very special features. At very low speed, every single bit counts, and therefore bit-synchronous protocols were the norm in the beginning of remote control. Some of these old protocols still survive and have to be interfaced in new systems today. The disadvantage of bit-synchronous protocols is that they require special hardware and special interpretation routines. In modern systems and in new protocol standards, byte-oriented protocols are used. These protocols have more overhead (more frame bits per "true" information bits), but they can be handled with normal (and much cheaper) line cards and modems.

The same basic design that formed SCADA from the beginning holds true today, even if much higher bandwidths based on modern PLC and fiber-optic cables are available. The only difference today is that the mirrored image of the process is much "closer" (in time) to the real process. It is now unusual to find point-to-point communication to the substations with lower than 2400 baud, and an increasing use of wide area network communication with much higher communication speeds has taken place.

In contrast, an industrial control system has the completely opposite design compared to that for network control, because of the limited geographical distribution of industrial processes. In this case, it is usually no problem to obtain high-bandwidth communication all the way down to the controllers and measurement points in the process. The central database is only a complication for the presentation functions and the higher-level applications because they access the process data directly from the measurement points — from the process itself.

Based on the above discussion and for historical reasons, it is easy to understand that specifications have concentrated on the calculation of data acquisition response times and bandwidth requirements. Such assessment will demonstrate how accurately the system is able to mirror the process and how precise the result of the higher-level applications will be, taking into account the configuration of the data acquisition* system. The data acquisition can be configured in a number of ways, which will be described in the following sections.

2.6.7 Polling Principles

Two main types of polling of RTUs are found in network control systems: cyclic and report-by-exception.

Cyclic. The measurands and indications are allocated to different polling cycles (scan rate), typically on a number of seconds level, e.g., every 2–4 seconds for high-priority data and each 10 seconds for less-important points. The front end will request information from the RTUs in these cycles, and the RTUs will answer with all data allocated to this level. The central SCADA system will check if the data are changed from the last cycle and, if so, update the real-time database and start other dependent applications. The communication channel is idle between polling cycles.

Commands and set points are sent from the SCADA servers when requested by the operators. If command requests are given high priority; they will interrupt any sending of telemetered data from the RTUs, otherwise, the sending of commands will have to wait until the line is free.

Cyclic polling will, of course, give very consistent response times that are independent of how much the process really changes; i.e., the data acquisition response times are always, even during big disturbances, the same as under normal conditions. The SCADA servers will be more heavily loaded during disturbances because change detection and all event activation takes place in SCADA for cyclic polling schemes.

Report-by-Exception (RBE). In the report-by-exception principle, the RTU only sends information when a telemetered value has changed (for measurands over a deadband). The front end polls the RTUs continuously, and the RTU will answer either with an empty acknowledgment when no data are available or with data if a data point has changed. Because not all data points are sent, in each telegram, data points under report-by-exception schemes require identifiers.

In report-by-exception schemes, the line is immediately polled again after the RTU has answered. This means that the communication line is always 100% loaded. With higher communication speeds, the lines are still 100% loaded, but the response times are improved.

Commands and set points will be sent on request from the SCADA system as soon as a new poll is initiated; i.e., the command replaces the next poll. This means that commands and set points are sent more consistently in a report-by-exception system because the waiting time for ongoing inward information is shorter.

Priority schemes can also be applied by report-by-exception. Objects can be allocated different priorities depending on their importance. Normally, indications and very important telemetered values, e.g., frequency measurements for automatic generation control (AGC) and protection operation, are allocated to priority

---

* As described in sections 2.6.5 and 2.6.6.

1, normal telemetered values like active and reactive power flows to priority 2, and sequence of events data to priority 3. The polling scheme is designed to first request highest priority information on all RTUs before priority 2 is requested and so on. This is important in multidrop configurations in order to achieve good response times for important information, e.g., status for breakers, from all RTUs on the same line.

Report-by-exception polling gives much faster response times for telemetered information in almost all circumstances. During high-disturbance situations, the report-by-exception will be marginally slower than cyclic polling because more overhead in the telegrams is required to identify the data.

Unbalanced polling is when the polling request, i.e., the initiative for communication, always comes from the Front-End computers — the master. With balanced protocol, the RTU can send a request to the front end to poll when something has changed in the RTU. Balanced protocols are typically used in networks with low change rates, many small RTUs, and low requirement on response times, e.g., in distribution medium-voltage networks, for Feeder Automation, and with dial-up connections.

Some protocols allow the downloading of settings to RTUs, thus avoiding the need to visit the site to conduct such modifications. Downline loading is normally handled in the file transfer part of the protocol. The format of the downloaded file is vendor specific and no standardization is proposed for these parts.

Use of Wide Area Networks for Data Acquisition. There is now a clear trend in SCADA implementations for the use of wide area network (WAN) and TCP/IP communication for data acquisition from RTUs. The communication principle applied on the WAN is packet switched communication. Standard RTU protocols based on TCP/IP have been defined, e.g., IEC 60870-5-104 for these types of networks. The reason for this trend is that the customers are installing much more communication capacity all the way to the substations, e.g., by installing fiber optics in the power line towers and direct laid with new cable networks. Packet-switched technologies use this additional communication capacity more efficiently, and the spare capacity can be used for many other purposes by the utilities, e.g., selling telecommunication services.

In a packet-switched network, the communication routes are not fixed. The individual packets of data search for the best possible communication path. This means that different packages can take different routes even if they belong to the same logical telegram, and the complete information is only assembled at the receiving end of the communication. Precise response times in such networks are not possible to define. However, with stable communication and enough spare capacity on the WAN, the response times will be sufficient for all practical purposes.

One result of packet-switched communication is that time synchronization of RTUs over the communication lines will not be accurate because of the unpredictability of transmission times. Time synchronization of RTUs in these types of networks is normally done locally, usually with GPS.

Sending of set points for closed loop control applications might also be a problem in packet-switched networks. The closed loop regulation characteristics will be influenced when the delay times vary. In these applications, special consideration has to be taken to keep data sending and receiving times constant, e.g., by defining a certain number of predefined and fixed spare routes in the network that the communication can be switched between if communication problems occur on the normal route, e.g., through the breakdown of a communication node.

WAN communication does not require front ends and uses a standard commercial router directly connected to the (redundant) LAN of the control center. The router will put the data together based on the received packages and send the RTU telegrams to the SCADA servers. The same principles for polling apply to WAN as for point-to-point connections.

2.7 OUTAGE MANAGEMENT

Outage management is one of the most crucial processes in the operation of the distribution network, having the goal to return the network from the emergency state back to normal. This process involves three discrete phases:

1. Outage alert
2. Fault location
3. Fault isolation and supply restoration

Various methods have been developed to assist the operator and depend on the type of data available to drive this process (Figure 2.10).

Utilities with very limited penetration of real-time control (low AIL) but good customer and network records use a trouble call approach, whereas those with good real-time systems and extended control are able to use direct measurements from automated devices. The former solution is prevalent in the United States for primary networks (medium voltage) where distribution primary substations are smaller. Except for large downtown networks, the low-voltage (secondary) feeder system is limited with, on the average, between 6 and 10 customers being supplied from one distribution transformer. This system structure makes it easier to establish the customer-network link, a necessity for trouble call management systems if outage management is to yield any realistic results. In contrast, European systems with very extensive secondary systems (up to 400 consumers per distribution transformer) concentrate on implementing SCADA control; thus, any MV fault would be cleared and knowledge of the affected feeder known before any customer calls could be correlated. In this environment, to be truly effective, a trouble call approach would have to operate from the LV system, where establishing the customer network link is more challenging. In these cases, trouble call response was aimed at maintaining customer relations as a priority over fault location, which is achieved faster through a combination of system monitoring applications (SCADA, FA, and FPIs*) and advanced applications.

---

* Fault passage indicators.

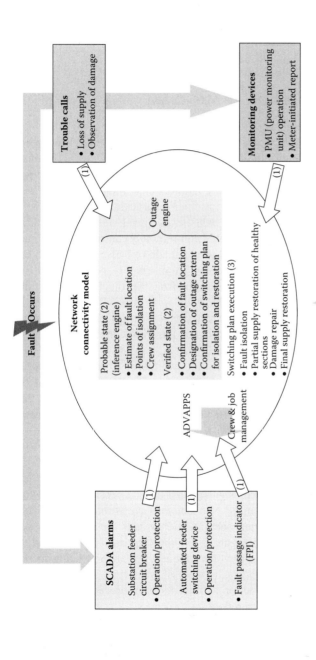

FIGURE 2.10 Outage management flowchart.

DMSs are now combining the best of these approaches to achieve real-time solutions on the network with customer-oriented feedback. It is, though, worthwhile to describe the principles of the various methods to understand how their combination can improve the overall solution. The fault location and restoration function of outage management can be both a local solution (reclosers and auto sectionalizers) or a centrally based process. The latter will be described here, with the local approach being covered in later chapters on feeder automation.

2.7.1 Trouble Call-Based Outage Management

Trouble call-based outage management was the first approach for including customer information in network operations. Through an IT approach, it improved operations where SCADA was limited to large substations and was effectively non-existent in the distribution network. Limited SCADA provided the operator with little information of the actual network failure, unless seen by the substation protection, until a customer called the customer information department complaining of loss of supply. Loss of supply, whether as a result of a known SCADA operation or the operation of a self-protected non-remotely controlled device such as a fuse or recloser, frequently resulted in a cascade of calls, all of which had to be managed for high customer satisfaction. Trouble call management systems are designed by extracting maximum information from the call itself, to determine the fault location; to providing the caller with up to date information on the outage; to monitoring the progress of the restoration process and finally to maintain statistics per customer of quality of supply ensuring correct assessment of penalties. The whole process is shown in Figure 2.11.

Trouble call systems must be designed to have fast response during storms when call bursts result as the first storm effects are felt by customers. As the number of circuits affected rises, the tendency is for the calls to peak and then fall off, even though the number of affected circuits has not peaked, because customers start to realize that the storm has had its maximum effect and will pass. However, sustained outage duration will result in delayed bursts as customers start to lose patience. The call history versus affected circuits profile of a typical storm is shown in Figure 2.12.

Fault Alert. The first trouble call signifies that there is potentially a network failure; however, in some cases it may be an isolated fault within the customer's premises. This is quickly confirmed once additional calls are received. A call entry screen of a typical TCM, in Figure 2.13 shows the data captured and the customer relation activities, such as a callback request that are now being accommodated in modern systems.

Fault Location. The determination of fault location proceeds through two steps, inference and verification, to reflect the two possible states of an outage. The core of the process is often called the outage engine. It automatically maintains the status of different outages and the customers (loads) associated with each outage state by processing line device status and trouble groups. The method relies on a radial connectivity model of the network, which includes a

Central Control and Management

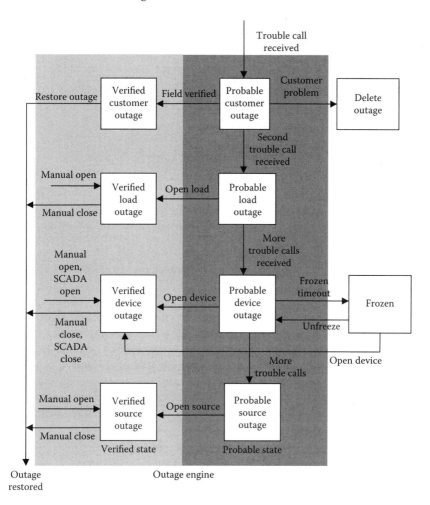

FIGURE 2.11 Trouble call management sequence of events from first call entry to restoration. (Courtesy of ABB.)

customer-network link pointing every customer in the CIS to a location on the network. As mentioned previously, in U.S.-type distribution systems, associating customers with a distribution transformer is less complex than in European systems, where the secondary (LV) networks are more complex and the number of customers per transformer greater. Various hybrid assignment methods (such as postal code) in addition to GIS* methods have been used to check early mains records. An outage is defined as the location of an operated protection device or

---

* GIS methods refer to classical geo-surveys and mapping. In urban underground networks established before GIS implementations, rastering of paper mains records and accepting their accuracy sometimes was the only cost-acceptable method, and thus additional public domain data was included to improve or verify location.

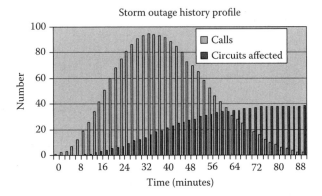

FIGURE 2.12 Duration profiles of the number of outage trouble calls and affected circuits for a typical storm.

open conductor and the extent of the network de-energized as a result of the operation including the affected consumers.

Probable Outages. Most outage engines analyze trouble calls that are not associated with a known or verified outage, and group them into probable outages. During each pass, all new trouble calls and all trouble calls previously grouped into probable outages (which are not yet assigned to verified outages) are noted. This new set of trouble calls is used to trace the network to infer a new set of probable locations of outages. New probable outages are identified, previously probable outages moved, and previously probable outages deleted, so that there is a probable outage at each location in the new set of probable locations. A location is a protective device that could have operated.

Typical algorithms predicting outages use a "depth first" algorithm with post-order traversal of the feeder network graph. Empirical rules are developed to cater for branching of the radial network, and the branch may include protective devices. For example at each node of the network, the ratio of trouble calls to the number of customers is calculated then compared to the accumulated ratio for all nodes in the subgraph below that node. Lists of trouble calls and probable outages below the outage at this location are calculated. After visiting a node, a decision is made whether or not to create a probable outage at the node or to defer the decision until concluding examination of an upstream node. The following rules are typical for deterministic inference:

- Each trouble call must be assigned to some probable outage.
- Before an outage is created, at least two loads must have trouble reports (customer calls).
- An outage should be placed at the lowest protection device or load in the network, above the point where the ratio of reports to customers is below a defined threshold.
- Determination of logical "AND" together or "OR" together results of affected load counts and percent of customer calls.

Central Control and Management

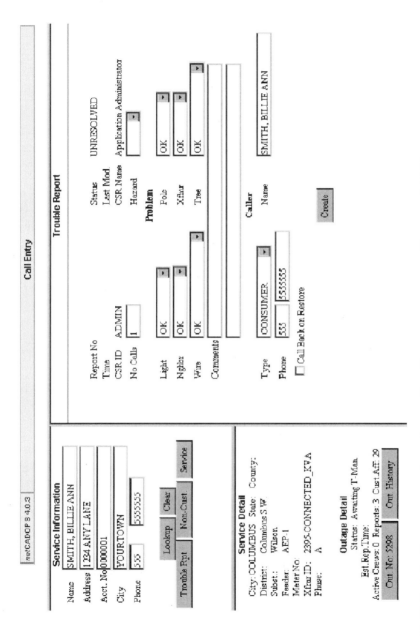

FIGURE 2.13 Typical call entry screen of a modern trouble call management system. (Courtesy of ABB.)

The trouble call approach is used predominantly in networks where SCADA deployment is very limited, with the consequence that the system is not observed. Methods using fuzzy logic are now being investigated to extend outage determination accuracy. Traditional methods predominately rely on calls and network topology to find a common upstream device. The proposed fuzzy logic method extends the input information to include protection selectivity. A fuzzy logic-based decision process screens this information [6]. The process combines fuzzy sets that reflect equivalent logical operations. For example, a feeder with two branches will have a fuzzy set "A" for branch "a" reflecting the devices that could operate for a call at the end of the branch.

$$A_a = \{(x_{ia}, \mu_{ia}(x_{ia}))/x_{ia}\},$$

where upstream devices $\alpha = x_{ia}$ with set membership values $\mu_{ia}(x_{ia}) = i_a/i_a$.

A second set is formulated reflecting branch "b" and each set if combined using the Hamacher principal to account for the common devices in each set upstream of the branch point to form one fuzzy logic-based decision algorithm for the feeder. Comparison of the fuzzy logic-based extension to the traditional approach outage detection (OD) using Monte Carlo simulation suggests dramatic improvements on the prediction of outage location for multiple faults.

	Fuzzy OD (% Accuracy)	Traditional OD (% Accuracy)
Case A (single fault)	98.6	99.1
Case B (multiple fault)	83.7	3.7
Case C (multiple fault)	80.4	2.8

Fast and accurate prediction of the location of a probable outage is vital because it is used to direct the field crews to confirm the fault location and points of isolation. The faster the crew reaches and confirms the extent of the damage, the more efficient the restoration process.

Verified Outages. An outage is verified by confirmation by the field crew of the operation of a manually controlled switch or open circuit or the notification of a SCADA operation. Once the outage is verified, the outage engine repeatedly analyzes switching events and other connectivity changes (phased restoration) to update customers associated with the outage. Different events in this category are typically processed as follows:

- For "open" operations at a probable outage, the outage is verified.
- For "open" operations where there is no probable outage, a new verified outage is created.
- For "open" operations to a de-energized device, no outages are created or modified.
- For "close" operations at a verified outage, the outage is removed.

Central Control and Management 57

- For "close" operations that energize de-energized loads, the outage is partially or totally removed.
- For "close" operations of a de-energized device, no outages are modified.

A "close" operation from SCADA is only accepted after a certain time delay to allow for closing onto a fault resulting in a new trip.

Supply Restoration. Emergency switching plans and field crew actions are iteratively undertaken to isolate and then restore supply. Supply restoration is often partial where normally open points or alternate feeds are used to back feed to the healthy parts of the system isolated from the primary supply by the fault. As the manual actions are completed, the operator, having received confirmation from the field, enters the connectivity changes into the DMS (TCM) OMS network model. The outage engine automatically keeps track of the changes and the event. Thus at all times the utilities call taker is aware of the customers still without supply and the prognosis of the situation.

2.7.2 Advanced Application-Based Outage Management

As different outage management methods use more and varied input data for decision making, the difference becomes one of emphasis and origin. Although the method described in the preceding section is able to benefit from the use of SCADA data, the emphasis is still on an IT-based solution. The advanced application-based method, in contrast, is triggered directly by real-time input from the operation of SCADA data collection devices, with trouble calls adding additional information after the event. The information from the measurement devices, usually a protective relay or FPI, is delivered directly to the topology engine within the real-time system network model and is used to determine the fault location. Trouble calls usually add little extra information for a highly automated network. This fault location, isolation and supply restoration (FLIR) divides the process into three discrete but linked stages:

- Fault localization (FL)
- Fault isolation (I)
- Service restoration (SR)

The generalized nature of the FLIR function enables it to operate across many voltage levels should the distribution system represented in the DMS be an integrated multivoltage level network model.

Fault Localization. Faults on the distribution network are sensed by protection and isolated by circuit breakers installed in primary (source) substations. With the exception of reclosers, there is seldom any other primary switchgear installed on the feeder network capable of breaking fault currents;* therefore,

---

* Fuses at distribution transformers or on single phase laterals will clear faults but will not provide information directly to the control room. If fitted with blown fuse indication, they will assist in fault location.

there is little other information to guide in the location of the fault down the feeder. The only exception is the application of ring main units with vacuum circuit breakers, including integral protection, which are now becoming cost-effective. The implementation of feeder automation, either in the form of FPIs associated with remotely controlled line switches or as communicating FPIs, immediately provides the chance for improved resolution of the fault location. Nondirectional FPIs are the most common and are applied to radial circuits, whereas the directional varieties are used on meshed networks. FPIs are available for short circuit and ground (earth) fault application. Central fault localization methods in operation today use two data sources: (1) the operation of non-directional and directional FPI and (2) information from communicating protective relays (distance or overcurrent). The first pass of the analysis is to determine the area of the network below (radial) the FPI that operated farthest from the source yet above the next FPI that has not operated or the network between a number of directional FPIs that have operated (mesh). In general, the analysis divides the network and determines subnets containing FPIs that have not operated and those that have. The former are usually termed "faultless," whereas the other areas are classified as "in fault." The localization resolution is limited to the density of FPI deployment but can be enhanced in the presence of certain protective relays by using network circuit parameters and protective relay values to determine the distance to the fault from the protective device location. Although this calculation relies on the accuracy of the network model impedance data, it often enables distinction between alternative branches of the network that lie in the "in fault" subnet. It also allows calculations to be limited to the "in fault" or "fault suspicious" subnet.

Fault Isolation. Fault isolation requires fast determination of the best points of isolation knowing the location of the fault. In a radial network, this determination is fairly trivial, requiring the selection of the nearest upstream device from the faulted section. In a meshed environment, the task is to determine those switching devices that form the boundaries of the portions of the network classified as "faultless." In this process, advanced algorithms should distinguish between manual and remotely controlled switches, including whether any device has restrictions. Devices with operating restrictions should be eliminated from consideration. These algorithms determine the necessary switching sequence for isolation and usually present the suggested switching plan for operator approval and execution.

Service Restoration. Open switches are identified that can be closed to restore supply to the isolated subnet. In the case of radial systems, the closing of the normally open point (if one exists) will be necessary to restore supply beyond the lowest point of isolation. The ability to test the feasibility of supplying load from an alternate feed by running a load flow calculation between restoration steps is a function available in advanced restoration algorithms. The final restoration sequence usually reverses the isolation steps, but in doing so the switching policies of operation must be met. As in the isolation function, most systems

Central Control and Management

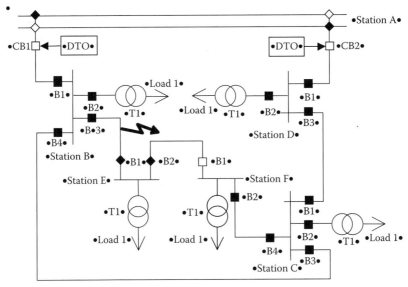

FIGURE 2.14 Sample FLIR analysis. (Courtesy of ABB.)

present the operator with a recommended sequence for approval and implementation, which is confirmed one step at a time.

FLIR Example. The operation of the isolation and restoration functions of a typical FLIR demonstrated by the following example using a combined radial and mesh network, shown in Figure 2.14.

A fault occurs on the line between stations B and E, which is cleared by overcurrent protection operating circuit breakers CB1 and CB2 in station A. The information from FPIs distributed in the network allows the location of the fault to be confirmed on the line between stations B and E. The fault isolation function determines that the fault is in the radial circuit supplied from station B switch B3 down to NOP B2 in station E. Switch B3 was determined as the favored remote-controlled switch for fault isolation and switch B1 of station E as the preferred manually operated switch. Once the fault is isolated by opening remote-controlled switch B3, operation of the restoration function, once applied, would suggest a two-stage restoration. First, restoring supply by closing CB1 and then, after field crews had verified the outage and opened manually operated switch B1 (station E), they could proceed to close manually operated NOP B2 (station E) in preparation for closing remote-controlled switch B1 (station F) to complete the second and final restoration step in reconnecting customers supplied from station E.

2.7.3 GIS-Centric versus SCADA-Centric

There exist two approaches to the deployment of trouble call-based outage management systems:

- GIS-centric
- SCADA-centric

GIS-centric TCM systems are attached more closely to the GIS environment, having a separate connectivity model to achieve satisfactory response times for fault location inferencing from trouble calls. Their HMI is either shared with the GIS or a dedicated facility specially designed for the application when the TCM is a stand-alone system. Any real-time network change has to be passed from the SCADA system to maintain real-time synchronization of the TCM connectivity model with the SCADA model. This is mandatory if the results are to be valid. This type of trouble call management system was the first to be implemented, mostly in North America where little SCADA or feeder automation existed. It provided an IT solution to customer satisfaction and outage management. In Europe, call-taking systems were implemented with basic customer location algorithms linked to the LV networks to improve customer relationship management while avoiding costly LV connectivity models.

A SCADA-centric deployment implies that the TCM application is tightly embedded under the SCADA HMI, network connectivity model and real-time process. Even though the TCMS may use a dedicated simplified connectivity model, this connectivity is continuously updated within the real-time system. Any data changes, whether temporary or permanent, are synchronized by the same data maintenance/data engineering mechanism. This type of deployment is essential where the AIL is high and increasing, thus placing ever-demanding pressure on real-time performance as more real-time device status changes with all their associated alarming occur within the network. This approach also allows full advantage to be taken of the advanced application-based outage management functions. The advantages are now being realized, and the industry is seeing successful implementation of hybrid TCM/advanced application outage management systems.

2.8 DECISION SUPPORT APPLICATIONS

Included in advanced applications, in addition to the FLIR function, are other decision support applications that can be employed independently or as support within the FLIR function. Although these are designed to work in real-time environments, their most frequent mode is as operational planning tools in which

Central Control and Management

the operator plans for contingencies, changes in normal switching configurations, and planned switching actions. When used in the real-time environment, they are usually triggered in one of three optional ways:

- Manually — on operator demand
- Cyclically — in the background at a predefined cycle (say, every 15 minutes)
- Event driven — on a change to the network configuration as the result of a switch operation or significant load change

In order to reduce the computation load in a vast distribution network, the topology engine is set to identify only the parts of the network that have changed since the last calculation. The calculations are then restricted to the changed areas of the network.

2.8.1 Operator Load Flow

This function provides the steady-state solution of the power network for a specific set of network conditions. Network condition covers both circuit configuration and load levels. The latter is estimated for the particular instance through a load calibration process that adjusts static network parameters (as built) with as much real-time data as possible available from SCADA. The calibration methods typically operate in a number of discrete steps. The first step is a static load calibration. Input for this calculation is the static information, like load profiles, number of supplied customers, and season. The results are static values for the active and reactive power consumption. The second step is a topological load calibration. This function uses the static results of the power consumption, the latest measurement values, and the current topology of the network to determine dynamic values for the active and reactive power consumption. The topology engine determines, for meshed network, islands of unmetered load points supplied from metered points, whereas for radial systems, the analysis is a fairly trivial tree search for the upstream metered point. These values are used as pseudo measurements together with the real measurements as input for a state estimation that comprises the third step. This final step adjusts the nonmeasured values and any missing values to represent the state of network loading with losses included.

Fundamental to the load calibration process is the load model employed. Different power utilization devices (lighting, air conditioners, heating load, etc.) exhibit particular load characteristics. These characteristics are represented by one of three types, constant power, constant impedance (lighting), and constant current, each exhibiting a different voltage dependency (Table 2.2).

At any distribution load point such as an MV/LV substation, there is a high probability that different types of load will be connected. These load types may be inherent in the load classification or customer class (commercial, residential, industrial). Thus, all loads can be described as a combination of the three characteristics by assigning a percentage factor to each characteristic.

TABLE 2.2
Typical Nonlinear Load Types Showing Voltage Dependencies

Behavior	Factor Name	Voltage Dependency
Constant power (SCP)	F_{CP}	None
Constant impedance (SCI)	F_{CI}	Quadratic
Constant current (SCC)	F_{CC}	Exponential with exponent P, Q where $0 \leq P, Q \leq 2$

$$P_{load} = P_{CP} + P_{CI} + P_{CC} = P_{load} \times F_{CP} + P_{load} \times F_{CI} + P_{load} \times F_{CC}$$

$$Q_{load} = P_{load} \times (\sqrt{(1-(\cos\varphi)^2)})/\cos\varphi$$
$$= Q_{CP} + Q_{CI} + Q_{CC} = Q_{load} \times F_{CP} + Q_{load} \times F_{CI} + Q_{load} \times F_{CC}$$

The default values are often assigned to these models for both the percentage contribution and the exponential components of the constant current exponent.

In addition to voltage dependency, loads at the distribution level [2] vary over time and are described by daily load profiles. Loads peak at different times, and when combining loads, the difference between the addition of all the individual peaks compared to the peak of the sum of the load profiles is called load diversity. Diversity exists between loads of the same class (intraclass diversity) and between different classes (interclass diversity). The higher up the system that the loads are accumulated, the less the diversity. Thus, to model loads accurately for analysis at different times of the day, the application of load profiles is recommended. Load profiles for typical load classes are developed through load research at an aggregation level of at least 10 loads per load point. Thus, loads can be represented by one class or a combination of classes. Load classes also vary from season to season and from weekday to holidays to weekends.

The load models for operator load flows must be able to represent loads with limited or more expansive information. Three different ways of modeling the active and reactive power values P and Q follow:

Single Point Value Based. Typical load based on installed KVA or other predefined magnitude, say, peak value. A fixed typical load S_{typ} and a power factor $\cos\varphi$ (each defined per individual load) is used to determine the active and the reactive power consumption.

$$P_{load} = S_{typ} \times \cos\varphi \quad \text{and} \quad Q_{load} = S_{typ} \times \sqrt{1-\cos^2\varphi}$$

Central Control and Management

A single point load description does not reflect the time-dependent consumption patterns of loads at the distribution level nor load diversity.

Profile Customer Class Based.* The 24-hour normalized load profiles for different seasons are available for the load point when used in conjunction with the single load value; this allows some representation of diversity and a more accurate value for calibration at times different from that of single load value, thus incorporating time dependency into the load model.

$$P_{load} = S_{typ} \times L_p (season, day\ type, hour) \text{ and}$$

$$Q_{load} = S_{typ} \times L_Q (season, day\ type, hour)$$

This can be extended for the values of the active and the reactive power for each type of voltage dependency (P_{CP}, P_{CI}, P_{CC}, Q_{CP}, Q_{CI}, and Q_{CC}), which are calculated individually first. Additionally, the customer classes (CC) are distinguished:

$$P_{XX} = \sum_{CC} S_{CC}(season) \times F_{XX,CC} \times n_{CC} \times L_{P,CC}(season,...)$$

$$Q_{XX} = \sum_{CC} S_{CC}(season) \times F_{XX,CC} \times n_{CC} \times L_{Q,CC}(season,...)$$

$$XX \in \{CP, CI, CC\}.$$

Billed Energy Based. The metered (billed) energy E_{Bill} and the number of supplied customers n for one transformer is used:

$$P = \Sigma(E_{billed}.LF_c)/DF_c \text{ for all customers 1 to n,}$$

where LF_c = load class load factor and DF_c = load class diversity factor. In a second step, the value of S_{Bill} is used as typical load in the formulas of method 1.

There is a general priority list, which defines the preferred calculation method. For each load, which input data are available is checked. If input data for more than one calculation method are available, the method with the highest priority is used. The output of the static load calibration is used as input for the topological calibration.

2.8.2 Fault Calculation

There are two categories of fault type — balanced or symmetrical — three phase faults and asymmetric faults when only two phases or ground is involved.

---

* See Chapter 3, Figures 3.21 and 3.23, for examples.

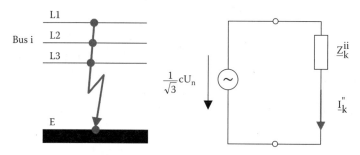

FIGURE 2.15 Three-phase balanced fault. (Courtesy of ABB.)

The symmetrical short circuit analysis function is defined by IEC 909 in which initial values of node/bus voltages provided by the load flow calculation (LFC) are neglected.

The fault calculation function for a three-phase balanced fault (L1-L2-L3-E) remote from a generator simulates a fault on every bus in the electrical power system (Figure 2.15). For each fault case, the initial symmetrical short circuit current at the bus and the currents in the connected branches are calculated. Based on Thevenin's theorem, the current is determined by

$$I_k^" = -\frac{c \cdot U_n}{\sqrt{3} \cdot \underline{Z}_k},$$

with the nominal voltage U_n, the voltage factor c, and the short-circuit impedance of a three-phase system $\underline{Z}_k$. The factor c depends on the voltage level. In addition, the initial symmetrical short circuit apparent power and the fault voltages at the neighboring buses are determined.

Asymmetrical Short Circuit Analysis. The following types of unbalanced (asymmetrical) short circuits are typically calculated by the asymmetrical short circuit analysis functions:

- Line-to-line short circuit without earth connection (Figure 2.16a)
- Line-to-line short circuit with earth connection (Figure 2.16b)
- Line-to-earth short circuit (Figure 2.16c)

The calculation of the current values resulting from unbalanced short circuits in three-phase systems is simplified by the use of the method of symmetrical components, which requires the calculation of three independent system components, avoiding any coupling of mutual impedances.

Using this method, the currents in each line are found by superposing the currents of three symmetrical components:

Central Control and Management

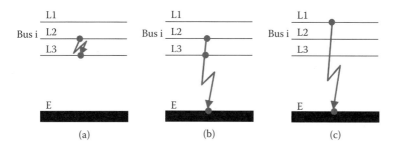

FIGURE 2.16 Three-phase unbalanced faults. (Courtesy of ABB.)

- Positive-sequence current $\underline{I}_{(1)}$
- Negative-sequence current $\underline{I}_{(2)}$
- Zero-sequence current $\underline{I}_{(0)}$

Taking a line L1 as a reference, the currents I_{L1}, I_{L2}, and I_{L3} are given by:

$$\underline{I}_{L1} = \underline{I}_{(1)} + \underline{I}_{(2)} + \underline{I}_{(0)}$$

$$\underline{I}_{L2} = \underline{a}^2 \underline{I}_{(1)} + \underline{a}\,\underline{I}_{(2)} + \underline{I}_{(0)}$$

$$\underline{I}_{L3} = \underline{a}\,\underline{I}_{(1)} + \underline{a}^2 \underline{I}_{(2)} + \underline{I}_{(0)}$$

$$\underline{a} = -\frac{1}{2} + j\frac{1}{2}\sqrt{3} \; ; \; \underline{a}^2 = -\frac{1}{2} - j\frac{1}{2}\sqrt{3},$$

where $a = 120°$ operator.

Each of the three symmetrical components systems has its own impedance that reflects the type, connection, and grounding (transformers) of the network equipment and must be entered into the distribution model database.

The method of symmetrical components postulates that the system impedances are balanced, as represented by balanced line geometry and transposed lines, although the absence of the latter gives insufficient error in distribution networks.

Breaker Rating Limit Check. The main use of the fault calculation in a DMS is to determine whether a circuit breaker will be operated above its rating, and thus an alarm can be created to notify the operator if an operating state would be a violation.

2.8.3 Loss Minimization

The loss minimization application provides a comprehensive method for investigation of the reduction of radial distribution network real losses through network reconfiguration within specified operating constraints. The loss minimization application has the following features:

- The ability to identify switching changes for reduction of distribution losses.
- The ability to calculate the necessary reallocation of load among feeders to reduce distribution losses.
- The capability to verify that the proposed optimized system condition is within the allowable operating limits (capacity and voltage).
- The capability to run cyclically or to run at a specified time each day and to execute on operator request. Similar to the load flow, computations will only be run on the portion of the network affected by changes (switching, incremental load adjustments, etc.).
- The capability to restrict the optimization to use remotely controllable switches only.

The function always starts from a feasible system base case. In the event that the as-operated base case is infeasible (constraints violated), the loss minimization function will first determine (advise) a feasible state. Starting with this state, a list of switching actions and the associated reduction in losses will be listed. Each successive switching action will provide increased loss reduction. The sequence of switching actions, which may involve multiple switching of the same switch, is specifically prepared to ensure that at no time during the sequence will any equipment be overloaded or voltage limit exceeded.

2.8.4 VAR Control

The VAR control function is designed for the control of MV capacitor banks located at HV/MV substations and on MV feeders. The function determines radial systems capacitor configurations that reduce reactive power flows into the MV system, while maintaining the system within user-defined voltage and power factor operating limit conditions. The resulting capacitor configuration reduces MV feeder voltage drops and losses under varying system load conditions.

VAR control functions usually produce MV network local capacitor control strategies on an individual HV/MV substation service area basis. A typical calculation sequence is as follows:

Step 1. Determine the service area operating state regarding VAR compensation. Three scenarios are possible:

- The service area is in a normal state. This is determined by comparing the actual total service area power factor (as measured by SCADA)

Central Control and Management

with the target power factor and by the nonexistence in the MV feeders of high/low voltage and power factor limit violations (as calculated by state estimation).
- The service area needs VAR compensation. This is determined by comparing actual total system power factor with the target and by the existence in the MV feeders of low voltage and power factor limit violations.
- The service area is in an over-VAR compensated state. This is determined by comparing actual total system power factor with the target and by the existence in the MV feeders of high voltage and power factor limit violations.

Step 2. The target total system power factor is user defined.

User-defined deadbands and threshold values are used in the determination of the service area operating state to reduce the number of VAR control application runs.

Operating states 2 and 3 represent candidates for VAR control. The availability of controllable capacitor banks in these service areas also determines whether the VAR control application is executed or not on the particular service area.

Certain assumptions are made to simplify the calculation yet achieving results within the system operating tolerances. Typical assumptions include that all MV-level VAR adjustments will not be reflected in the higher voltage networks and that all tap changing equipment (transformers and line regulators) will be held constant. Further, that common practice of employing a limited number of capacitor banks on feeders will be adopted.

2.8.5 Volt Control

The volt control function is designed for the control of on-line tap changers (OLTCs) associated with substation transformers and line voltage regulators with the objective to reduce overall system load. Typically, this function is used at times when demand exceeds supply or flow of power is restricted due to abnormal conditions. The function calculates voltage set points or equivalent tap settings at the OLTCs to reduce the total system served load.

The calculation method supports two load reduction control strategies:

Target Voltage Reduction. The first method reduces the system load in such a way that a lowest permissible voltage level is not violated. The maximum load reduction is, therefore, a function of the user-defined lowest permissible voltage level.

Target Load Reduction. The second method determines the system voltage level required to achieve a user-defined load reduction target.

Calculation Method. Calculation methods follow a similar approach to VAR control. An area is first selected where load reduction is required. For example, consider Figure 2.17, where a sample system of three radial feeders is described. Each of the three feeders has different lengths (hence impedance), total load, load distribution, and number of line regulators.

The first step is to determine which HV/MV substation service areas are candidates for OLTC volt control. The criteria used reflects

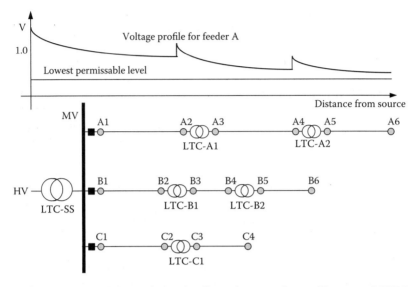

FIGURE 2.17 Example network showing line voltage regulators. (Courtesy of ABB.)

- The availability of OLTCs for control as determined by SCADA
- User-defined active power limits

Each candidate substation service area is analyzed one at a time using one of two possible calculation methods.

Again, corresponding assumptions similar to the VAR control function allow realistic performance of the calculation for control room use.

2.8.6 Data Dependency

The validity of results from all advanced applications are entirely dependent on data availability and its quality. The first level of data is that of the network topology. The connectivity of the network incorporating the status of switches must be correct, because fundamental operating decisions and maintenance of safety use this as the basis. Maintenance of remotely controlled switch status is automatic within the SCADA system but all manually operated switch statuses have to be manually input by the operator from information communicated from the field. The normal state of network configuration must be stored during temporary switching to enable restoration to the normal state as defined by NOPs. As extensions are made to the network, all changes added to the DMS have to be synchronized with the actual field state. The data modeling becomes more complex for North American networks where each phase needs to be represented due to the mix of three- and single-phase supply and where switching devices can open one phase independently. Outage management and basic switching plan generators within the advanced applications portfolio depend entirely on correct topology (connectivity) for valid results.

Advanced applications that perform calculations on the network such as power flow, fault level, loss minimization, and optimization require another level of data describing network parameters such as circuit impedances, capacity limits, and loads. The latter are the most difficult to determine not only because the time variation of consumption differs from customer class to customer class, but also there is voltage/current dependency. Few real-time measurements of individual customer loads or composite customer load groups exist except at the substation feeder sources. The most commonly available value at customer substations is an annual peak load value that is manually collected after the system peak has occurred and is seldom time-stamped. The implementation of automatic meter reading, even though not widespread, is improving load behavior information. However, the automatic readings are normally not real-time values but historical energy consumption values. Devices within the network with local control such as line voltage regulators and capacitor controllers require additional data of their control behavior such as set point, and deadbands to describe their operation.

The issue of data accuracy and sustainability must be considered when setting the expectations and benefits derived from the advanced applications performing network analysis calculations. It is not just the initial creation of the DMS database that requires considerable effort, but also the cost of maintaining its accuracy in an ever-changing environment.

It can be concluded that DMS advanced applications should be divided into two categories: those that operate satisfactorily with topology only and those that in addition to topology require network parameter data as shown in Table 2.3. The former will give results that are unconstrained by network capacity limitation and must rely on the operator's knowledge, whereas the latter consider the constraints of the network capacity and voltage regulation.

In utilities with good planning practices where a detailed network model of the MV system has been created and verified, this model provided of sufficient granularity in terms of switch and feeder representation, and can be used as the foundation of the DMS connectivity model.

The concept of value versus data complexity is illustrated in a subjective manner in Figure 2.18, where a data complexity and sustainability factor is assigned a value between 1 and 10. Similarly, the value to the operator of the results from a constrained advanced application can also be assigned between 1 and 10. The ratio of the two values, the Value to Complexity Ratio (VCR) illustrates subjectively the potential benefits and distinguishes between unconstrained and constrained applications.

2.9 SUBSYSTEMS

2.9.1 SUBSTATION AUTOMATION

Traditional SCADA systems acquire the majority of data on the power system from substations. This has been achieved by installing RTUs, hardwired into the

TABLE 2.3
Categorization of Advanced Applications

Application	Topology-Based (Unconstrained)	Parameter-Based (Constrained)
Network coloring	✔	
Switch planner	✔	✔[a]
FLIR	✔	✔[a]
Operator load flow		✔
Fault current analysis		✔
Volt/VAR control		✔[a]
Loss min/optimal reconfiguration		✔[a]

Note: Some functions will operate without the need for the power flow to give meaningful results such as the case of a switch planner and FLIR that could be implemented initially as unconstrained applications, thus reducing the need to collect and verify additional complex data.

[a] Applications operating under constraints require the load flow as an integral part the application.

protection relay and switch auxiliary contacts, as the communications interface to the central control system. This approach was costly and restrictive, not using the full capability of modern communicating relays. Substation automation (SA) describes the concept that takes advantage of the configurable communication IED technology in implementing a local multilayer control hierarchy within a substation. Logically there are two levels, the lower or bay (feeder) and the higher or station level. Suppliers have adopted different architectures for SA. These architectures use either a two or three information bus configuration — the difference being where the protection intelligence is located. The three-bus approach retains the segregation of the protection logic and device at each bay, whereas the two-bus architecture uses a common central processor for all protection. The former approach has been favored because it reflects traditional protection philosophy.

The main varieties of SA platforms being marketed that attempt to integrate the distributed processing capabilities of the many IEDs are as follows:

1. **RTU-Based Designs.** These systems evolved from conventional SCADA where RTUs were modified to provide the communications interface with, and utilize the distributed processing capability of, modern IEDs. Some RTUs have added limited PLC functionality and de facto standard sub-LAN protocols such as Echelon LONWorks or Harris DNP 3.0. The main disadvantage of this approach is the inability to pass through instructions to the IEDs in a one-to-one basis because they effectively appear as virtual devices to the central control room.

Central Control and Management

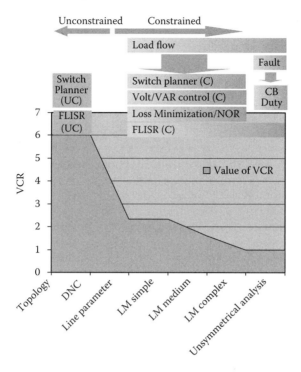

FIGURE 2.18 Subjective benefits versus data creation effort for different data categories and associated DMS advanced applications where DNC = dynamic network coloring, LM = load model with simple being a one-value model based on installed distribution transformer capacity, medium is represented by an estimated load value and estimate of voltage dependence, and complex requires in addition to the first two model types load profiles of typical days. Unsymmetrical analysis refers to applications requiring unbalanced loads and sequence impedances.

2. **Proprietary Designs.** These are fully functional, modular, distributed systems provided by one supplier using proprietary architectures and protocols. The systems are not open because the full protocols are not published and the HMI is vendor specific to their IEDs and architecture. Any extension of the substation is confined to the original SA supplier's equipment.
3. **UNIX/PLC Designs.** These systems use RISC workstations operating under UNIX integrated with PLCs to give very high-speed multi-tasking solutions. The resultant cost is higher than other platforms.
4. **PC/PLC Designs.** The design of these systems is based around a sub-LAN with the PC providing the HMI and integrated substation database. The PLC supports customized ladder logic programs designed to replace conventional annunciators, lockout relays, and timers. Because the sub-LAN communication protocol is usually dictated by the selection of PLC, the PC protection IEDs require special gateways

TABLE 2.4
Summary Comparisons of the Main Features of the SA Alternatives

Feature	RTU Based	Proprietary	UNIX/PLC	PC/PLC	Black Box
SA Design Type					
HMI operating system	Windows, DOS	Proprietary	UNIX	Windows	Windows
HMI software	Limited	Proprietary	HP, SI, US Data	Many	Proprietary
Sub-LAN protocol	LONWorks, DNP	Proprietary	PLC based	PLC based	VME bus
IED support	Protocol conversion	Proprietary	Protocol conversion	Protocol conversion	Self-contained
Comparative functionality	Medium	Limited to products from a single vendor	High	High	Medium
Cost	Middle	High	Middle to high	Middle	Low
Major suppliers	Harris, SNW, ACS	ABB, Schneider, GE, AREVA	HP, SI	Tasnet, GE K series, Modicon, Allen-Bradley	RIS

and interface modules. The resulting integration issues are not trivial if devices from different vendors are used but the approach has a degree of openness.

5. **Black Box Designs.** These systems are designed to integrate, within a single PC framework, selected SA functions. All functions such as programmable protection, ladder logic, input/output (I/O), and front panel annunciation are achieved within one common PC server. The main disadvantage is that protection is centralized, which deviates from the conventional protection philosophy of individual circuits having independent protection.

Comparison of the different SA types described is summarized in Table 2.4. The benefits in flexibility with primary and secondary configurations can be substantial.

2.9.2 SUBSTATION LOCAL AUTOMATION

The most common architecture in use today, whether proprietary or not, is that where each bay has independent protective devices, thus it is worth expanding on more of the details of this design. The protection and control units act

Central Control and Management

independently in cases of a disturbance, cutting off the faulty portion of the network. The same protection relay and control units have a communications bus and thus act as data transfer units to local systems or telecontrol systems, replacing the conventional RTUs. Communicating annunciators can also be connected to the overall system. All data acquisition devices are connected through a common communications bus that is used to transfer the data to the local or telecontrol system through a data communications unit, if needed. The protection relays, control units, and alarm centers provide the operating systems with the following:

- Time-stamped event data
- Measured electrical quantities (directly measured and derived)
- Position data for switching devices (circuit breakers and disconnectors)
- Alarm data
- Digital input values
- Operation counters
- Data recorded on disturbances
- Device setting and parameter data

The local or telecontrol (SCADA) system can send the following to the units:

- Control commands
- Device setting and parameter data
- Time synchronization messages

Using the communications bus provides several technical and economic advantages compared with conventional signal cabling. The need for cabling is considerably less when much of the necessary information can be transferred through one bus. Intermediate relays are not needed, either. On the other hand, the feeder-specific current transducers and circuits can be discarded, because the measurement data are obtained through the protection circuits. As less cabling and a smaller number of intermediate relays are needed, the fault frequency of the substation is reduced. Protection relays can be used for condition monitoring of secondary circuits, such as tripping circuits. The actual transfer of messages is also monitored, and communication interruptions and faults can be instantly located. Systems are easier to extend, because new units can be easily added thanks to this communications principle.

Each device provides a time-stamped message on events (starting, tripping, activation, etc.) through the bus. These events are sorted by time and transmitted to the event printers or to the monitoring system. Data covered include maximum fault current data, the reason for starting or tripping, and fault counters. Device setting and parameter data can also be conveyed by control units, and commands from the control system can also be transmitted to the switching devices in the same way. Microprocessor-based relays and control units store a lot of data on

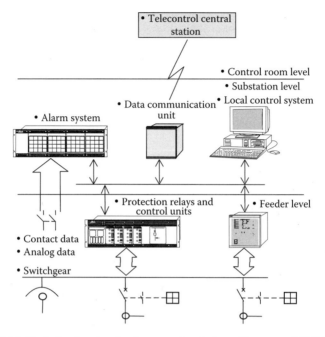

FIGURE 2.19 Typical substation local automation design. (Courtesy of ABB.)

faults occurring in various registers. Also, the devices can provide current and voltage measurements. The devices may also have calculation routines providing the power, energy, and power factor derived from the measured current and voltage values. These data can be used locally or by the telecontrol system.

Each device in the system has its own internal clock, which must be synchronized with the other clocks in the system. Events and other important messages are time-stamped in the secondary device. The messages can be sorted by time on the basis of this time-stamp. To keep the clocks synchronized, a synchronization message aligning the clocks to millisecond accuracy is sent out at regular intervals.

The substation level has supervision or control systems to perform centralized automation functions at the substation level. These local control systems (Figure 2.19) are based on the same concept and technology as the SCADA system. However, they are more basic in terms of equipment and software than the SCADA system, and are scaled for use at the substation level. Functions typically performed at substations include the following:

- Schematic mapping of substation and switch position indication
- Presentation of measured electrical quantities
- Controls
- Event reporting

Central Control and Management

- Alarms
- Synchronization
- Relay settings
- Disturbance record collection and evaluation
- Processing of measured data, trends, power quality data, and so on
- Recorded data on faults and fault values
- Network primary device condition monitoring
- Substation level and feeder level interlocking
- Automatic load disconnection and reconnection
- Various regulations (voltage regulation, compensation, earthing coil regulation)
- Connection sequences on bay and substation levels (e.g., busbar or transformer change sequences)

Technically, it is possible to integrate similar functions in a single device. One protection package can contain all the protection functions needed for the feeder in question. A protection relay can also incorporate control, measurement, recording, and calculation functions. Protection relays can provide disturbance records, event-related register values, supply quality and consumption measurements, and counter and monitoring data. The number of inputs and outputs available in the protection packages are increasing, and thus all the data related to a feeder can be obtained centrally through the relay package. The more data there are to be processed, the more economical it is to link the data through one sufficiently intelligent device. Also, communications devices are becoming smaller and smaller. Linking of occasional data (door switches, temperature, etc.) to the system via small I/O devices will become economically feasible. Thus, this type of data will also be available centrally in local systems and at the master station SCADA systems.

Control Units. Necessary protection and control functions have been integrated into the feeder protection packages. A feeder-specific control unit provides position data on the switching devices of the feeder, and the unit can be used for controlling the motor-operated switching devices of the switchgear locally. Control units can also be used to perform bay-specific interlocking. Position data and measurement, calculation, and register data of the control unit can also be transferred to the local or telecontrol systems through the communications bus, and the local or telecontrol system can control the motor-operated switching devices of the feeder through the control unit.

Annunciator Units. The annunciator collates contact alarm signals from all over the substation or distribution process. The data, obtained as analog messages, are sent to the analog annunciator, which generates alarms according to preprogrammed conditions. The purpose of the annunciators is to facilitate management of disturbances. Typically, an annunciator broadcasts first-in alarms showing the origin of the disturbance. The annunciators can be programmed with alarm delay times, alarm limits, alarm blinking sequences, or alarm duration, for instance.

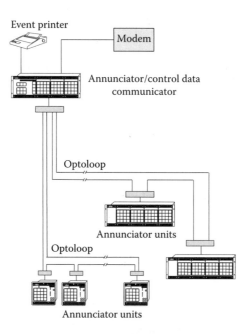

FIGURE 2.20 Typical alarm system. (Courtesy of ABB.)

The contact functions of the annunciators can be opening or closing. In addition, alarms can be programmed to be locked to each other in order to avoid unnecessary alarms caused by the same fault. For remote supervision, the annunciators have two or several programmable group alarm outputs, which can indicate whether an alarm demands immediate action or is merely given for information. These groups can be programmed as desired. The annunciators can also be linked to the substation communications bus, thus providing information for local or telecontrol systems. In small substations, a control data communicator integrated into the annunciator can be used to collect data from the entire substation. For event reporting, an event report printer can be directly connected to such an annunciator (Figure 2.20).

Disturbance Recorders. Disturbance recorders have become more popular in fault diagnosis and post-fault analysis. Disturbance recorder functions are, nowadays, integrated into protection devices. Disturbance recorders provide curves of analog values such as currents and voltages before and after a fault, and digital data such as autoreclose event sequences before and after a fault. The channel number, sampling rate, and signals monitored can be programmed separately for each purpose. A disturbance recorder can be triggered on set conditions in the signal monitored, or by a triggering message received over the communications bus. Disturbance recordings can be downloaded through the communications bus and analyzed in a separate computer program. Data from the disturbance recorders can also be fed into standard programs such as spreadsheet and calculation programs for further analysis.

Substation-level Communication. Microprocessor-based relays, control units, and annunciators communicate with one another, so data related to these devices are made available through the communications bus to other devices linked to the system. A substation-level communications bus interconnects all the feeder-level devices and alarm units, and the local supervision and monitoring system of the substation. The usual communications medium used on the substation level is optical fibers, so the data transfer is not susceptible to electrical interference. A data acquisition unit connected to the substation-level communication bus collects data from devices linked to the bus and transfers it to a higher-level system, such as a telecontrol or process monitoring system (Figure 2.21a).

Communication from the data acquisition unit to the telecontrol system is called telecontrol communication. At its simplest, the data acquisition unit is a gateway between the substation level and the telecontrol communication. The data acquisition unit can also function as a reporting unit, and then an event printer can be connected to it. In polling systems (e.g., those using an SPA bus), the data acquisition unit is the master of the substation communication system, whereas in spontaneous systems (e.g., those using a LON bus), the devices can send data autonomously to the bus and poll other devices for data. In a spontaneous system, event data are transmitted instantaneously as they occur, whereas in a polling system, they are sent only on request. On the other hand, a polling system is simpler to implement and manage, because the master node can freely determine when and from which node to request data. An early proprietary bus such as SPA bus is a communications standard for the ABB SPACOM/PYRAMID series relays developed in the 1980s. It has evolved into a general substation-level communications standard. SPA is polling by nature. Protection relays, control units, and alarm centers are linked under one master device with opto loops. All nodes, i.e., slaves, have unique numbers. The master device queries each slave in turn for the desired data, and the slaves respond to this query. The system response time depends on the number of devices and the amount of data polled. Important data can be given priority and polled more frequently than other data. The SPA bus is an asynchronous bus with a maximum speed of 9.6 kbit/s.

The LON bus is a widely used open communications standard developed by ECHELON. It supports a variety of communications media, from optical fibers to distribution line carrier (DLC). The maximum data transfer rate for a fiber-optic LON bus is 1.2 Mbit/s. In substations, the LON bus is configured as a radial system with star-couplers. The LON bus is a spontaneous system where all devices (nodes) can repeat a change of state (Figure 2.21b).

2.10 EXTENDED CONTROL FEEDER AUTOMATION

Extended control as discussed earlier is a general term for all remote control and automation of devices outside the substation and includes all devices along distribution feeders such as switches, voltage line regulator controls, feeder capacitor controls, and devices at the utility customer interface such as remotely read

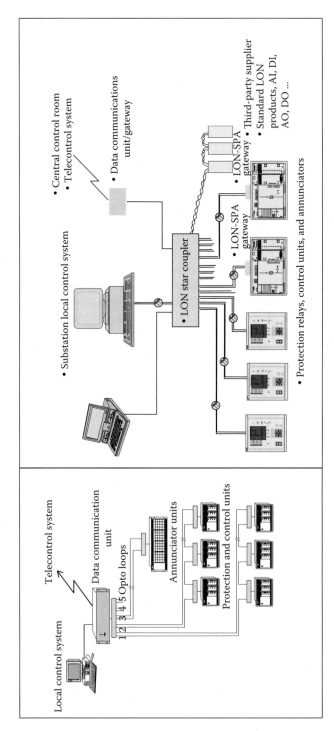

FIGURE 2.21 (Left) Proprietary bus layout and (right) LON bus layout. (Courtesy of ABB.)

Central Control and Management

intelligent meters. All line devices fall under the feeder automation umbrella, which is the major topic of this book and thus is treated in more detail in subsequent chapters once the foundation of distribution system type and protection has been covered. Feeder line devices are controlled either directly from the DMS master station or via the normal supplying primary substation RTU or SA server. Much depends upon the communications infrastructure established and whether the architecture is based on distributed data concentrators such that the feeder device IEDs become virtual devices when viewed from the DMS master through the concentrator. Whatever the configuration, all devices will eventually be controlled from a central location.

2.11 PERFORMANCE MEASURES AND RESPONSE TIMES

2.11.1 Scenario Definitions

Distribution management systems are required to operate in both normal and abnormal conditions. It is especially important that the system works properly in abnormal situations when the operator is dependent on the system to precisely follow the rapid-changing state of the network and associated information it is crucial that no data is lost in times of high activity.

Response time is a combination of the transaction times of the central system, communication, and data collection systems. The response times are defined in most specifications by categorizing the functions into critical (core) and noncritical (other) as typified below.

Critical	Noncritical
All SCADA applications	Database modification and generation (DE)†
All DMS advanced applications	Display modification and generation (DE)
Processing of requests from operator consoles	Training simulators
Information storage and retrieval system[a]	Program development system

[a] Sometimes called utility data warehouse or historian, † DE — data engineering.

The critical functions are considered available when operating as specified at the scheduled periodicity and within the specified execution time. Their availability is also dependent on hardware, which in turn is considered available when sufficient processors, peripheral, and remote devices together with interfaces to external enterprise IT systems are running.

The required performance of these functions is based on a specific configuration and a number of scenarios that depict the state of the distribution network under a variety of conditions around a base condition. The configuration scenarios are defined as follows:

TABLE 2.5
Hardware Configuration Matrix Indicating Redundancy and Quantities for a Typical SCADA/DMS System with Main and Backup Control Centers (MCC and BCC), Information Storage and Retrieval (IS&R) System, Distribution Training Simulator (DTS), and Program Development System (PDS)

Device	MCC Redn	MCC Qty	BCC Redn	BCC Qty	IS&R Redn	IS&R Qty	DTS Redn	DTS Qty	PDS Redn	PDS Qty
Processor and auxiliary memory	Y	1[a]	N	1	Y	[b]	N	1	N	
Database	Y	1[a]	N	1	Y		N	1	N	
LAN[a]	Y	1[a]	N	1	Y		N		N	
Printer	N	2	N	2	N	1	N	1	N	1
Operating consoles	N	5	N	3	—		N	2	N	2
Support consoles	N	2	N	1	—		N	1	N	
Communication front ends[a]	Y	1[a]	Y	1	—		—		N	

[a] Redundant pair
[b] IS&R uses the MCC SCADA/DMS servers in this example; however, some implementations call for separate redundant servers.

Configuration: This requires and establishes that the specified setup of the hardware has been established and software is operational (Table 2.5).

In terms of software, the normal display and command windows have to be running at all operator consoles and showing certain displays such as the network overview and alarm list. The complete database for the entire network has to have been loaded. It defines system activities and condition on which the steady-state and high-activity state are layered.

Steady-State or Normal Condition: This represents the normal system operating conditions and is measured over a typical 1-hour period where specific functions are invoked.

High-Activity State: The network is in a stressed situation, many events occur, and the operators are busy restoring the network to normal condition. A high-activity state can last for hours.

A typical definition of these two states appears in Table 2.6.

Emergency Condition: A power system disturbance has occurred. A big burst of events is received at the control center. This represents specific conditions during a 15-minute period typical of outage situations.

Table 2.7 shows a typical 10,000-event burst defined as a number of events received by the central system each minute during the burst in addition to the high-activity state definitions. It is assumed that the high-activity state precedes and follows the burst.

Central Control and Management   81

TABLE 2.6
Activity Definitions for Two Performance Scenarios

Activity	Steady-State	High-Activity
Percentage of total analog points that change sufficiently every 10 seconds to require complete system processing	10%	50%
Number of alarms per minute to be generated and processed (50% status, 50% analog)	30 per minute	600 per minute
Display activity in terms of one new display requested per operator workstation per time interval	1 every 60 seconds	1 every 10 seconds
Data point entry frequency measured as one data point per operator console per time interval	1 every 60 seconds	5 every 60 seconds
Supervisory control sequence frequency (device open or close) per operator console per time interval	1 every 5 minutes	1 every 60 seconds
Advanced applications (operator load flow) in terms of amount of network analyzed per time interval	10% of network every 15 minutes	50% of network every 5 minutes
Information storage and retrieval system response in terms of the number of query and report requests from one console and the number of query items	5 queries of each having 500 items	1 queries of each having 500 items

TABLE 2.7
Typical Alarm Burst Profile

Minute	Alarms (50% Status, 50% Analogues)
1	5000
2	2000
3	1000
4	500
5	500
6	400
7	300
8	300

2.11.2 CALCULATION OF DA RESPONSE TIMES

The calculation of response times depends on the response times of the central system (retrieval and display from the real-time database), the communication* time, and the remote data collection system (RTU). The latter two elements

---

* See Chapter 7 for calculation of communication speeds for different communication media.

comprise the data acquisition system. Response times are normally calculated for a variety of signals, the most common being as follows:

- An indication change
- A measurand change (outside limits, within limits, maximum, and average times)
- Time to send a command and receive back an indication as confirmed

Indications. Response times for one indication change from only one of the RTUs in a multi drop configuration to reach the Front-End database in a report by polling scheme (worst case) is given by

$$T_{TOT} = T_{MEAS} + (N-1) \times (T_{POLL} + T_{NACK}) + (T_{POLL} + T_{IND}),$$

where T_{MEAS} is the time required to send one measurand telegram. If one measurand is just being sent from one of the RTUs on the line, this telegram has to be completed before a new poll request is issued. T_{POLL} is the time to send one poll request. If a multidrop line is used the poll request has to be sent to all other RTUs on the line before the RTU with the indication change is polled. The worst case is that all RTUs minus one have to be polled before the actual RTU is reached. Therefore, the poll request has to be multiplied by the number of RTUs on the line minus one, $(N-1)$. Observe that in normal polling schemes, priority 1 signals, i.e., normal indications, are polled before measurand (priority 2) changes are polled. T_{NACK} is the time to acknowledge and indicate that no information is available for sending. T_{IND} is the time to send one indication response.

Each type of telegram has a certain number of bytes. Each byte is 11 bits including a start bit, a parity bit, and a stop bit plus the byte itself with 8 bits. The frame header is 5 bytes and the trailer is 2 bytes of an IEC 60870-5-101 telegram. We will assume for the calculations that a poll telegram and an acknowledge telegram both require 12 bytes, an indication telegram requires 15 bytes, a command and check back telegram 17 bytes, and a measurand telegram requires 20 bytes. The exact number of bytes is dependent on the RTU protocol cand can be found in the respective protocol definitions.

The baud rate for a communication line defines how many bits per second can be transmitted. With this information, it is now possible to calculate the total time (worst case) to receive an indication change to the communication board of the front end on a multidrop 1200-baud line with 5 RTUs.

$$T_{TOT} = 20 \times 11/1200 + (5-1) \times (12+12) \times 11/1200 + (12+15) \times 11/1200 = 1573/1200 \text{ seconds} = 1.31 \text{ seconds}$$

Central Control and Management

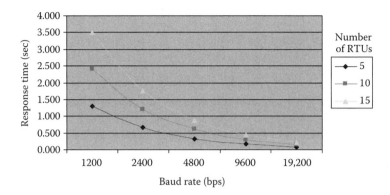

FIGURE 2.22 Response Times for different communication baud rates and number of RTUs per channel.

From this calculation, the effect of baud rate and the number of RTUs on response time can be determined as shown in Figure 2.22 for baud rates from 1200 to 19,600 and RTUs in groups of 1, 5, 10, and 15.

This calculation shows the communication time between RTU and front end. Before the indication is shown on the screen for the operators, the following times must be added:

- T_1, time for treatment in Front-End and to send to Master system. A front end system normally buffers telegrams coming in on many RTU lines and builds a buffer to be sent to the master system. T_1 is normally 0.1 seconds including buffering time.
- T_2, time to update the master system real-time database in the database update program. T_2 is normally 0.01 seconds.
- T_3, time to do demand update of screens. T_3 is normally 0.2 seconds.
- T_4, period for cyclic update of displays. T_4 is normally a maximum of 5 seconds (average 2.5 seconds).

For indications, a demand update scheme is used. This gives, if all times are added, a response time for one indication under the same condition as above between RTU and operator screen to be approximately 1.62 seconds.

Metered Values. The corresponding time for one measurand response for only one RTU on a multi-drop line with the measurand on priority 2 polling scheme and no indications changing:

$$T_{TOT} = T_{MEAS} + (N - 1) \times (T_{POLL} + T_{NACK}) + (T_{POLL} + T_{MEAS})$$

With the same values as above we get

$$T_{TOT} = 20 \times 11/1200 + (5 - 1)(12 + 12)(12 + 12) \times 11/1200 + (12 + 20) \times 11/1200 = 1628/1200 = 1.36 \text{ seconds.}$$

For a measurand change that does not exceed a limit, a cyclic update scheme is normally used. This then gives a maximum time from a measurand change in the RTU until it is displayed on a single communication line of 6.36 seconds (1.36 + 5) and an average time of 3.86 seconds when a cyclic update of 5 seconds is used.

For a measurand that changes above a limit, a demand update is used. This will then give a total time of 1.67 seconds using T_1, T_2, and T_3.

Output Command with Check-Back. The corresponding formula for an output command with check-back under the same conditions as above is given below. The first two parts of the formula, $T_{CHB} + T_{COM}$, are the time until the output relays are set in the RTU. T_{OPR} is time to operate the primary device and T_{TOTIND} is the time for the indication response, as calculated above. Observe that check-back and command message will interrupt polling but cannot interrupt ongoing telegrams.

$$T = T_{CHB} + T_{COM} + T_{OPR} + T_{TOTIND}$$

where

$$T_{CHB} = T_{MEAS} + T_{CHBS} + T_{RTU} + T_{CHBR} + T_{FE}$$

(Wait for possible measurand telegram + sending of check-back + RTU treatment time + receiving check-back response + the front-end treatment time.)

The RTU treatment time for check-back or command message will be approximately 0.1 seconds. The Front-End treatment time will be approximately the same time.

$$T_{CHB} = 20 \times 11/1200 + 17 \times 11/1200 + 0.100 + 17 \times 11/1200 + 0.100 = 594/1200 + .100 = 0.495 + 0.2 = 0.695 \text{ sec}$$

and where

$$T_{COM} = T_{MEAS} + T_{COM} + T_{ACK} + T_{RTU}$$

(Wait for possible measurand telegram + sending of command + RTU treatment time + acknowledgment.)

$$T_{COM} = 20 \times 11/1200 + 17 \times 11/1200 + 12 \times 11/1200 + 0.100 = 539/1200 + 0.100 = 0.449 + 0.100 = 0.549 \text{ sec}$$

T_{OPR} is the operating time of the primary equipment. This will vary substantially dependent on device; e.g., a breaker will operate within 0.5 seconds, and an isolator may take up to 10 seconds to operate.

Central Control and Management

We can now calculate the whole time from requesting a check-back command until the response indication is shown on the single-line display. It is assumed that the primary device has an operating time of 2 seconds.

$$T = T_{CHB} + T_{COM} + T_{OPR} + T_{TOTIND} = 0.695 + 0.549 + 2.0 + 1.62$$
$$= 4.86 \text{ sec.}$$

2.11.3 RESPONSE TIMES

General. System response times are normally defined for the following areas of system interaction. The tables below define the response times as 90% values. The 90% value means that the response time is less than the stated value at least 90% of the time.

Data Acquisition and Processing. These response times are measured in terms of the elapsed time from when an event occuring at the input of a RTU until it is displayed to an operator who has a picture containing that object on one of console VDU. The times shown include allowance for the 5-second picture update cycle for measurands inside limits. All other events cause demand-driven picture updates. The tests are conducted by making the specified data changes on one of the physical (not simulated) RTUs.

Values taken from the calculations in the previous sections are now summarized.

Indication Change

	Steady-State 90% (s)	High-Activity State 90% (s)
Time RTU to screen	1.62	Steady-state plus 50%

Measurement Change

	Steady-State 90% (s)	High-Activity State 90% (s)
No limit violation Time RTU to single line	6.36	Steady-state plus 50%
Limit violation Time RTU to single line	1.67	Steady-state plus 50%

Command, Regulation

	Steady-State 90% (s)	High-Activity State 90% (s)
Control Command		
The time from when OK is clicked until back indication is displayed	4.86	Steady-state plus 50%

Human-Machine Interactions

The time for picture call-up is measured from when the key is pressed until the complete picture is shown.

Picture Call-Up Local Consoles

Type of Picture	Steady-State 90% (s)	High-Activity State 90% (s)
World map	3	4,5
Single line	1	1,5
Alarm list	1	1,5

2.12 DATABASE STRUCTURES AND INTERFACES

One of the challenging issues within a DMS is the resolution of different data structures that potentially exist between the various applications, both internally within the SCADA, the advanced applications, and outage management function and externally with the enterprise IT functions such as GIS, work management (CMMS, ERP),* and customer information management (CIS/CRS).

2.12.1 Network Data Model Representations

Data models for representing the electrical network with varying degrees of detail have developed in the industry as appropriate to the application using the data. These are described pictorially in Figure 2.23.

All the models have a node branch relation model because this most efficiently describes connectivity. The most detailed representation is required for the operations model where all devices and operational constraints have to be represented as unique entities not only for network analysis but also to fit into the SCADA data model for the control and monitoring task. Even this complexity of model is simplified compared to the asset model usually held in GISs or in CMMSs where every asset and its details are required for inventory and maintenance purposes. In particular, within a GIS every cable size and joint for underground networks is recorded, and similarly for overhead systems, points where conductor sizes and pole geometry changes are identified. This level of detail is not required for the real-time DMS network model; hence, if data are to be provided from a GIS system, some form of model reduction must be available to extract a composite branch representation. This concept permeates throughout the asset database depending on the particular granularity of the GIS data model implemented.

---

* CMMS — computerized maintenance management system, ERP — enterprise resource planning.

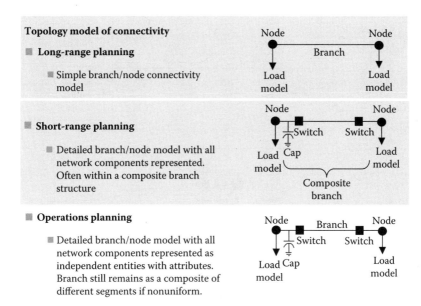

FIGURE 2.23 Three levels of network model, the operations planning model being the most complex with highest device resolution.

2.12.2 SCADA Data Models

Traditionally, SCADA systems applied to power systems employ a hierarchical structure of the power system consisting of station, substation, bay, and terminal. This is needed to explicitly model all the components that affect network operation. A simple switch within a substation must be part of a bay, and a line is a terminal interconnect between bays in different substations. Topology for network models must be interpreted from this structure. In contrast, network application functions (advanced applications) only require a simplified equivalence of the structure. In fact, to be effective this equivalence must be specifically based on a branch-node structure, where the branches connecting the nodes represent nonzero impedance elements such as lines and transformers. Methods are now coordinating these two requirements by generalizing the two levels of network topology nodes and branches into vertices and edges.

A vertex or terminal is defined as a fixed point in the network that is described by the SCADA hierarchy — each single layer not being allowed to overlap. An edge is an arbitrary connecting element that can include an impedance-bearing element such as a line or a zero impedance element such as a switch or busbar. This method permits, in the simplest form, rapid analysis of the structure's connectivity because only the status of the edge needs to be considered. The analysis can be extended in complexity by considering the type of edge; for example, treating a transformer as an open edge breaks a multivoltage level network to its discrete networks of the same voltage level. This architecture allows

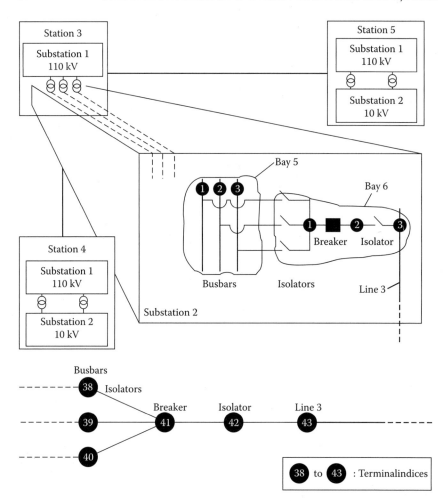

FIGURE 2.24 Diagram showing development of vertex/edge model from traditional SCADA data structure. (Courtesy of ABB.)

very fast real-time assessment of the connectivity state and de-energization, which is also used for dynamic network coloring.

The SCADA system real-time database is described in terms of point and data acquisition data (RTU and ICCP*), where every point must be defined at input in a very flat structure. This flat structure with no inherent relationships has to be built into a process model as discussed previously in the SCADA section.

The SCADA data model concept in simplified form is shown in Figure 2.25. Point data comprises either a measurement or an indication. These data have no practical meaning unless linked to a picture, so in the most simplest form of pure SCADA, these two categories of point data have to be linked to a position in a

---

* ICCP — inter control center protocol, a standard used to pass data from one control center to another.

Central Control and Management

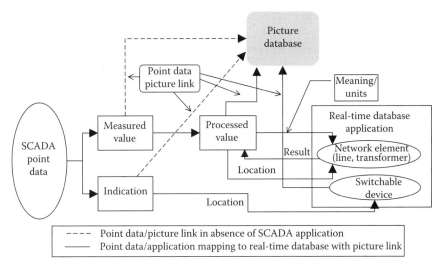

FIGURE 2.25 Simplified structure of SCADA real-time data model.

SCADA picture (dotted line in Figure 2.25). Practical SCADA systems, though, have applications that run in the real-time database, so an additional mapping to the network model is required before linking to the picture.

Schematic diagrams available in GISs are generally not satisfactory for use in a control room for monitoring and controlling the distribution network. Therefore, SCADA systems require additional picture data to describe their graphical displays. These data describe all picture objects in terms of text, value (measured value and its location to be displayed), and symbol or drawing primitive. Picture data and object data must be linked to the same SCADA device or element associated with the point data.

In addition, the data acquisition system connection through point addressing to the RTU also has to be described at input.

All this data input is coordinated and verified for consistency as part of the data engineering process.

2.12.3 DMS Data Needs, Sources, and Interfaces

SCADA/DMS data are input via the data engineering process, which is composed of creating pictures and linking the corresponding point data with the picture data. Integrated graphical tools that ensure data consistency are employed in modern systems, with a checklist to guide the user in completing all required data elements before populating the real-time database. Many data (Table 2.8) are resident in other enterprise IT applications, and it is natural to assume data can be transferred through standard interfaces between applications.

Few standard interfaces exist at present in the industry, and even if emerging standards are applied, any differences between the data models used by the

TABLE 2.8
Typical Data Maintained in Enterprise IT Systems

Enterprise IT Application	Data
CIS/CRM (customer information/customer relationship management system)	Consumer data Account number Telephone number Customer class/rate class Consumption (kWH) Customer-network link locator/ID (needed for trouble call/OMS), etc.
GIS (geographical information system)	Power system element data Network parameters Load values Connectivity Geographic maps (electrical network/background street maps) Single-line diagrams (not always maintained), etc.
CMMS/WMS (computer maintenance management system/work management system)	Asset maintenance and performance records Manufacturer, type, serial number Date of commissioning Belonging to hierarchical structure Reference to product documentation Maintenance requirements modeling Dependency of network on device for RBM[a] Maintenance history Planned maintenance Work order and job scheduling Costing Maintenance management, etc.
PMS (personnel management system)	Personnel records and details (needed for crew management) Field workforce skills/authorization levels Contact details Vacation schedules Overtime limits, etc.

[a] RBM — risk-based maintenance.

respective application must be resolve in implementing the interface. The data transaction frequency and performance requirements will differ between the applications. For example, the provision of consumer changes to a trouble call/OMS system from the CIS are such that a batch update once in 24 hours is sufficient, whereas the maintenance of correct topology within the same application is crucial and will require real-time data transfer from the SCADA system

Central Control and Management

for all network connectivity changes. These will become excessive during storm conditions and for systems with a high AIL.

The most complex interface is that between the SCADA/DMS and GIS systems, because not only do data model inconsistencies have to be resolved but the level of interface has to be decided in not only the design of the DMS but, more importantly, the degree of data modeling and extent of data population within the GIS. The previous remark applies to the connectivity and parameter model for electric network applications. In addition to the network data, are the two other data categories covering the following:

- Picture data (symbology, coloring, text placement, measurement display, etc.)
- SCADA data (point data, data acquisition system addressing, etc.)

The data maintenance responsibility within the utility organization has to be defined in terms of not only the master database for the as-built and as-operated network but also the data change process for updating each of the views. Such a decision is crucial to the master direction of data flow of the interface, plus the frequency and type of data to be transferred. Different levels of interface have been established within the industry often dictated by a legacy implementation, particularly for GISs where the database had been populated for asset management application only and sometimes expanded to include supporting engineering applications. The level of interface possible is dictated by the exact implementation of the various IT systems due to the data availability within the implemented data model. Differences predominate even if the applications are implemented at the same time but by different autonomous departments without completion of an enterprise wide data architecture design study. At present, without standards and guidelines within the industry across all the issues above, every GIS/SCADA/DMS interface has to be customized using common IT data transport mechanisms (CSF, ASCII, XML). Seldom is any detailed specification made prior to customization, setting interface levels, data availability and requirements of interfacing applications, and (of vital importance) the naming conventions and element definitions being employed.

Typical levels of interface are given in Table 2.9 as examples on which detailed designs can be based.

As the GIS delivers less and less data, the missing data have to be added by the data engineering application of the SCADA/DMS. In cases where load and customer data are not stored in the GIS database as network attribute data this has to be imported or obtained from a different source such as CIS using a batch transaction base interface.* The above discussion is highly simplified but serves

---

* This is also usually the source from which the GIS would have obtained it if included in the GIS data model.

TABLE 2.9
Typical GIS/SCADA/DMS Interface Levels Summarizing Data Provided by Each Application, and Data Transfer to and from Each Application

Level	Description	Master	Data Added with SCADA Data Eng Tool	GIS/SCADA/DMS	SCADA/DMS/GIS
1	GIS database is enhanced to add all required SCADA data as attributes to the asset data including RTU data All operating diagram displays are implemented and maintained in the GIS	GIS	None	SCADA point data Picture data Network parameter data Connectivity	Optional Device status changes Selected measurands
2	GIS database of network attribute data and graphic displays maintained, including attribute data required for SCADA operation of the network	GIS for all network model data and displays	All point data, RTU information, linking of point data, and network model data and picture linking	Picture data Network model data, including parameter data Connectivity	Optional Device status changes Selected measurands
3	GIS database of connectivity and GIS-native[a] network attribute data	GIS for connectivity and GIS native network parameter data SCADA for remainder	All point data, RTU information, complementing network model data for SCADA operation, operating diagrams, and picture linking	Network data Connectivity and GIS-native parameter data	Optional Device status changes Selected measurands
4	GIS database of native network parameter data	GIS database of network attribute data SCADA for remainder	All data for operations with exception of GIS-native parameter data	GIS-native network parameter data	Optional Device status changes Selected measurands
5	GIS without network parameter or connectivity data or no GIS implementation	SCADA	All data from diverse sources	None	None

[a] GIS-native data: Data that has been entered into the GIS solely for the need of GIS-resident applications and without reference to the requirements of the SCADA/DMS.

TABLE 2.10
Responsibilities of Each IEC Working Group

IEC TC 57 Working Group	Topic	EPRI UCA2	EPRI CCAPI
WG 3,10,11,12	Substations	✔	
WG 7	Control centers	✔	
WG 9	Distribution feeders	✔	
WG 13	Energy management systems		✔
WG 14	Distribution management systems		✔

to illustrate that not only must the roles of each application in terms of data maintenance responsibility by type of data be made available, but the data model must also be defined at the outset of implementation of the two functions if complex or native interfaces are to be avoided.

2.12.4 DATA MODEL STANDARDS (CIM)

The industry is, through IEC Technical Committee (TC) 57, developing standard data models and business structures, starting with transmission systems that will be extended to cover distribution networks. IEC 61970-301 in Working Groups (WG) 13 (EMS) and 14 (DMS) are developing the common information model (CIM).* The principal task of the overall standard is to develop a set of guidelines, or specifications, to enable the creation of plug-in applications in the control center environment, thus avoiding the need to customize every interface between different applications from different suppliers. There are a number of WGs that are either directly or indirectly associated with the development of this standard and EPRI† is also contributing through two major projects (Table 2.10).

Although at present, this part of the standard, IEC 61970-301, defines a CIM that provides a comprehensive logical view of energy management system (EMS) information, the standard is a basic object-oriented model extendable for distribution networks. The CIM is an abstract model that represents all the major objects in an electric utility enterprise typically contained in an EMS information model. This model includes public classes and attributes for these objects, as well as the relationships between them. The CIM is part of the overall EMS-API‡ framework. The purpose of the EMS-API standard is to facilitate the integration of EMS applications developed independently, between entire EMS systems developed by different vendors, or between an EMS system and other systems concerned with different aspects of power system operations, such as generation or distribution management. This is accomplished by defining standard application program interfaces to enable these applications or systems to

---

* IEC 61970 bases its concepts on the EPRI Control Center API (CCAPI) research project (RP-3654-1).
† EPRI — Electric Power Research Institute, United States.
‡ API — Applications program interface.

access public data and exchange information independent of how such information is represented internally. The CIM specifies the semantics for this API. Other parts of this standard specify the syntax for the API.

The objects represented in the CIM are abstract in nature and may be used in a wide variety of applications. As stated earlier and of importance to the content of this book, the use of the CIM goes far beyond its application in an EMS. This standard should be understood as a tool to enable integration in any domain where a common power system model is needed to facilitate interoperability and plug compatibility between applications and systems independent of any particular implementation. It provides the opportunity of common language between different data structures of different applications, wherein each data structure has been optimized for the performance of that application.

CIM Model Structure. The CIM is defined using object-oriented modeling techniques. Specifically, the CIM specification uses the Unified Modeling Language (UML)* notation, which defines the CIM as a group of packages. Each package in the CIM contains one or more class diagrams showing graphically all the classes in that package and their relationships. Each class is then defined in text in terms of its attributes and relationships to other classes.

The CIM is partitioned into a set of packages. A package is just a general-purpose grouping mechanism for organizing the model. Taken as a whole, these packages comprise the entire CIM.

The CIM is partitioned into the following packages for convenience:

Wires. This package provides models that represent physical equipment and the definition of how they are connected to each other. It includes information for transmission, subtransmission, substation, and distribution feeder equipment. This information is used by network status, state estimation, power flow, contingency analysis, and optimal power flow applications. It is also used for protective relaying.

SCADA. This package provides models used by supervisory control and data acquisition applications. Supervisory control supports operator control of equipment, such as opening or closing a breaker. Data acquisition gathers telemetered data from various sources. This package also supports alarm presentation.

Load Model. This package provides models for the energy consumers and the system load as curves and associated curve data. Special circumstances that may affect the load, such as seasons and data types, are also included here. This information is used by load forecasting and load management packages.

Energy Scheduling. This package provides models for schedules and accounting transactions dealing with the exchange of electric power between companies. It includes megawatts that are generated, consumed, lost, passed through, sold, and purchased. It includes information for transaction scheduling for energy, generation capacity, transmission, and ancillary services. It also provides information needed for OASIS† transactions.

---

* The UML notation is described in Object Management Group (OMG) documents and several published textbooks.
† OASIS — Organization for the Advancement of Structured Information Standards.

Central Control and Management

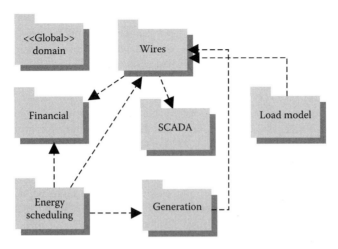

FIGURE 2.26 CIM package relationships diagram.

This information is used by accounting and billing for energy, generation capacity, transmission, and ancillary services applications.

Generation. The generation package is divided into two subpackages: production and operator training simulator (OTS).

Financial. This package provides models for settlement and billing. These classes represent the legal entities who participate in formal or informal agreements.

Domain. This package provides the definitions of primitive datatypes, including units of measure and permissible values, used by all CIM packages and classes. Each datatype contains a value attribute and an optional unit of measure, which is specified as a static variable initialized to the textual description of the unit of measure. Permissible values for enumerations are listed in the documentation for the attribute.

Figure 2.26 shows the packages defined above for the CIM and their dependency relationships. A dashed line indicates a dependency relationship, with the arrowhead pointing from the dependent package to the package on which it has a dependency.

CIM Classes and Relationships. Within each CIM package are classes and objects and their relationships. These relationships are shown in CIM class diagrams. Where relationships exist between classes in other packages, those classes are also shown identifying the ownership package.

A class is a description of an object found in the real world, such as a transformer, switch, or load that needs to be represented as part of the overall power system model. Classes have attributes, each attribute having a type (integer, floating point, boolean, etc.), which describes the characteristics of the objects. Each class in the CIM contains the attributes that describe and identify a specific instance of the class. CIM classes are related in a variety of ways given below, which describe the structure and type of relationship.

Generalization. A generalization is a relationship between a general and a more specific class. The more specific class can contain only additional information. For example, a transformer is a specific type of power system resource. Generalization provides for the specific class to inherit attributes and relationships from all the more general classes above it. In the schemas, the relationship is depicted as an arrow pointing from the subclass to the general class.

Simple Association. An association is a connection between classes that can be assigned a role. For example, there is a Has A association between a transformer and a transformer winding. In the schemas, this is shown as an open diamond pointing to the higher class.

Aggregation. Aggregation is a special case of association. Aggregation indicates that the relationship between the classes is some sort of whole-part relationship, where the whole class "consists of" or "contains" the part class, and the part class is "part of" the whole class. The part class does not inherit from the whole class as in generalization. Two types of aggregation exist, composite and shared. The Consists Of, Part Of labels are used in the schemas.

Composite Aggregation: Composite aggregation is used to model whole-part relationships where the multiplicity of the composite is 1 (i.e., a part belongs to one and only one whole). A composite aggregation owns its parts (e.g., a topological node could be a member of a topological island).

Shared Aggregation: Shared aggregation is used to model whole-part relationships where the multiplicity of the composite was greater than 1 (i.e., a part may be a part in many wholes). A shared aggregation is one in which the parts may be shared with several aggregations, such as a telemetry class may be a member of any of a number of alarm groups. The Member Of label is used in the schemas.

In the schemas showing the association and aggregation relationships, the possible extent of the relationship is given as one of the following:

- (0..*) from none to many
- (0..1) zero or one
- (1..1) only one
- (1..*) one or more

These rules are shown diagrammatically in Figure 2.27, illustrating some of the relationships in the wires and SCADA packages.

CIM Specification. Each CIM class is specified in detail by attributes, types, and relationships. Building on the nomenclature in the above section, an example is given introducing not only the connectivity attribute but also the cross package relationship that a wires package class would have with the SCADA class. Connectivity is modeled by defining a terminal class that provides zero or more (0..*) external connections for conducting equipment. Each terminal is attached to zero or one connectivity nodes, which in turn may be a member of a topological node (bus). Substations that are considered subclasses of a power system resource must have one or more connectivity nodes. The Connected To association

Central Control and Management

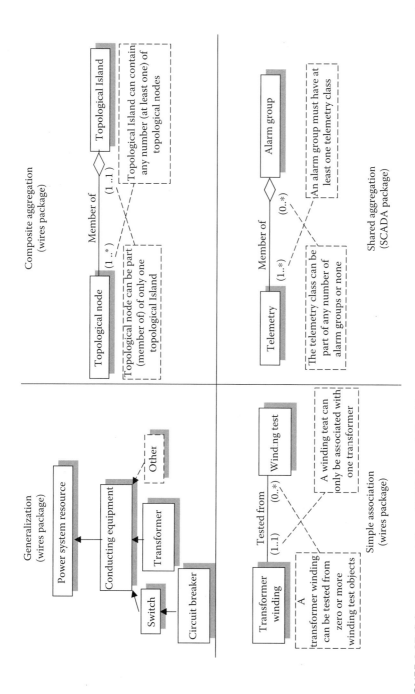

FIGURE 2.27 Schema representation of CIM class relationships.

uniquely identifies the equipment objects at each end of the connection. The relationship to the SCADA package is established through the terminal association with a measurement that can be zero or many. The complete schema of this subset within the wires package is shown in Figure 2.28.

Each object in the CIM model is fully defined by a standard set of attributes and relationships either unique (native) to the class or derived from the whole or superior class. An example is shown for a breaker class as follows:

Breaker attributes
- Native attributes
  - Fault rating (amps)
  - Breaker type (oil, SF6, vacuum, etc.)
  - Transit time (sec) from open to close
- Inherited attributes from:
  - Conducting equipment class
    - Number of terminals
  - Power system resource
    - Name of power system resource
    - Descriptive information
    - Manufacturer
    - Serial number
    - Location: X,Y coordinate or grid reference
    - Specification number if applicable
  - Switch
    - Modeling flag designation of real or virtual device for modeling purposes
    - State open or closed
    - Switch on count number of operations since last counter reset

Breaker associations
    Native roles
        (0..*) Operated by (0..*) IED breakers can be operated by protective relays, RTUs etc.
        (1..1) Has a (0..*) reclose sequence
    Roles inherited from:
        Conducting equipment
        (0..1) External connection for (0..*) terminal
        (0..*) Protected by (0..*) protection relay
        (1..1) Has a (0..*) clearance tag
        Power system resource
        (0..1) Measured by (0..*) measurement
        (1..1) Has a (0..1) outage schedule
        (0..*) Member of role A (0..*) power system resource
        (0..*) Member of role B (0..*) power system resource
        (0..*) Member of company (0..*) company PSR may be part of one or more companies

Central Control and Management

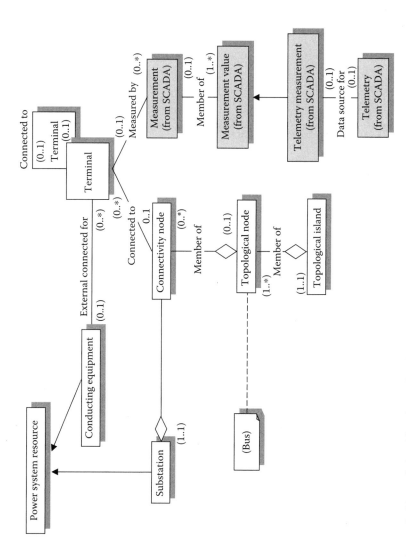

FIGURE 2.28 Class and object relationship schema for a portion of the CIM Wires Package showing treatment of connectivity and association with SCADA Package.

(0..*) Operated by (0..1) company PSR may be operated by one company at a time
Switch
(0..*) Has a (0..*) switching schedule A switch may have a switching schedule

A typical CIM model structure for the majority of wire package classes is shown diagrammatically in Appendix 2A at the end of this chapter. This sample is for illustration purposes only and should not be taken as a comprehensive CIM model for distribution. The reader interested in the details of the entire standard is recommended to study the technical committee and working group publications, because any further detail is outside the objectives of this book.

2.12.5 Data Interface Standards

Development of standards to achieve both vertical and horizontal integration in a plug-and-play manner for DMSs is another of the goals of WG 14. The WG has defined the business activity functions within a distribution utility for which enterprise application integration (EAI) is required. These business segments or departments, though, are supported by more than one IT application as mapped in Table 2.11 below. The interface architecture will form the part of the eventual IEC standard that will rely on the definition of an interface reference model (IRM) and messaging middleware accommodating this business segmentation, where wrappers or common interfaces attach each application to the message bus.

The IEC work has defined the above business activity segments in more detail in its publications and reports.

2.13 SUMMARY

This chapter is aimed at giving an overview of the major functions that comprise central control activities. The treatment is far from exhaustive, because the whole area is worth a volume on its own. The evolution of DMS systems endeavors to explain the configuration of many legacy systems, whereas the real-time components of most present DMSs are now tightly integrated internally. Data requirements of different applications that play in the DMS arena are explained to highlight where effort has to be applied before implementing a DMS. External integration with other enterprise IT systems is a developing requirement with the potential to deliver benefits not only from process efficiency but also from improved information allowing better policy setting. Until standards are accepted and implemented across all enterprise applications, the interfaces between any two applications (e.g., SCADA/GIS, SCADA/CIS) should be defined at the outset of a project before individual procurement of the packages to ensure a correctly working solution. At present, even though standard IT data transfer protocols are commonplace, data and data modeling inconsistencies between applications dictate that interfaces be customized for most implementations. The need for stan-

TABLE 2.11
Business Segmentation Treatment of DMS Integration by IEC TC 57 WG 14

	Systems				Metering and Load Management	Billing	Work Management
Business Functions	SCADA/DMS	Network Calculations	GIS	CIS			
Operations planning	X	X	X	X			X
Records and asset management			X				
Network operations	X	X	X	X			X
Maintenance and construction			X	X			X
Network expansion planning	X	X	X	X			X
Customer inquiries	X		X	X	X	X	X
Meter reading and control			X		X	X	X
External systems	X	X	X		X	X	X

dardization within EAI has been recognized by the industry, and the IEC is working towards such a standard in TC 57 WG 14. The integration of the many functions and the use of value-adding data will enhance the ability to improve the management of the utilities network assets. This work is introduced in summary here for completeness and to highlight the role of integration in any modern DMS implementation. The reader with further interest in these standards should consult the more extensive reports from the IEC working groups.

APPENDIX 2A
SAMPLE COMPREHENSIVE CIM STRUCTURE

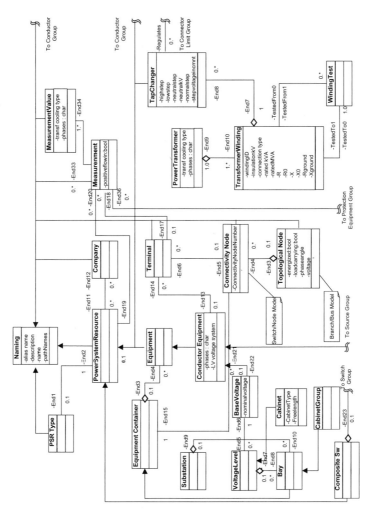

Partial diagram for a distribution network based on CIM principles. (Courtesy of ABB.)

REFERENCES

Ackerman, W.J., Obtaining and Using Information from Substations to Reduce Utility Costs, ABB Electric Utility Conference, Allentown, PA, 2000.

Antila, E., Improving Reliability in MV Network by Distribution Automation.

Apel, R., Jaborowicz, C. and Küssel, R., Fault Localization in Electrical Networks — Optimal Evaluation of the Information from the Electrical Network Protection, CIRED 2001, Amsterdam, June 18–21, 2001.

Apostolov, A., Distribution Substation Protection, Monitoring and Control Systems with Web Browser-Based Remote Interface, Distributech Europe, Berlin, Nov. 6–8, 2001.

Becker, D., Falk, H., Gillerman, J., Mauser, S., Podmore, R. and Schneberger, L., Standards-Based Approach Integrates Utility Applications, *IEEE Computer Applications in Power*, 13, 4, Oct. 2000.

Bird, R., Substation Automation Options, Trends and Justification, in *DA/DSM Europe Conference Proceedings, Vol. III*, Vienna, Oct. 8, 1996.

Cegrell, T., *Power System Control Technology*, New York: Prentice Hall, 1986.

ESRI and Miner & Miner, *Electric Distribution Models, ArcGIS™ Data Models*, Redlands, CA: ESRI.

Harris, J., Johnson, D., et al., Mobile Data for Trouble/Outage Restoration at Reliant Energy — HL&P, Distributech, Miami, Feb. 2002.

IEC 61970-301, Energy Management System Application Program Interface (EMS API), Part 301 Common Information Model (CIM), Working Group Draft Report.

Kaiser, U., Kussel, R. and Apel, R., A Network Application Package with a Centralized Topology Engine, IEEE Budapest, Aug. 29–Sep. 2, 2001.

Kussel, R., Chrustowski, R. and Jakorowicz, C., The Topology Engine — A New Approach to Initializing and Updating the Topology of an Electrical Network, *Proceedings of 12th PSCC*, Dresden, Germany, Aug. 19–23, 1996, pp. 598–605.

Lambert, E. and Wilson, R.D., Introducing the First Part of a New Standard for System Interfacing for Distribution Management Systems (DMS), CIGRE, Paris, 2002.

MacDonald, A., *Building a Geodatabase — GIS by ESRI™*, Redlands, CA: ESRI.

Nodell, D.A., Locating Distribution System Outages Using Intelligent Devices and Procedures, Distributech, San Diego, Jan. 2001.

Rackliffe, G.B. and Silva, R.F, Commercially Available Trouble Management Systems Based on Workstation Technology, ABB Electric Utility Conference, Raleigh, NC, 1994.

Robinson, G., Key Standards for Utility Enterprise Application Integration (EAI), IEC TC 57 Report, http://www.wg14.com/.

Schulz, N., Outage Detection and Restoration Confirmation Using a Wireless AMR System, Distributech, San Diego, Jan. 2001.

Skogevall, A., S.P.I.D.E.R. Basics, ABB Network Partner, Vasteras, Sweden, 1995.

Sumic, Z., Fuzzy Logic Based Outage Determination in Distribution Networks, Distributech, San Diego, Jan. 2001.

Vadell, S., Bogdon, C., et al., Building a Wireless Data Outage Management System, Distributech, Miami, Feb. 2002.

Varenhorst, M., Keeping the Lights On, *Utility Automation*, July/August 2000, 19–22.

3 Design, Construction, and Operation of Distribution Systems, MV Networks

3.1 INTRODUCTION

Distribution system design varies in philosophy from country to country. The resulting standards employed in any country are dependent on the original ways that electrification was introduced either as a pioneering country or as a result of engineering sources adopted by countries electrifying at a later date. There are many texts published on the design and planning of electric distribution networks [1, 2, 3] that cover all aspects. There, are, however specific aspects of distribution network design that directly affect the implementation of effective distribution automation. Specifically, the network structure and therein the flexibility for reconfiguration together with the grounding (earthing) methods that impact the determination of fault location. These two areas, as they affect selection of distribution automation systems, will be thoroughly examined in this chapter.

This chapter only looks at the aspects of network design where there is an interaction with automation and provides a basic knowledge of this interaction, specifically:

- That the performance of a network is a function of its switchgear content, its structure and level of automation
- That the time an operating crew spends on a network will depend on it structure and level of automation
- That system reinforcement might be delayed if automation is used to provide a fast-acting method of load transfer
- That automation can have a significant effect on control of system voltage
- That, because most customer outages are caused by the medium-voltage (MV) network, utilities tend to examine this first but are now recognizing that attention to low-voltage (LV) networks can bring benefits

A distribution network is a network that distributes. To distribute is to spread something, and in this context, that something is electrical power. A network is a system of interconnected lines. So, a distribution network is a system of interconnected electric lines, which spreads electrical power from a source to a number of load points. A network normally operates at one voltage, for example, medium voltage or low voltage, and is connected to different voltage networks using transformers. For example, a 20 kV network may be connected to a 110 kV network at a 110/20 kV substation, and the 20 kV network may be used to supply a low-voltage network using 20/0.4 kV transformers.

A very simple distribution network of one feeder is shown in Figure 3.1. At the top of the diagram is a busbar at the source substation operating at 20 kV, and one feeder is shown leaving this busbar. The feeder supplies 11 distribution substations, each with a transformer to step down to 400 volts and then to feed a low-voltage network. A number of different types of switchgear might be contained within this feeder, which will be dealt with later.

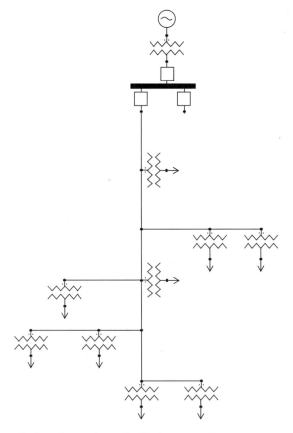

FIGURE 3.1 Typical medium voltage distribution network.

In practice, a network may consist of more than one feeder connected to the source busbar, and the resulting network can be quite complex, as shown in Figure 3.2.

This network provides 50 MVA of supply to a small town in Germany and operates at 20 kV. It is supplied from two 110/20 kV transformers, one in each of two source substations and each feeding a part of the network, but never operating in parallel. Two of the busbars are actively supplying the network; the remainder may have been supply points in the past, but without transformers, they are now just switching points that enable the network to have some operational flexibility. In fact, this network has a third transformer, normally left connected to the 110 kV system but with its 20 kV circuit breaker left open, often called "on standby," to provide an alternative supply if one of the other transformers were to fail.

It has been quoted that to design a network that just works is quite easy, but to design one that works well is quite difficult, and to design one that works extremely well is extremely difficult. But, because the distribution network is the prime physical asset of the distribution company, it makes engineering and economic sense to design the network to be the best possible to meet the owner's requirements.

The optimum design of a distribution network is a complex process that depends on many variables, and sometimes the relevance of these variables depends on the utility concerned. For example, the level of harmonic distortion may be extremely important to one utility but not to another. Also, its relevance may vary between different networks within one utility.

As time goes by, the demands on the network, and hence its design, will change. Sometimes, it is pressure from the end customer that forces change, at other times it is the controlling effect of the regulation authority in deregulated utilities that precedes change. At other times, it is the development of new technology, for example, the introduction of the recloser that causes the design engineer to reappraise existing policies. The development of improved network planning software has improved the accuracy of design and has given the design engineer more options, which at first sight is a great benefit, but it should be remembered that the software is a tool and only gives the expert results when used by an expert engineer.

Each country will have its own local legislation as to how power networks should perform. The legislation is often complex, but the basic engineering requirements are that the network should be fit for its required duty and that correct protection should be applied to prevent dangerous operation. There are normally limits on voltage and frequency. Increasingly, as utilities become deregulated, there is legislation regarding the quality of supply that customers should expect to receive in terms of interruption frequency and duration.

3.2 DESIGN OF NETWORKS

Network design becomes extremely complicated because there are many variables, some of which interact with one another. The most important variables are perhaps:

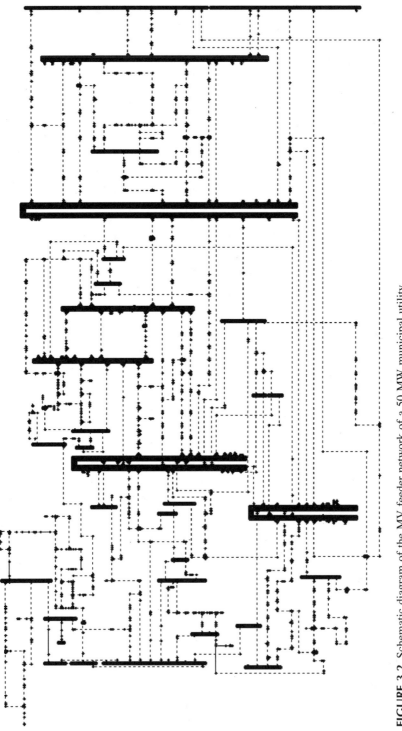

FIGURE 3.2 Schematic diagram of the MV feeder network of a 50 MW municipal utility.

Distribution Systems, MV Networks

1. The selection of voltage
2. The choice between overhead or underground networks
3. The sizing of the distribution substation
4. The structure of the upstream network
5. The required performance of the network
6. The complexity of the network
7. The requirements for voltage control
8. The requirements for current loading
9. Load growth
10. The selection of the method of system earthing
11. Losses
12. The country where the network is located
13. The cost of installation of the selected design
14. The cost of owning the network after construction

Some of these factors affect the capital cost of the network, whereas others affect the revenue cost. However, the two can be combined if the total life cycle cost is estimated as the design process is made. In this context, the total life cycle cost can be defined as the capital cost plus the net present value of the revenue costs for the life of the network. Clearly, the capital cost and revenue cost can be dependent on each other, an example being electrical losses, where an increased capital cost may decrease the value of the losses to the point that the increased capital spending is economically justified.

3.2.1 SELECTION OF VOLTAGE

In most cases, the network design engineer does not have freedom to select the voltage that a network will work at. This may be because the utility has already standardized voltage levels. For example, 11 kV is the normal distribution voltage in the U.K., and in most cases, the operating voltage of the low-voltage network is already chosen, either by law in the relevant country or by some other established practice.

The selection of voltage for the supply to industrial customers provides more flexibility, and there is a general rule that the higher the load, the higher the voltage. Some industrial customers create disturbances that will affect other customers, and this is normally alleviated by connection at the highest suitable voltage. For example, single-phase 25 kV is a common global voltage for railway traction supplies, and the supply substations are normally fed from 115 kV or higher.

There is much discussion regarding the selection of voltages between 6.6 and 33 kV for secondary distribution. The decision will depend on many factors, including the cost of the materials for the network. In Europe, voltages around 20 kV have become common, which ensures that there is a good supply of competitively priced hardware available, which in turn tends to reinforce the selection of 20 kV as the secondary distribution voltage. In contrast, voltages in the United States range between 13 and 25 kV.

But one golden rule is that, for the transfer of a given amount of power, the current needed decreases as the voltage increases. Because electrical losses are proportional to the square of the current, then it follows that longer distribution circuits could be more efficient at higher voltages.

3.2.2 Overhead or Underground

The choice between overhead networks and underground networks has, historically, been made on cost. In general terms, the overhead network is cheaper than the underground network, especially as the voltage rises. Some utilities will estimate the underground MV network as costing four to seven times the cost of an equivalent overhead network.

But there are other factors. There is an environmental lobby that lobbies for underground systems based on the aesthetics, especially in areas of outstanding natural beauty such as national parks. Also, the reliability of an underground network is improved because it is not subject to weather-related faults and outages.

The selection of voltage and overhead lines can sometimes be traced to a decision made many years ago when the situation was different. An interesting case relates to a network in Asia. At the time that the installation was made in a semirural area, the load density was low and the circuit length was relatively high. Hence, an overhead system at 33 kV was chosen, and new load could be added with a pole-mounted transformer. As time has progressed, however, the load has increased more than expected, and urbanization has taken place. The utility now wants the network to be undergrounded in the urban areas, which would mean substations with 33 kV distribution switchgear. Although at the time of original installation, 33 kV was ideal, the changed circumstances would suggest that 20 kV would be more economic nowadays.

3.2.3 Sizing of Distribution Substations

We have already seen that a low-voltage network will be supplied from a medium voltage network via the MV/LV distribution transformer, and it makes engineering sense to examine the complex design process at the distribution transformer. If we consider a model load scenario, then we can consider the effects of different transformer sizes and how this affects both LV and MV network design. We can make some preliminary investigations on the economics and the network performance.

We will choose a model consisting of 1600 equal loads of 5 kVA, laid out in geographical street map format, such that the matrix is 40 by 40 loads. Each load is 25 meters from the next in each perpendicular direction. The total load on the network is, therefore, 8000 kVA. Figure 3.3 shows the top left-hand corner of the model network, and we will use it to consider the options for location of each distribution substation and how each is connected to the MV source.

Distribution transformers can be manufactured in any size, but are commonly sold according to a range of standard sizes. In this example, we will start with one of the smallest ratings, 25 kVA and, if each load is 5 kVA, then we will need

Distribution Systems, MV Networks

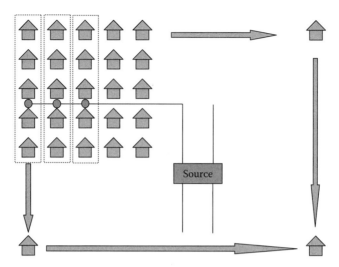

FIGURE 3.3 LV service area for 25 kVA transformer.

one transformer for each block of five loads, as shown in the diagram. We can quickly calculate the length of LV conductor needed to supply the five loads from the distribution substation.

If we have blocks of five loads, then we can see that there will be 320 substations, and we can connect these to the MV source substation, which, logically, would be in the center of the load matrix. If each outgoing circuit at the source substation were capable of 2 MVA, then we would need four outgoing circuits. We could supply 80 distribution substations from each source feeder, and if we assume that the MV circuits are laid out in a particular way, then it is relatively easy to calculate the length of MV circuit needed. Assuming cost information for the MV circuits, the LV circuits and the distribution substations, then we can evaluate the cost of this design of network.

The process could be repeated using a 125 kVA transformer (Figure 3.4), in which case each distribution substation would supply 25 loads. The length of LV circuit would be greater but the length of MV circuit would be less. Because the 125 kVA transformer will cost more and MV and LV circuits are different in cost, this solution will have a different total cost.

In the same way, we can work with 250 kVA transformers, 500 kVA transformers and 1000 kVA transformers. Table 3.1 shows the equipment quantity required. Using typical budget figures, the cost of each option has been calculated for an overhead line system and an underground cable system.

From the curve of cost per kVA of load supplied, Figure 3.5, a number of rules can be seen:

- For underground systems, as the transformer size increases, the cost per kVA load decreases to an optimum point at about the 500 kVA transformer rating.

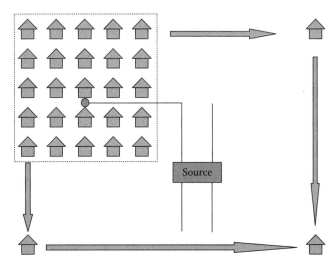

FIGURE 3.4 LV service area for 125 kVA transformer.

TABLE 3.1
Cost of Network to Supply Different Load Densities

Transformer Rating (kVA)	Transformer Count	LV Length (km)	MV Length (km)	Cost per kVA $USD	
				Overhead	Underground
25	320	32.0	9.7	128	670
125	64	38.4	8.9	80	338
250	32	39.2	5.1	68	275
500	16	39.6	4.5	77	248
1000	8	52.0	3.5	79	294

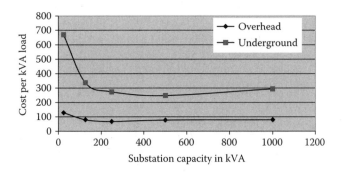

FIGURE 3.5 Plot of cost per kVA of load supplied.

Distribution Systems, MV Networks

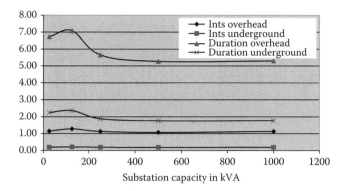

FIGURE 3.6 Outage frequency and duration for different substation capacities for a given voltage level.

- For overhead systems, as the transformer size increases, the cost per kVA load decreases to an optimum point at about the 200 kVA transformer rating.
- Underground systems are more expensive than overhead systems.

The network is probably the most basic that can be designed. The MV circuits are purely radial, connecting the source to each distribution substation only, and there are no alternative sources of supply. Hence, for a fault on any MV circuit, all customers on that circuit would be off supply until repairs had been made. The circuit breaker on each circuit at the source is fitted with basic protection only, whereas for the overhead line option, a reclosing circuit breaker would help for temporary faults. Similarly, the LV circuits only have a fuse at the distribution substation and there is no LV alternative supply, so for a fault on any LV circuit, all customers on that circuit would be off supply until repairs had been made. Now, the number of faults increases as the circuit length increases, so as the transformer size increases, then the number of LV faults will increase and the number of MV faults will decrease. We can make some calculations as to how each network design option will perform in terms of reliability, specifically the number of interruptions per year seen at the average load point and the annual outage duration seen at the average load point. These are summarized in Figure 3.6.

These curves introduce some new rules that can be added to those given above:

- The average annual outage durations for the overhead options are worse than those for the corresponding underground options.
- The average annual number of interruptions for the overhead options are worse than those for the corresponding underground options.
- For overhead systems, the average annual outage duration decreases as the transformer size increases. This statement may be misleading

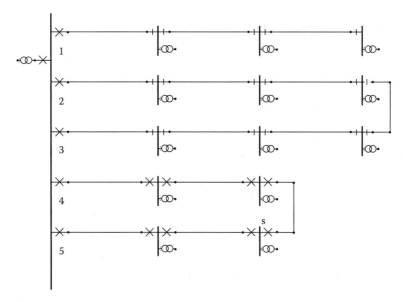

FIGURE 3.7 Typical underground MV network design.

because it is the changes in the relative lengths of MV and LV circuit that affect the outage durations.

3.2.4 Connecting the MV (The Upstream Structure)

The design of the MV network, by which the distribution substations are connected to the network source substation, plays a significant part in the cost of the overall network. It also plays a significant part in the way that the network will perform. Figure 3.7 shows some typical MV networks for underground systems, each feeder starting at the MV busbars on the left of the figure.

Feeder number 1 is a radial feeder because it starts at one end and has no method of connecting to any alternative supply. For a fault on any one section of cable, customers downstream of that fault must, therefore, wait for repairs to be made before supplies can be restored. The switchgear on this radial feeder is typically the ring main unit, described in Chapter 4.

Feeder 2 and feeder 3 form an open loop network. Each is similar to the radial feeder but they are able to mutually support each other in the case of a cable fault. For any cable fault, the customers downstream of that fault do not need to wait for repairs because they can be reconnected to supply by switching the network around. This will mean that, after downstream disconnection, the normally open point (NOP) switch between the two feeders will need to be closed, and we can see that each distribution substation can be supplied from two directions, or that each has a switched alternative supply. In this example, the switched alternative supply is made from the same source substation, but this need not be the case. If

Distribution Systems, MV Networks

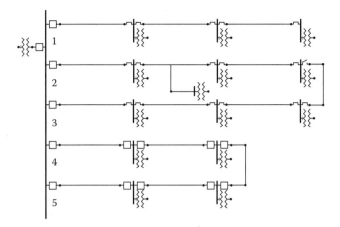

FIGURE 3.8 Typical underground MV network designs, including tee off substations.

two feeders from different source substations meet at a normally open switch, then we will still have a switched alternative supply but from two different source substations. This configuration is actually very common in practice.

Feeder 4 and feeder 5 form a closed ring and each distribution substation is fitted with a circuit breaker. The circuit breakers at each end of any cable are fitted with a unit protection scheme such that any cable fault will trip the circuit breaker at both ends of the cable. In this way, the load at any distribution substation is not disconnected for a single cable fault at any position on the closed ring. This arrangement may also be described as giving each distribution substation a continuous alternative supply, which therefore provides higher reliability.

Some utilities use a tee off substation between ring main units, which is shown added to feeder 2 in Figure 3.8.

This type of substation does not have in-line switchgear to help cut the ring into sections but only the local protection for the distribution transformer. It is, therefore, somewhat cheaper to provide, but it suffers from a disadvantage in that a cable fault in the three-way cable that connects this tee off substation to the ring will mean that the load on the tee off substation cannot be restored from the MV system until repairs have been made. Of course, it is possible to have a switched alternative supply available for the LV network supplied from this substation, which can be used for MV cable faults in the zone; it is a balance between MV switchgear savings and additional LV network costs.

Figure 3.9 shows some typical MV network arrangements for overhead networks, the source circuit breaker for each feeder now being fitted with autoreclose. Autoreclose is often fitted to overhead networks to deal with temporary faults on the overhead lines caused by weather. This is dealt with in more detail in Chapter 5. Feeder 1 is again a radial feeder but note that there is much less switchgear; in fact, there is only one in-line switch disconnector. Compared to the underground network, this reduces the cost but degrades the performance.

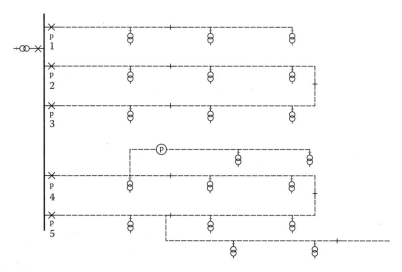

FIGURE 3.9 Typical overhead MV network designs.

Feeder 2 and feeder 3 again depict an open ring with the normally open point on the right-hand side of the diagram. The reduced switchgear content should again be noted.

Because the continuous alternative supply arrangement for overhead lines is not common, this has not been shown in Figure 3.9. Instead, feeder 4 and feeder 5 show a more typical arrangement that has both open ring and a radial components.

Feeder 4, apart from being part of an open ring, also has a radially connected spur line, in this case fitted with a pole-mounted recloser downstream of the point where it connects to the main line. Feeder 5, also part of the open ring, also has an additional line connected, but this additional line is part of an open ring to another substation via the normally open point on the right of the diagram.

3.2.5 THE REQUIRED PERFORMANCE OF THE NETWORK

We have seen that the performance of any distribution network is affected by aspects of its design such as whether it is overhead or underground and the switchgear content. The question that has to be asked now is exactly what level of performance is required for any particular network. This can be divided into two categories, meeting the requirements of the customer and meeting the requirements of national legislation. It must, of course, be pointed out that in many cases, the legislation is designed to offer a certain standard to the customer; hence, the two categories tend to merge together.

Most utilities have preformulated policy regarding the level of investment for new networks or extensions to existing networks, and these often try to take account of the needs of the customers supplied by that network. Sometimes, this

TABLE 3.2
U.K. Restoration Standards

Amount Load Affected by Fault	Restoration Standard
Less than 1 MVA	Restoration following repair
1 to 12 MVA	Restoration of customers less than 1 MVA within 3 hours
Greater than 12 MVA	No supply interruption from design perspective

is load related, whereby loads fitting into predefined ranges will have different policies — for example, a hospital over 10 MVA might be treated differently than an office with 30 kVA of load. The hospital might be supplied at MV from a continuous alternative supply, whereas the office might be supplied from the LV network. This approach has some advantages but can become complicated when different categories of load are fed from the same network. In addition, it assumes that the utility is fully aware of the needs of its customers. However, if the utility is flexible enough, then it might be able to provide a higher level of performance than that initially thought appropriate if the customer were to pay the additional costs, both capital and revenue based.

In a later chapter, the concept of customer outage costs is discussed, and it is these values that the utility needs to consider if making investment decisions on behalf of the customer. Apart from internal guidelines, many utilities are subject to national guidelines, sometimes legislation, that define security standards. For example, in the U.K., there is a national requirement that for a first circuit outage, where the load affected by a fault, restoration should be on a timescale related to the group demand initially interrupted.

In addition, regulation in the U.K. provides for guaranteed standards of service that can result in utility payments to customers in the event the standards are not met. These standards have been in place for a number of years and have limited effect (due to the size of fine and the relatively small number of events). (See Table 3.2 and Table 3.3.)

In addition to the guaranteed service standards, there are additional performance measures against which the utilities must report. There is no fixed penalty for these additional items; however, the regulator takes these performance items into account at the next price review. (See Table 3.4.)

3.2.6 THE NETWORK COMPLEXITY FACTOR

So far, we have only examined straight feeders, but of course most real networks comprise an assortment of teed connections and, sometimes, multiple open points. Software is the only practical way to evaluate the reliability performance of these designs, but it is helpful to look at a simple way of defining the overall effects of more complex networks. The complexity of the network not only influences the resultant performance but also the crew travel times.

TABLE 3.3
U.K. Guaranteed Service Standards

Service	Performance Guarantee	Penalty/Fine
Service fuse failure/replacement	3 hours (normal) 4 hours (outside normal hours)	£20 per customer
Network fault restoration	18 hours	£50 per domestic customer £100 per nondomestic customer plus £25 for each subsequent 12-hour period
Notice of service interruption	2 days' notice for planned interruptions	£20 per domestic customer £40 per nondomestic customer
Voltage complaints	Appointment within 7 days Written reply in 5 business days	£20 £20
Appointments	Attendance per agreement	£20
New/altered service estimates	Simple (no extension) — 5 days Extension required — 15 business days	£40

TABLE 3.4
U.K. Typical Regulatory Standards

Performance Category	Required Performance
Restoration of supply after failure	95% of supplies restored within 3 hours 99.5% of supplies restored within 18 hours*
Voltage complaints	Resolve within 6 months
New supplies	Provide new supply** within 30 days for domestic, 40 days for nondomestic
Written inquiries	Respond within 10 days

* Targets differ for each utility based on geography.
** Where no extension is required.

Figure 3.10 shows five feeders that are more complex, or less simple, than the straight feeder shown at the top of the figure.

As the feeders become more complex, we can see that the number of teed connections increases, also that the positioning of the tees before and after the midpoint varies. Suppose that we count the number of tees and their positions and then define the straight feeder as having a network complexity factor (NCF) of 1.0. Because the other feeders are more complex, they will have a higher NCF and we have assumed that the most complex feeder here has an NCF of 3.5. Then, by applying linear regression and rearranging the five less simple feeders

Distribution Systems, MV Networks

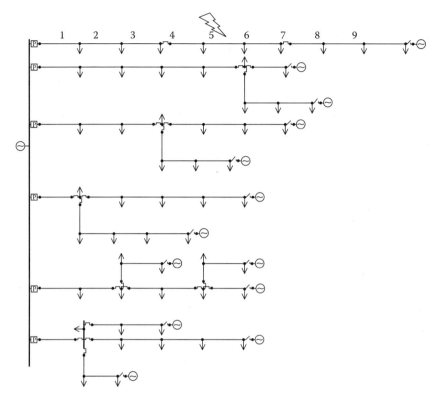

FIGURE 3.10 Networks of varying complexity.

in different orders in the diagram until the best regression fit is obtained, we find an empirical formula that

$$NCF = 2 \times N - 0.5 \times (N1 + 3 \times N2) - 1$$

where N is the number of ends in the feeder (excluding the source circuit breaker end), N1 is the number of tees in the first half, and N2 is the number of tees in the second half, as illustrated in Table 3.5.

This formula is based simply on the geographical arrangement of the feeder and could be expanded to take account of other differences in the feeders. One such parameter is the number of in line switches on the feeder but it has been found that the effect of such switches has a very small effect on the network complexity factor. It should be noted that, for a lateral connection to be correctly regarded as a teed connection, there should be some form of inline switchgear in the connection, even if only one switch at the tee point as shown in Figure 3.10. In this context, a lateral connection that does not have any form of inline switchgear could have all the loads replaced by one equivalent load, rather similar

TABLE 3.5
Values of Network Complexity Factor

Network	Number of Ends in Circuit (N)	Number of Tees in First Half (N1)	Number of Tees in Second Half (N2)	Value of NCF By Definition	Value of NCF By Calculation
1	1	0	0	1.0	1.0
2	2	0	1.0	1.5	1.5
3	2	0.5	0.5	2.0	2.0
4	2	1.0	0	2.5	2.5
5	3	1.0	1.0	3.0	3.0
6	3	2.0	0	3.5	4.0

to basic electrical network theory, and therefore cannot be counted as a teed connection.

This is amplified in Figure 3.11, which shows a circuit in Sweden and illustrates how the NCF can be calculated by using the formula and data from the feeder.

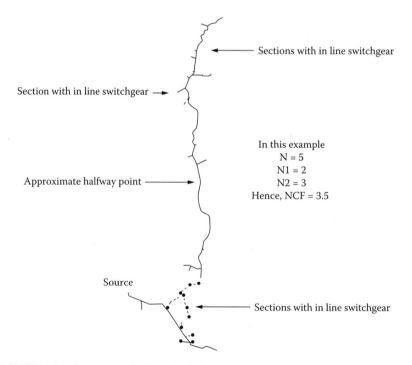

FIGURE 3.11 Illustration of NCF calculation.

3.2.7 VOLTAGE CONTROL

Because of the impedance of the cable or overhead line conductor of a distribution feeder, the load current will cause a drop in voltage along that feeder. As the load increases, either because of the variations of the daily (or seasonal) load curve or as customers apply additional loads, the voltage drop will increase correspondingly. Also, the power factor of the load has a significant effect on the voltage drop — the worse the power factor, the worse the voltage drop. However, the voltage delivered to the customer must remain within certain limits that are normally specified in the law of the country concerned. For example, European Standard EN 50160 requires that the voltage at the customer be within 10% of the declared voltage, and this standard is adopted into the national laws of each member of the European Union.

The utility operator has four major tools to control voltage on the distribution network:

- The source transformer tap changer
- The distribution transformer tap changer
- The inline voltage regulator
- The power factor correction capacitor, switched or unswitched

Voltage Control by the Source Transformer Tap Changer. The voltage ratio of the source transformer that feeds an MV busbar is typically 110/20 kV. The incoming 110 kV is applied to the whole primary winding, and the secondary voltage is taken off the secondary winding. However, one winding can have a small number of additional turns at one end, such that, by switching the incoming supply to one of the additional sets of turns, a change in the voltage ratio is achieved in proportion to the additional turns. These additional turns are known as transformer taps and are usually fitted on the primary winding simply because the tap switching mechanism then operates on the lower current of the primary winding, at least for voltage stepdown transformers. Similarly, the taps are usually located on the "earthy" end of the winding so as to reduce the voltage capabilities of the tap changing mechanism.

Suppose a transformer had a main winding with 10 additional taps at one end and that the voltage ratio of the transformer was set to the nominal 110/20 kV if the fifth tap were selected. If there were a 1% change in the voltage ratio by moving from one tap to the next, then this transformer would be able to change its secondary voltage between 95 and 105%. This selection of taps either could be made while the transformer is off load and disconnected from all sources of supply (off load tap changer), or it could be made while the transformer is live and carrying load current up to its full rated value (on load tap changer). Both options have their advantages and disadvantages.

The on load tap changer needs contact selector switchgear that can carry and switch the full load current and, usually, a motorized device for operating the contacts. It is, therefore, more expensive than the off load tap changer, but its major

FIGURE 3.12 Typical on load tap changer. (Courtesy of ABB.)

advantage is that load is not disconnected during the tap changing operation. The on load tap changer, illustrated in Figure 3.12, is commonly applied to the source transformer, whereas the off load tap changer is applied to distribution transformers.

Now, the MV substation busbar voltage will vary according to the load flowing through the secondary winding resistance of the transformer, also any variations in the incoming voltage to the primary winding. If the busbar voltage can be measured using an appropriate VT, and fed into a voltage control relay, then a motorized on load tap changer can be adjusted to control the busbar voltage to within the limits of the tap settings, in this case to within 1%. This type of control scheme is known as automatic voltage regulation or AVR.

An early AVR would have been based on a small, suspended weight that was free to move either upwards or downwards in the relay cabinet. The downwards force arising from gravity was counteracted by an upwards force that depended on the system voltage. If the voltage were too low, then the weight would be allowed to fall slightly, thereby making an electrical connection that would operate the tap changer to raise the voltage. In the opposite direction, if the voltage were too high, then the weight would be raised until it operated a contact to control the tap changer to lower the voltage. A time delay of up to two minutes would be provided to ensure that the tap changer only responded to sustained voltage exercusions.

A typical microprocessor-based voltage control relay is shown in Figure 3.13, which can be used either with hardwired auxiliary systems or with substation automation. There are three basic controls that need to be applied to the on load tap changer for correct operation. The first is the nominal voltage level of the input to the regulating relay, typically 110 VAC. The second is the bandwidth outside of which the regulating relay should operate, and the third is the time delay after which the relay should function. These are illustrated in Figure 3.14.

In this example, the busbar voltage is decreasing with respect to time until it exceeds the bandwidth, at which time the time delay starts. If the voltage is still outside the bandwidth at the end of the time delay, the regulating relay operates

Distribution Systems, MV Networks

FIGURE 3.13 Typical voltage control relay.

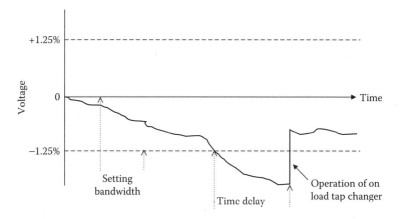

FIGURE 3.14 Principle of automatic voltage control.

the on load tap changer to restore the voltage within the required bandwidth. The time delay is needed to stop a voltage dip, say, caused by a large motor taking 20 seconds to start, from causing a tap change operation that would be cancelled by an opposite tap change after the motor has come up to speed.

This type of control corrects for the voltage that is measured at the MV busbar and can be supplemented by a system called line drop compensation, which inputs the line impedance to the regulating relay, hence taking account of the voltage drop that will occur on the distribution feeders.

Many source substations have parallel operated transformers, each fitted with voltage control. In these cases, one transformer is designated as the master and is controlled by the regulating relay, while the other(s) is the slave and always follows the commands of the master. Another system is known as reverse reactance. Transformer tap changers normally have a nonautomatic setting that facilitates the prevention of circulating currents when paralleling of distribution networks that may be fed from different sources.

Voltage Control by the Distribution Transformer Tap Changer. Distribution transformers up to a value of typically 1000 kVA can be fitted with a tap changer, although it is normally an off load device. Five taps, with ratios of −5%, −2.5%, nominal, +2.5% and +5%, are common. This tap changer is used to provide a basic control over the voltage on the LV side of the transformer, where the loads may still vary considerably over time, but where the on load tap changer would be too expensive.

However, the combination of the source transformer on load tap changer and the distribution transformer off load tap changer are capable of giving reasonable control of voltage. It should be remembered that the requirement for a limit to the voltage deviations normally applies to the customer metering point. Therefore, on networks where the customers are all on the LV system, there is no need to limit the voltage drop on the MV feeders to the appropriate limit, say 10%. The distribution transformer could be set to give a boost of 5%, which means that the MV drop can be higher.

Voltage Control by the In Line Voltage Regulator. The inline voltage regulator, also known as a booster, is a small device that is commonly used on MV overhead lines. It is a transformer of, nominally, 1:1 ratio, fitted with an on load tap changer with five taps, each of 1.25%, which is used to boost or lower the voltage as load current increases and decreases. It can be controlled by a voltage regulating relay much the same as for the source substation transformer, but for reasons of economy, it usually responds to a simple current transformer in one or more phases of the network. Some MV networks have been built without a source transformer on load tap changer because, at the time of construction, the variations in voltage caused by load cycles were insignificant. As load has grown on certain feeders, voltage problems have arisen, and one solution is to add an in line voltage regulator to the start of those feeders.

Voltage Control by the Power Factor Correction Capacitor. Capacitors have been used to correct for low power factor for many years, partly because customers with a low power factor have been charged for the supply of reactive power. But because the capacitor reduces reactive power, which gets worse as load increases, it has the effect of improving the system voltage. Locating the capacitor has always been the subject of many different views. For low power factor loads, there is an argument that correction capacitors should be located at each load point but this will lead to a large number that, because of economies of scale, could be replaced at less cost by a smaller number of larger capacitors. The opposite view is that one capacitor should be applied at the source substation busbars. It is then easier to control because it can be connected to the busbars by a circuit breaker, and it remains under the control of the utility. The circuit breaker connection to the busbar is for protection purposes, and the capacitor is rarely switched in and out.

With the advent of remotely controlled network switches, there is now an argument to apply the capacitor on each feeder. As a general guide, the best

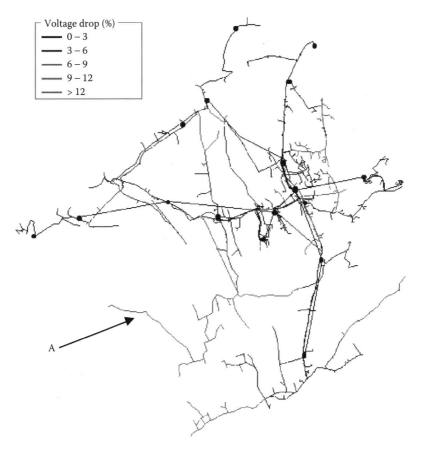

FIGURE 3.15 Application of voltage control by capacitor.

economics are obtained when a capacitor that provides two thirds of the feeder reactive needs is located two thirds of the length of the feeder.

In the network shown in Figure 3.15, the voltage drop at A has been calculated to be 17.5% at full load and 3.1% at 25% load. Hence as the load varies over the day between these two limits, the customer will see a range of voltage drop of 14.4%. Following the "two thirds" rule would place a 3 MVAr capacitor on the feeder and the voltage drop has been re-calculated as 11.3% at full load and a voltage rise of 3.1% at 25% load. The range of voltage seen by the customer remains the same at 14.4% but the actual values have changed.

What we could also do is switch the capacitor into service at times of heavy load and switch it out of service at times of light load. The voltage drop would then range from 11.3% at full load to 3.1% at low load, from which we can see that the range has decreased. This is illustrated in Figure 3.16 where the voltage seen by the customer is shown by the path "a, b, c, d, e, f."

If the off load tap changer at distribution transformers at the further end of the feeder were set to give 5% boost to the voltage, then the customer voltage,

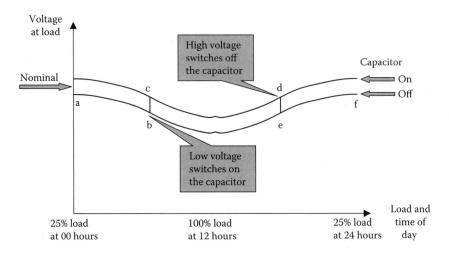

FIGURE 3.16 Variation in voltage with switched capacitors.

with switched capacitors, will range from 6.3% to 1.9% rise, which is well within the expected range.

Economic Effect of Adding Capacitors and Loss Reduction. We have already seen that the addition of capacitors to the distribution system can have a useful benefit in terms of voltage control. This is because, for a given load, the phase current is reduced by improving the power factor and, hence, bringing the current and voltage nearer to being in phase with each other. This reduction of current also has an effect on the (technical) electrical losses of any power system, and when considering the economics of power systems, it is useful to consider the economic value of these losses.

This is perhaps best illustrated by an example, which we can select as being the same real power system in Europe as used in Section 3.2.6, but, for reasons of clarity, simplified to be just one of the five feeders found on the total system. The feeder comprises 58.3 km of overhead line, part operating at 20 kV and part at 10 kV, 13.4 km cable operating at 20 kV, and 1049 kVA of installed load, all load being of 0.9 power factor. In addition, there is a single-transformer 20/10 kV substation at about the midpoint of the feeder, and this transformer is not equipped with any automatic voltage control. The feeder is supplied from a 110 kV source comprising a single 110/20 kV transformer with on load automatic tap changer, and we have assumed that the load is at the maximum demand value.

Four alternatives have been considered:

1. The basic condition with no capacitors added to the system.
2. The addition of a 140 kVAr capacitor at the 20 kV busbars of the source substation; the value of 140 kVAr being chosen to bring the system power factor from the 0.9 found on the basic system to a suggested

TABLE 3.6
Variation in Losses as a Function of Installed Capacitance

	Basic Condition, No Capacitance	140 kVAr at Source Busbar	100 kVAr at Two-Thirds Location	150 kVAr at Two-Thirds Location
Line losses, kW	23.10	23.00	20.20	19.80
Cable losses, kW	2.20	2.20	2.00	1.90
Core loss, kW	8.50	8.50	8.50	8.50
Winding losses, kW	1.40	1.30	1.20	1.20
Total losses, kW	35.20	35.00	31.90	31.40
Busbar volt drop, %	0.42	0.32	0.34	0.31
Feeder end volt drop, %	9.52	9.41	8.40	7.85
Busbar power factor	0.90	0.95	0.93	0.95

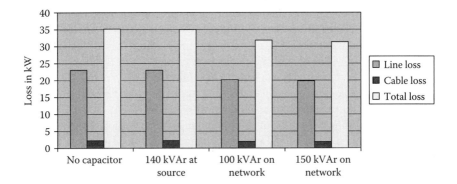

FIGURE 3.17 Variation in losses as a function of installed capacitance.

value of 0.95 — this value of 0.95 has been suggested as being suitable to optimize the tariffs for the utility purchase of power at that substation.
3. The result of the application of the two-thirds rule* whereby a 100 kVAr capacitor would be added two-thirds of the way down the feeder.
4. The addition of 150 kVAr at the two-thirds point, this value being chosen to improve the power factor at the source substation busbars to 0.95.

The results of the calculations are given in Table 3.6 and partially displayed in Figure 3.17.

---

* The two-thirds rule is a quick way of estimating the optimum value of capacitors to be added to a power system on a feeder basis. It suggests that a capacitor to compensate for the reactive power flow at the source end of the feeder should be of two-thirds of the value of that reactive flow and be placed at the two-thirds of feeder length position.

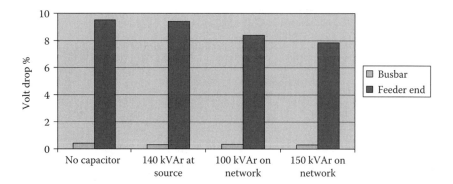

FIGURE 3.18 Range of voltage drops for different capacitor installations.

We can see that there are significant reductions in the total losses (caused primarily by the reduction in line losses) when power factor correction capacitance is added to the network rather than to the source busbar. The losses in this example have been calculated at the maximum demand of the feeder and, of course, as the daily (annual) cyclic load curves are applied, then the actual value of the savings will change, but the percentage change for each alternative will not vary. Whether these savings in losses are significant to the utility will depend on how electrical losses are evaluated by the utility, but it is recommended that the savings should be estimated as a matter of routine.

Figure 3.18 shows the plot of the variation in voltage drop, from which it can be seen that the range of change in voltage seen by the customer is reduced for the condition of capacitance being added to the network.

3.2.8 Current Loading

Each and every item of an electrical plant will have defined maximum current that it can safely carry. For example, a particular design of 185 mm^2 33 kV single-core cable with XLPE insulation, when laid in trefoil, bonded, and earthed at each end, can safely carry 560 amps when laid in air, 450 amps when laid in the ground (under specified conditions) and 400 amps when laid in ducts. If load higher than the ratings is applied, then the associated rise in temperature will degrade the cable insulation and may reduce its useful service life. Because network load varies on both a daily basis and an annual basis, many utilities apply an acceptable overload factor to their cables, for example, 10% continuous and 20% for three hours. The rationale behind an overload factor is that, for much of the life of the cable, it will operate significantly below its rated current, so for a short period, maybe during an abnormal system condition lasting for a few days, it is acceptable to take the small risk of longer term overload damage.

As for all plant items, the transformer has a defined load rating that is usually linked to the normal method of cooling. A typical oil-filled transformer might have its insulating oil circulating by natural convection and the cooling air passing

Distribution Systems, MV Networks

over the radiators, again by natural convection, and be assigned a given rating based on these two methods of cooling. There are two ways of increasing the rating of the transformer, first by pumping the oil through the tank and radiators, and second by using air fans to pass more air over the radiators. Both will have the effect of removing more heat from the transformer, thereby allowing an increased load without increasing the temperature of the transformer windings. Of course, some transformers have their rated power defined assuming that oil pumps and air fans are operating all the time, and in these cases, there is no easy way of increasing the rating. Some utilities will assign two ratings to their transformers, one being the continuous rating without pumps and fans and the other corresponding to using both pumps and fans and having a maximum transformer temperature that will cause a known rate of accelerated deterioration of the transformer insulation. Such cases can permit a doubling of the rating.

Great care must be taken to ensure that the load flow in all components of a distribution system takes account of the following:

- The assigned rating of that component
- The increase in loads that will be connected to the system during the life of the project, or at least until some form of reinforcement is expected
- All reasonably foreseeable abnormal operating conditions
- Any assigned overload margins

However, we shall see now that extended control can take advantage of the short-term thermal rating of plant to be able to accept an increase in load flowing for the corresponding time period.

3.2.9 LOAD GROWTH

In most electricity networks, there is a steady increase of load with respect to time, and rates of maybe 5% per year are not uncommon. At the same time, a given network may be subject to a more sudden and specific load growth, for example, if a new industrial plant, maybe requiring 20 MVA of supply, is proposed by industrialists. The important point is that this growth of load can, within limits, be foreseen and appropriate action taken. It would be wise for a utility to have a general reinforcement plan based on, perhaps, a five-year period and to coordinate this with a two-year plan that is more specific and would lead to a firm construction plan.

But as the load on a substation increases, another factor will become paramount a long time before the plant suffers from continuous overload. The firm capacity of a substation is normally based on the capacity that remains after the first circuit outage, commonly known as the "N-1" approach.

Figure 3.19 shows two substations, each fitted with two transformers of continuous rating 10 MVA and of emergency rating 11 MVA supplying a load. Let us suppose that the load on each substation is 9 MVA but increasing at a steady annual rate. Based on the N-1 rule, the firm capacity of each substation

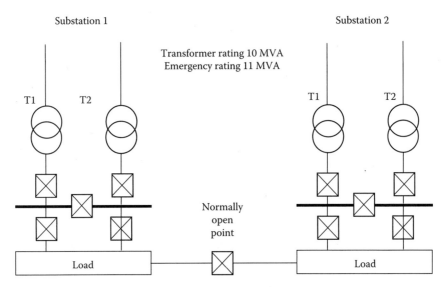

FIGURE 3.19 Typical source substation arrangement.

is 11 MVA, and with the 9 MVA load, the situation is satisfactory. But as the load increases to 11 MVA and above, then, for a first outage, the substation is not firm and the utility would reinforce the substation, either by installing two larger transformers or by installing a third transformer. Let us suppose that the annual increase in load prompts reinforcement 4 years from now.

We have already seen that it is common to have one or more normally open switches on the network supplied by a substation to provide alternative supplies for fault or maintenance situations. The diagram shows the normally open point(s) as being between the two substations. We shall now consider that the normally open connection between the substations could be used to transfer 4 MVA of load from substation 1 to substation 2 following the loss of one of the transformer circuits in substation 1, and this situation is explained in Figure 3.20.

Starting at Point A, we can see the load growing until it reaches Point B, where it matches the substation firm capacity with two transformers of 11 MVA. At that point, in year 4, the addition of the third transformer is one of the required options. But the other option is to operate the normally open connections as a load transfer scheme, which is shown as the line B to C in the diagram. After load transfer becomes available, the load continues to increase along the line C to D, and at Point D the substation again becomes unfirm. At Point D, the third transformer will be needed. But we can see that the load transfer scheme has delayed the expenditure on the third transformer from year 4 to year 10 or 11. This deferral of expenditure is, of course, of great value to any utility.

This type of load transfer can easily be provided by using the utility crews to travel to the network switches and operate them as and when required. There is a serious restriction, though, in that the plant ratings of the circuits left in use after the first circuit outage will dictate that the load transfer be completed before

Distribution Systems, MV Networks

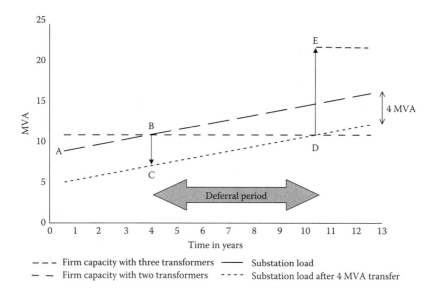

FIGURE 3.20 Deferral of capital expenditure arising from load transfer.

thermal damage occurs, which would typically be two or three hours. What is far better than using utility crews is to employ extended control of the normally open connections to carry out the load transfer.

We can say that by increasing the firm capacity of a substation, load transfer will permit the deferral of capital expenditure, and that load transfer by extended control will improve the speed of the transfer and increase the savings with respect to crew time. The deferral period can be calculated from

$$\text{Deferral Period} = \frac{\text{Capacity of Load Transfer Scheme}}{\text{Annual Load Increase}}$$

A further application of load transfer by extended control can be on a regular, daily basis. Let us consider two substations where one has its peak load during the working day and the other has its peak load in the evening. This situation could arise for a daily industrial load, which decreases when the workforce travel home in the evening and then start to create demand from their homes. If the network capacity exists to provide the transfer of load, in one direction in the morning time and in the other direction in the evening time, then the load transfer scheme may be able to defer reinforcement of one of these two substations.

3.2.10 Earthing (Grounding)

One basic problem for the distribution network designer is the way that the neutral is earthed because it affects a number of technical and economic solutions. The most important items are the relay protection, operation principles, equipment earthing and power system component selection. The choice between methods is fundamentally between solid earthing, resistance earthing, compensated earthing and unearthed (isolated) neutral, although limited combinations can occur.

In solid earth systems, the neutral point, usually the source transformer star point, is connected directly to earth. Where the ground earth resistivity is good, this produces high earth fault currents, which are relatively economic to detect and clear quickly, but it can cause high step and touch potential in the vicinity of the fault. Because approximately 80% of faults on overhead lines are of a transient nature, the reclosers needed to trip and then restore supply can use relatively basic protection schemes. However, the fundamental principle of using a recloser* is that it converts what would be a sustained interruption (for example, two hours) into a transitory interruption (for example, 30 seconds). The earthing connection also provides a direct voltage reference for the neutral voltage.

In resistance earthed systems, the neutral point is connected to earth via a resistor, which may be the natural ground resistance in areas of high resistivity or a separate power system resistor. This resistor is able to control the earth fault current to a value selected by the network operator and may be associated with a reduced cost of earthing electrodes in the ground. The earthing connection also provides a direct voltage reference for the neutral voltage.

In compensated systems, the neutral point is connected to earth by a reactor that is tuned to balance, or nearly balance, the capacitive current. Compensation is widely used in the cases where the continuity of supply is important and where earthing circumstances are good enough so that a sustained earth fault can be allowed. The penalty of this arrangement is the protective relays and high accuracy demands for the voltage and current transformers. The disadvantage of both unearthed and compensated neutral systems is the overvoltages during earth faults. This is the main reason for these solutions being less popular in distribution systems with a lot of underground cables or in networks with electrical machines, as in industrial plants.

In unearthed systems, the neutral point is left completely disconnected from earth; hence, the voltage reference is provided by the phase capacitance to earth. As the network size increases, so does the network capacitance and the duty required of switchgear, both to clear faults and for routine network switching operations. In such cases, reactance is added to compensate for the capacitive current, and the network becomes a compensated network.

There is no single answer to the selection, but a variety of approaches exist in different countries and in the different power system cultures.

3.2.11 Lost Energy

Lost energy can be taken in two distinct areas, energy not supplied and electrical losses. Electrical losses can also be regarded as being in two areas, technical losses that arise from the operation of the distribution systems itself and nontech-

---

* The methodology of reclosers is considered in Chapter 4.

nical losses, otherwise known as illegal abstraction of electricity or theft, the latter being outside the scope of this book.

Energy Not Supplied. Unless there is a continuous alternative supply available, any fault on a power system that causes protection to operate will cause loss of supply to one or more customers. We will describe in Chapter 6 how this loss can be characterized in terms of reliability indices such as SAIFI and SAIDI. But the loss of supply can also be described according to the energy that is lost as a result of the protection operation. Some utilities use the expected loss of energy per year as a benchmark to compare the reliability performance of alternative network options and usually attribute a financial value to the kWh lost. Other utilities may be subject to a penalty payment to each affected customer that is based on the kWh lost (discussed in Section 8.10.1) — in this way, the penalty is proportional to the load as well as the duration of the outage, which many consider to be a fair method of compensation for the larger customers.

There is a very simple method by which the expected energy loss can be calculated — simply multiply the disconnected load (in kW) by the outage duration (in hours) to arrive at the kWh lost. Now, the outage duration can be estimated from the position of each fault on the network, the estimated time to operate switchgear, and the estimated repair time. These estimated figures can be compared to historical fault events and revised estimates made to ensure reasonable consistency between calculated and measured outage times.

However, the calculation of the load interrupted causes more of a problem because the load at the vast majority of customer sites varies according to other parameters. For example, most loads vary according to a profile with respect to time of day, day of the week, and month of the year, each customer class exhibiting a particular load profile as shown in Figure 3.21.

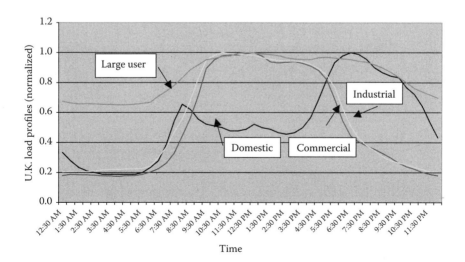

FIGURE 3.21 Standard load class profiles used by the supply industry in the U.K.

TABLE 3.7
Average Load Value over 24-Hour Period for the U.K. Standard Load Class Profiles

Type of Load	Minimum Value	Maximum Value	Average Value
Domestic	0.19	1	0.53
Commercial	0.18	1	0.51
Industrial	0.20	1	0.54
Large User	0.65	1	0.84

Figure 3.21 uses standard U.K. profile data for domestic, commercial, industrial and large user customers to illustrate how the load changes over the course of the 24 hours during the summer period, and the data have been normalized so that they represent the variation of a 1 kW load. It can be seen that the domestic load ranges from about 0.2 kW during the night to 1 kW during the early evening. The average value for the load over the 24-hour period can be calculated and is shown in Table 3.7.

To calculate the expected energy loss with complete accuracy, we would need to consider a fault at different times of the day and also how long the fault outage was (CAIDI). For example, we could consider a fault at 0600 to a domestic customer with an average daily load profile shown in the Figure 3.22.

From Figure 3.22, we can see that, if the outage duration were 2 hours, then the average load interrupted for that period would be about 0.41 kW, but if the outage were 12 hours, then the average load would be nearly 0.6 kW, from which it is clear that the expected energy loss depends on both the load when the outage starts and the duration of the outage.

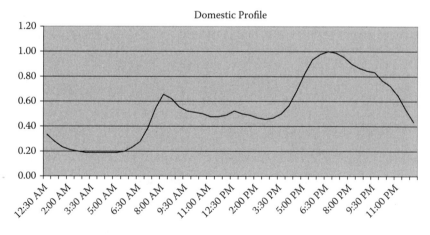

FIGURE 3.22 Standard U.K. class profile for a typical domestic load.

Distribution Systems, MV Networks

It is, of course, possible to calculate the actual energy loss accurately by taking all possible combinations of the time that the outage starts and the duration of the outage and then producing an average for the 24-hour period. This value can then be compared to the other methods of calculation to see which, if any, give comparable results. It has been found that, if we take the average load over the 24-hour period, we have seen in the table above and multiply that by the CAIDI, then the difference in results is very small and, for practical purposes, can be ignored. Because the average load over the 24-hour period is the same as the load factor, then the calculation can be made using the load factor.

Electrical Losses. Losses on an electricity distribution system represent the difference between the power purchased from suppliers and the power sold to customers. Losses can be classified according to whether they are technical losses, which cover resistance-based losses and magnetization losses, and nontechnical losses, which are losses from unmetered loads, for example substation environmental control and theft. In addition, many utilities charge a block rate for energy used for street lighting instead of placing a meter in every point of supply.

- Copper (resistive-based) losses apply to all conductors with a finite resistance and exist because of passing a load current through the resistance. They are proportional to the square of the load current and are subject to variations of the load current on a daily and annual basis.
- Magnetizing losses (also known as iron losses) apply to transformers and represent the energization of the magnetic circuit to operate the transformer core. They are of a constant value and are experienced for the entire time that the transformer is energized.

The calculation of the total loss in kWh over the lifetime of any project is relatively straightforward if load duration curves are known. But it is more difficult to estimate the financial value of these losses because it is very difficult to determine the price of electricity some years hence. Nevertheless, utilities use this process to compare the losses of, especially, different designs of transformers at the time of purchase.

It is impossible to achieve zero technical losses, although all losses can be monitored by an extended control system to at least permit the utility to understand where these losses take place.

The design of a distribution network will permit it to be operated in many different ways; for example, the actual location of a normally open point can, in theory, be at any switching device. It is possible to optimize the system losses on the circuits on each side of the NOP by careful selection of the location of the NOP. However, because the load current will vary on a daily and seasonal basis, so the losses will also vary. Monitoring of the load currents through an extended control scheme will permit the utility to vary the location of the NOP to match the variations in losses arising from the variations in loads.

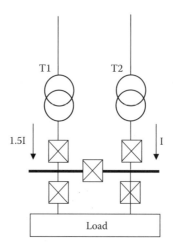

FIGURE 3.23 Loss balancing in substations.

Because resistive losses are proportional to the square of the current, it follows that, where possible, circuits should share loads equally. In Figure 3.23, the resistive losses in the secondary winding of T2 will be proportional to I^2 (or equal to kI^2) and in the secondary winding of T1 will be $2.25kI^2$. The total loss is, therefore, given by $3.25kI^2$. If we now equalize the currents to $1.25I$ by re-arranging the loads or closing the bus section circuit breaker, then the loss in each transformer will be $k(1.25I)^2$ (or $1.5625kI^2$), and the total losses will be given by twice this, or $3.125kI^2$, which will always be less than the initial condition. This principle of loss minimization applies to multitransformer supplies to, for example, a larger industrial customer.

Because operation of this substation with the bus section circuit breaker closed is one way to reduce the electrical losses, it is perhaps appropriate to consider the methods by which extended control can facilitate the operation of the transformer automatic voltage control (AVC) when the bus section breaker is closed. If two transformer AVCs are operated as in the independent automatic role, which is acceptable when the transformers are not in parallel, then there is a natural tendency for the taps to separate, that is, one transformer will go to top tap and the other to bottom tap, when the bus section is closed. This disparity in the taps will create a large circulating current that may cause overheating of the busbars and operation of protection. Although extended control/SCADA can be arranged to give an alarm if the taps are more than one or two different, in general independent automatic control is not viable for parallel transformer operation. However, the master-follower arrangement is normally acceptable because only one AVC relay, on the master transformer, is used, and any tap change operation initiated by the AVC will cause the second transformer tap changer to follow the master in sympathy.

Extended control of the AVC scheme can be used to select which control scheme is to be used and also to directly control the tap settings from a control

Distribution Systems, MV Networks   137

room to facilitate the leveling of taps between two substations to prevent excessive circulating currents between those two substations if the intersubstation normally open switch on the MV network is to be closed for any reason, for example, as part of a prearranged switching schedule for maintenance work.

Extended control of the tap changers can also be used to simulate an overvoltage, perhaps 6%, on the MV busbars of a substation, thus causing the tap changers to step down and reduce the load on the substation a little. This technique can be applied as a form of load reduction if needed by the utility, which, in turn, will have an effect on losses.

Some control of electrical losses can also be made by the extended control of MV system capacitors, which will improve the MV system power factor, in turn reducing losses.

3.2.12 COMPARISON OF U.K. AND U.S. NETWORKS

Distribution systems in the United States are designed and operated as four-wire multigrounded systems. Four-wire multigrounded systems include the three phase conductors as well as a fourth neutral conductor, which is carried with the phase conductors and grounded at numerous points along the length of the circuit. Overhead main feeders are typically constructed using large phase conductors (170 to 400 mm ACSR) and are typically tied to one or more other primary distribution circuits through normally open switches. Underground system construction practices vary more than those for overhead systems, but main feeder cable sizes typically range from 255 to 510 mm per phase. A concentric neutral of smaller individual conductors or a metallic tape typically surrounds the phase conductors. A variety of cable insulation materials and thicknesses are available for each voltage class. Installed cost is also greatly impacted by the type of underground construction utilized (e.g., concrete duct banks, PVC duct in earth, direct burial).

Urban and city networks in the U.K. are nearly always underground, although some overhead feeders will start at the edge of the town area and proceed to the rural area. Distribution systems constructed to typical U.K. standards are designed for operation as three-wire unigrounded systems. Three-wire unigrounded systems do not carry a neutral conductor along the length of the feeder, but are grounded at a single point at the source substation. Overhead three-phase MV main feeders are typically constructed using 75–90 mm^2 conductors. Underground systems utilize cable with somewhat larger conductor sizes (e.g., 185 or 240 mm). As with U.S. systems, a variety of cable insulation materials are available, having potentially large impacts on cost. Additionally, the type of underground construction will have a major impact on installed cost.

The different grounding practices between U.S. and U.K. primary systems have a number of operational and cost impacts. The four-wire multigrounded circuit configuration used in the United States may result in higher primary voltage system costs for a similar overhead line constructed to U.K. standards, because a neutral conductor is required along the length of the circuit. Additionally, the neutral conductor of a primary voltage system constructed to typical U.S. standards must be grounded through a driven ground rod at regular intervals, causing additional potential cost differentials. Underground system differentials are comparable for similar construction types, as the conductor represents a smaller portion of the total construction effort.

Primary voltage systems in the United States are configured radially, with power flowing from the source substation toward load under normal operating circumstances. Three-phase primary main feeders are typically 8–15 miles (13–24 km) in length and follow major thoroughfares. Three-phase and single-phase radial primary voltage laterals are tapped to the main feeder and can extend several miles from the main feeder. Primary systems tend to be overhead when possible and underground when required or when the cost differential between overhead and underground construction is covered by a third party. Typical utility practice includes two to six normally open ties to other circuits to allow for maintenance and contingency switching.

A typical urban underground system in the U.K. will have a number of branches, each terminated in a normally open switch, to provide a switched alternative supply for maintenance or fault outages. The switched alternative supply can be supplemented in a few special cases, notably the provision of high-reliability supplies to individual industrial customers and hospitals, where two cables will supply from each side of a source substation busbar. This requires circuit breakers with unit protection and is not common now.

The use of single-phase radial laterals in U.S. systems can result in cost savings because lesser amounts of primary cable/conductor are required. However, this potential savings is offset by the greater reach of the secondary/low-voltage systems used in U.K. systems and the necessity of a neutral conductor for grounding and return. (Secondary/low-voltage systems are dealt with below.) Underground construction is more expensive than comparable overhead construction and may result in increased initial capital costs for U.K.-style systems in urban areas (which tend to use a greater proportion of underground construction).

Distribution systems in the United States utilize a variety of switching (padmounted switchgear, line switches, sectionalizers) and protective devices (reclosers, fuses, fused cutouts) along the length of a primary distribution line. These devices isolate faulted line sections, allow service restoration or allow switching for maintenance and contingency purposes. Detailed discussion of these devices is outside the scope of this book; however, each of these devices can act as a switching point to minimize the duration of customer interruptions. Transformers are typically connected to the primary system through fuses, with the fuses being tapped into the primary line. In line switching devices are distributed along circuits in a manner to provide some measure of sectionalizing capability and switching flexibility.

Distribution systems in urban/suburban areas of the U.K. tend to focus on use of unit substations comprising a ring main unit and LV distribution board, both mounted on the MV/LV distribution transformer to form a single unit.

Distribution Systems, MV Networks

Ring main units to provide protective and switching functions are located at nearly all transformer locations in urban U.K. systems. The majority of distribution substations use a ring main unit comprising two 630 amp ring switches and a 200 amp tee off for the transformer connection and protection. The ring switches are designed as fault make (say, 350 MVA) and load break (say, 630 amp). The protection can be fuse based, arranged for all phases to trip if one ruptures, or circuit breaker based. Ring main unit switchgear is fully interlocked to permit switching operations in the correct sequential order. For example, a circuit main earth cannot be applied unless the associated ring switch is open. A past design practice has been to install tee off substations between ring main unit substations. These tee off substations have no line switchgear, simply a local transformer protection unit and a single cable, tee jointed to the ring main cable.

The ring main units isolate failed transformers (without resulting in customer outages when secondary networks are interlaced and fed from two separate transformers, as is done in some areas of the U.K.). Additionally, ring main units provide substantial switching flexibility because switches are located on either side of the transformer primary protective device (in line with the circuit). This configuration allows switching to isolate a faulted line section or restore service from either circuit source to which the transformer is connected. Finally, automation of a single ring main unit provides substantially more operating flexibility than offered by automation of a single switch in a U.S.-style system.

Switchgear employed on rural U.K. systems includes source reclosers, line reclosers, and disconnects and are virtually identical to U.S. practice except that U.K. devices are three-phase ganged devices. Unganged drop-out fuses are used for local protection of transformers and some lateral lines. Some utilities will not fuse each transformer, opting for a group fusing policy to provide quick restoration following lightning storms. Switchgear on the rural network is generally not interlocked, and grounds are usually applied externally.

The switching flexibility provided by ring main units allows restoration of all customers through switching actions only, allowing cable or system repairs to be handled during normal working hours, resulting in large potential operational savings. Systems in the United States that utilize load break elbows of loop through transformers as switching points can provide similar (but slower) switching flexibility to that provided by ring main unit application.

Service transformer characteristics vary greatly between U.S.- and U.K.-style systems. Residential service is typically provided by small single-phase transformers (10–50 kVA) connected to single-phase or three-phase laterals in U.S.-style systems. Transformers typically serve one to six residential customers. Larger three-phase transformers are used for commercial connections, and transformers are typically sized to meet the expected load. U.S. utilities typically stock a large number of single-phase and three-phase transformer sizes for both pole-mounted and pad-mounted applications. Typical single-phase transformer sizes include 10, 15, 25, 37.5, 50 and 100 kVA units. Typical three-phase transformer sizes include 150, 300, 500, 1000, 1500, 2000 and 2500 kVA units. Service transformers are either pole-mounted or pad-mounted in most cases. In urban areas, network transformers may be located in vaults below street level.

U.K. systems utilize larger three-phase service transformers for most applications (residential and commercial) and serve more customers from each transformer, partially as a result of the higher secondary voltages used. Additionally, residential and commercial customers may be served from the same transformer in cases, which is not typically possible in the United States due to different voltages required by commercial and residential customers. The larger transformer sizes utilized in U.K. systems also provide advantages with respect to coincidence of customer load as well as simply requiring installation of fewer transformers (i.e., capital and maintenance cost savings). Also, U.K. systems typically have much greater standardization of service transformer sizes (using far fewer transformer sizes when compared to U.S. utility transformer inventories). This standardization potentially results in inventory cost savings. Typical U.K. ground-mounted transformer ratings are 300, 500, 750 and 1000 kVA, but in accordance with IEC standards are now 315, 630, and 800 kVA. Service transformers in urban areas of the U.K. are typically housed in ground-mounted distribution substations that require that a site be purchased, typically 5×4 meters. In central areas of cities, these sites can be expensive, and the service substations are normally located in the building basement. Distribution substations are typically partially or completely enclosed by a fence or a building.

There are significant differences in the quantity of equipment and difficulty in connecting three-phase transformers to a three-wire ungrounded system and connecting single-phase transformers to single-phase or two-phase laterals. These differences arise from not only the different electrical characteristics of the systems, but also differing structural requirements. Single-phase transformers can have significant advantages over three-phase in cases where three-phase power is not required. The single-phase transformers are easier to install (two-phase or one-phase and one neutral connection are required on the primary of the transformer). Additionally, single-phase transformers consist of simpler core and winding designs, reducing their cost relative to three-phase transformers. Single-phase transformers utilized on multigrounded systems can be constructed with only a single high-voltage (MV) bushing and can utilize a lower insulation rating (BIL), resulting in further cost reductions.

When applicable, single-phase transformers (as used in U.S. systems) offer significant cost and labor advantages over three-phase transformers on a per-transformer basis. However, the advantages of single-phase transformers are offset at the system level, where a greater number of the smaller single-phase transformer connections are required and the higher secondary voltage used in U.K. systems allows a greater number of customers to be connected to each transformer.

3.2.13 THE COST OF INSTALLATION OF THE SELECTED DESIGN

The process for the electrical design of the distribution network must be made in conjunction with the process for the selection of the components from which

Distribution Systems, MV Networks

the network will be constructed; for example, the closed ring network must use unit-protected circuit breakers for the switchgear, and overhead networks must use switchgear that can be connected to an overhead line. Because the total capital cost of the system must be the sum of the capital cost of the components and the cost of installing those components, then it follows that the installation costs must be taken into account during the overall design process. Some examples of where the cost of installation may affect the component selection and the overall design include

- To deliver a large system transformer to the site may cost less if the transformer can be split into a number of subassemblies, for example, the main windings and the cooling radiators. This is because there is often a maximum weight that can be transported by road.
- Because a longer length of single-core cable can be stored on a cable drum than for three-core cable on the same drum, the number of cable joints to join the drum lengths into one cable length will be less if single-core cable is used.
- Cable terminating areas on switchgear and transformers must be matched to the type of cable used to connect the switch or transformer.
- The cost of a substation site and buildings will depend on the size of the switchgear chosen to the extent that a small extra cost of compact switchgear may be less than savings in land purchase and civil works.
- Modern communications, for example, fiber optics, used with digital communicating relays in a substation will be less expensive than traditional multicore connections within the substation. In addition, factory testing of this type of substation can reduce site commissioning tests.
- It is usually preferable to preassemble some components into subassemblies in the controlled environment of a factory rather than at thesubstation construction site. For this reason, some utilities are using complete substations that have been factory built before delivery to the site.

3.2.14 THE COST OF OWNING THE NETWORK AFTER CONSTRUCTION

We have already seen, in the section about losses, that we can capitalize a revenue cost by calculating the present value of the revenue cost over the lifetime of the project. It follows, therefore, that the revenue costs should be considered as well as the capital costs and the installation costs. The major revenue costs will include, but not in any order of priority:

- *The value of electrical losses, technical and nontechnical*
- *The cost of crew time for planned switching routines, including maintenance and construction work*
- The cost of maintenance of plant and equipment

- *The cost of crew time for unplanned switching routines, especially postfault switching*
- The cost of repairs following failure and third-party damage
- *The cost of energy not supplied for fault and other unplanned outages*
- *The cost of penalties for performance below required limits, sometimes paid to the customer and sometimes to a government-based authority*
- *The cost of energy not supplied to any customer who moves away from the utility supply in cases of unacceptable quality of supply*

When all three costs are considered, we can arrive at the life cycle costs, where

$$\text{Life cycle costs} = \text{capital costs} + \text{installation costs} + \text{present value of revenue costs},$$

which can then be minimized to provide the most economic solution in overall terms, and we can also investigate the ways that automation can influence the life cycle costs. We will show in later chapters that automation can influence the revenue costs in the list above.

3.3 LV DISTRIBUTION NETWORKS

For many customers, the LV distribution network is the final link in the chain of electricity supply because it connects between the distribution substation and the customer. Because most customer outages are caused by the MV network, utilities have tended to examine control of MV networks first but are now recognizing that attention to extended control of LV networks can bring benefits. As with the MV network, the LV network may be underground, overhead or mixed. Whereas the routings of the cables or overhead lines for the MV network are not always critical, because their function is only to interconnect between substations, LV circuits are normally routed along the streets in which the connected customers are located.

Depending on the load to be supplied, an LV distribution network can be single-phase or three-phase. If there are any single-phase loads (typically up to 10 kVA), then there must be a neutral return conductor, which is commonly provided by earthing one point of the secondary winding of the distribution transformer. The customer installation may earthed by a separate conductor from the neutral connection (separate neutral and earth, or SNE) or by one conductor (combined neutral and earth, or CNE). Three-phase LV networks are therefore normally four-wire (CNE system) or five-wire (SNE system).

3.3.1 UNDERGROUND LV DISTRIBUTION NETWORKS

An underground network is normally connected to a three-phase, ground-mounted, distribution transformer that is typically rated at between 300 and 1000 kVA and which would supply up to 500 customers. A typical network is shown in Figure 3.24.

Distribution Systems, MV Networks

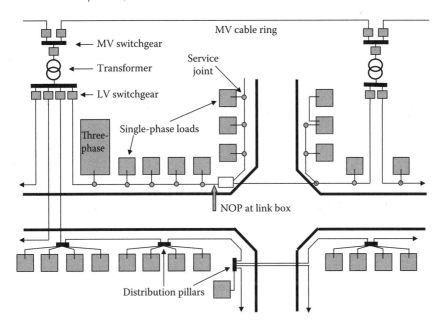

FIGURE 3.24 Typical underground LV network (European practice).

The diagram shows an MV underground cable ring supplying two ground-mounted distribution substations, each fitted with a ring main unit, one substation being fitted with a four-way LV distribution board and the other with a two-way LV distribution board.

The top part of the LV system shows the system where a service cable is jointed onto the LV distributor to connect with each load, and near to the right-hand substation, it can be seen that one load is loop connected to the next-door load. A underground link box is also shown, where one set of links has been left open to provide a normally open point between the two substations.

The lower part of the diagram demonstrates the application of small, pavement-mounted, distribution pillars to connect each load to the distributor. Although not shown in the diagram, a normally open point could be included in any of the distribution pillars.

3.3.2 Overhead LV Distribution Networks

An overhead network is normally connected to a pole-mounted distribution transformer, which is typically rated at between 50 and 100 kVA, and which would supply between 10 and 30 customers. Some connection arrangements are shown in Figure 3.25.

This type of substation is common on both rural networks in Europe and urban overhead networks in the United States where it would feed the customers through a radial network, normally without any interconnections to an adjacent network. A typical network is shown in Figure 3.26.

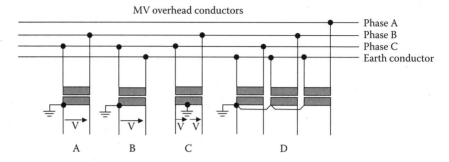

A Single-phase transformer connected phase to phase
B Single-phase transformer connected phase to earth
C Single-phase transformer with center earth, connected phase to phase
D Three-single-phase transformers connected phase to earth (MV) and in star (wye) (LV)

FIGURE 3.25 Earthing arrangements for distribution transformers.

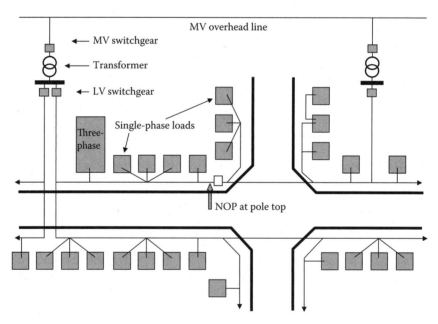

FIGURE 3.26 Typical overhead LV network (European and U.S. practice).

The diagram shows an MV overhead line ring supplying two pole-mounted distribution substations, each fitted with local MV switchgear, one substation being fitted with a two-way LV distribution board and the other with a single-way LV distribution board.

The diagram shows the system where an overhead service span is connected to the LV overhead distributor to supply each load, and near to the right-hand

Distribution Systems, MV Networks   145

substation, it can be seen that one load is loop connected to the next-door load. One service connection to the three-phase distributor line typically supplies three single-phase loads. A normally open point between the two substations has been created by using a set of LV disconnecting links at the pole-top level that have been left open.

3.4 SWITCHGEAR FOR DISTRIBUTION SUBSTATIONS AND LV NETWORKS

Switchgear needs to be provided on both the MV side of the transformer and the LV side of the transformer. The MV switchgear is to provide protection and a means of disconnection for the transformer, and the LV switchgear is to provide overload protection for the transformer and a means of disconnection between the transformer and the LV network. For ground-mounted substations, the MV switchgear is commonly a ground-mounted switchfuse or circuit breaker within the substation itself, which, if operated, will only affect the one transformer. For pole-mounted substations, the switchgear can be, typically, drop-out fuses mounted at the substation, which, if operated, will only affect the one transformer or be located at a remote site to control a group of several substations. Operation of the latter case will, of course, affect all the downstream substations. MV switchgear is covered in Chapter 4.

The LV switchgear for use at distribution substations can be classified as fuse-based or circuit breaker-based, and each has some advantages according to Table 3.8.

TABLE 3.8
Comparison of Fuses and Circuit Breakers for LV Application

Parameter	Fuse-Based	Circuit Breaker-Based
Current limiting action	Yes	No
After operation on fault	New element required	Close circuit breaker
Life expectancy of contacts	Renewed when single operation fuse is replaced after operation	Depends on fault level but may be limited, at full fault level, to relatively few operations. Risk if this number is exceeded
Prospective fault current	Typically up to 80 kA	Can be 80 kA but depends on design of circuit breaker chosen
Ability to open and close by extended control	Only if integrated with switchfuse assembly	Power mechanisms are commonly available
Remote indication of operation	Available on certain types of fuse element	Auxiliary contacts on circuit breaker are commonly available

FIGURE 3.27 Fastline-type LV distribution pillar for application ground-mounted (left) and pole-mounted (right) substations (Courtesy of ABB).

In Figure 3.27, we can see two different LV pillars made up from the same range of components. The left-hand pillar has an incoming connection from the transformer in the center of the busbars (not visible here) and six 400 amp outgoing fuse disconnectors, while the right-hand pillar has just two outgoing fuse disconnectors, each rated at 160 amps. This particular system can be mounted, as shown here, in a cabinet, or wall mounted in a ground-mounted distribution substation. For the ground-mounted installation, it is possible to mount the complete LV distribution pillar directly onto the distribution transformer, thereby simplifying the installation process. The fuse disconnectors shown here take the widely available IEC/DIN-type fuse NH. In common with a number of other products, the type of LV switchgear can accept circuit breakers if they are mounted on a special adaptor plate, which then replaces the outgoing fuse disconnector.

3.5 EXTENDED CONTROL OF DISTRIBUTION SUBSTATIONS AND LV NETWORKS

Historically, there has been very little control and monitoring of the LV distribution switchgear, because utilities have concentrated their efforts on improving the performance of their MV networks, and if the LV network supplies few customers, as in the United States, then control of the LV gives little practical benefit. However, where the LV network is larger, some form of control of the LV switchgear can be advantageous, normally as an adjunct to the control of the MV switchgear. The following control functions might be expected.

Remote Operation of Switchgear. Remote operation of switchgear to control loads or reclose after a trip operation requires some form of power actuator on the switching device. This is normally possible for circuit breakers but is normally not available for switch disconnectors. Remote operation would need, at the RTU in the distribution substation, two digital outputs (one for close operation and one for open operation) per switching device.

Remote Indication of Switchgear. Remote indication of switchgear can be provided to monitor the status of the switching device. For circuit breakers fitted with a set of auxiliary contacts, this poses no problem, but remote indication for fuses can be more difficult. This is because fuses can have two different types of local indication. These are local indication by using a visual indicator, for example, a disc contained within the body of the fuse that changes color if the fuse has ruptured, and local indication by using an external indicator, for example, a small rod that might move 2 or 3 mm when the fuse has ruptured. In the latter case, a small microswitch can be fitted to the fuse to provide the auxiliary indication and, if the switching device can be tripped by an electrical mechanism, to open the other poles of the switch.

However, it is possible to measure the voltage on both sides of a fuse and use the difference in voltage to indicate that the fuse has ruptured. This method could either use analog inputs to the substation RTU or take the form of an integrated device such as the OFM Fuse Monitor from ABB. The indication that the fuse has ruptured can be fed into a digital input on the substation RTU or arranged to operate a trip-all-phases arrangement. Remote indication for a fuse operation requires simply one digital input to the substation RTU, but a circuit breaker could use two digital inputs if the user wanted to have a "don't believe it" indication to show that the circuit breaker was neither correctly open nor correctly closed.

Voltage Measurement on LV Systems. A further application of extended control to the LV distribution system is the measurement of the LV system voltage using analog inputs to the substation RTU. The quantity of analog inputs can vary greatly with the user, with some users measuring just one voltage, usually on the busbars, whereas others may wish to measure more.

An interesting application can be derived if three phase-to-phase or phase-to-neutral voltages are measured. Suppose that the protection on the MV side of the substation comprises three fuses that do not operate as a ganged set; that is, if one fuse operates then the other two do not, unless they, too, have experienced enough current to operate independently. In this situation, the voltages on the LV system become abnormal, with two of the phase-to-neutral voltages becoming half their normal value. This can cause load connected between phase and neutral to be damaged, possibly resulting in a compensation claim against the utility. Also, the rotational field set up in three-phase motors will be lost and, unless loss-of-phase protection is provided at the motor, damage can occur. Because measurement of the three-phase voltages will indicate this type of single-fuse operation, then the utility can be made aware of the problem and take corrective action.

Automatic Changeover for LV Distribution Systems. Another option is to apply the circuit breaker-based LV distribution switchgear to provide a switched alternative supply to a customer who needs a high reliability supply. In this case, two LV circuits are taken to the customer load point, and one is run normally open using a normally open circuit breaker. Should the other supply fail for any reason, then the circuit breaker on that supply can be arranged to open, and by closing the normally open circuit breaker, customer supply can be restored.

3.6 SUMMARY

This chapter describes the most important design features of distribution systems, concentrating on those that interact with the provision of extended control. We can now use these descriptions to consider the application of extended control in more detail.

REFERENCES

1. Lakervi, E. and Holmes, E.J., Electric Distribution Network Design, in *IEE Power Engineering Series*, London: Peter Peregrinus Ltd., 1989.
2. Willis, H.L., *Distribution Planning Reference Book*.
3. Uwaifo, S.O., *Electric Power Distribution Planning and Development, The Nigerian Experience*, Lagos State, Nigeria: Hanon Publishers Limited, 1998.

4 Hardware for Distribution Systems

Electric distribution delivery systems comprise primary devices that deliver power and secondary devices that protect and allow control of the primary devices. The combination of these two elements creates an automation ready device, as explained in Chapter 1. This chapter will cover in more detail the selection of this primary and secondary equipment, starting with the primary switch.

4.1 INTRODUCTION TO SWITCHGEAR

Switchgear is a general term covering switching devices and assemblies of such devices with associated interconnections and accessories. The following classifications of switching devices are in general use within the utility distribution industry.

A *circuit breaker* is a switching device capable of making, carrying and breaking currents under normal circuit conditions and also making, carrying for a specified time and breaking currents under specified abnormal circuit conditions such as those of a short circuit.

A *switch* is a switching device capable of making, carrying and breaking currents that may include specified operating overload conditions and also carrying for a specified time currents under specified abnormal circuit conditions such as those of a short circuit. A switch, therefore, is not expected to break fault current, although it is normal for a switch to have a fault making capacity.

A *switchfuse* is a switch in which one or more poles have a fuse in series in a composite unit so that normal currents are interrupted by the switch and high-fault currents are interrupted by the opening of the fuse. The operation of one fuse may be arranged to cause the operation of the three-phase switch to ensure that all three phases of the load are disconnected.

A *recloser* is a circuit breaker equipped with relays in order to carry out a variable pattern of tripping and closing and is usually associated with overhead lines.

A *disconnector* is defined as a mechanical switching device that provides in the open position a specified isolating distance. It should be capable of opening and closing a circuit when negligible current is broken or made. It should be noted that disconnectors are unreliable for breaking capacitive current, with

TABLE 4.1
Switching Device Type, Capacities and Duty Cycle

Type	Fault Current Break	Fault Current Make	Rated Current Break	Rated Current Make	Duty Cycle
Circuit breaker	✓	✓	✓	✓	Single shot, fast
Switchfuse	✓	✓	✓	✓	
Recloser	✓	✓	✓	✓	Multishot, fast
Switch		✓	✓	✓	
Disconnector					Slow, no load breaking
Autosectionalizer			✓	✓	Fast
Switch disconnector		✓	✓	✓	

capacitive currents of less than 1 A causing damage to some types of disconnector. Although it should be capable of carrying normal load current and also carrying for a specified time currents under abnormal conditions such as for a short circuit, a disconnector is not capable of making or breaking short-circuit currents.

An *autosectionalizer* is a disconnector equipped with relays, or other intelligence, so as to operate within the dead time of a recloser.

A *switch disconnector* is a combination of the switch and the disconnector. Its main application is twofold — to switch off the electrical supply to a point of work and then create a safe working environment at that point of work.

The primary switchgear service duty is designated by (1) current breaking capacity, (2) current making capability, (3) normal current rating, (4) operating cycle and (5) voltage class and summarized in Table 4.1 for the seven categories described above.

Circuit breakers, reclosers and autosectionalizers in nonautomated (remote control) systems have integrated actuators operated by an appropriate secondary, whereas all the remaining devices are manually operated.

Secondary devices take the form of (1) protection relays for fault breaking devices such as circuit breakers and reclosers and (2) a simple intelligent electronic device for counting pulses when used with an autosectionalizer. These devices in various levels of functionality and input/output capability are integrated with the primary device to provide communication interface and logic for automation, thus producing an automation ready device.

4.1.1 Arc Interruption Methods

Primary devices depend on a method of interrupting the current. The higher the current, the more dependence on an efficient arc extinguishing method. Early interruption methods that relied on extinguishing the arc under insulating oil or air were used for lower voltages with special arc chutes to extend the arc beyond the distance for conduction to be maintained at the particular voltage level.

Hardware for Distribution Systems 151

Installations consequently absorbed significant space, and as space became a premium, new interruption methods were employed starting with minimum oil. In this case, the arc was contained in a small space, arc extinction being achieved through a series of baffles that built up extinguishing pressures as the arc formed. Oil soon fell into disfavor due to potential fire hazards, high maintenance costs and the significant power required to operate the switches [1]. Three basic interruption methods for switches are predominantly in use today.

Air Interruption. Air circuit breakers, although still used up to 15 kV, have in general been replaced by minimum oil devices, which in turn have been replaced by sulphur hexafluoride (SF_6). The air circuit breaker extinguishes the arc at current zero by ensuring that the contact gap will withstand the system recovery voltage by creating an arc voltage in excess of the supply voltage. This is achieved by forcing the arc through an arc chute having many paths, thus effecting cooling of the plasma and lengthening of the arc. Air break devices in their simplest form are inexpensive and are used where the current interrupting duty is restricted to no load, as disconnectors both indoors in cubicle assemblies, and outdoors as pole-mounted devices. The operation is manual; however, motor operation is possible when the speed of operation is not critical.

Sulphur Hexafluoride Interruption. SF_6 has been used both as an interruption medium and as busbar insulation. It has proved highly efficient in reducing the footprint of indoor metal-clad switchgear cubicles and is used extensively in such applications. As an interruption medium, it has again enabled smaller circuit breakers to be manufactured. SF_6 ring main units (RMUs) have now replaced oil gear as the preferred medium for both arc interruption and insulation.

In outdoor applications, it has been used for enclosed pole-mounted switches of high reliability that can be operated in all climatic extremes and require little maintenance. The distance between open contacts of the switch in most designs is sufficient to provide a recognized point of isolation. In the past, this has only been available with air-break switch disconnectors.

The SF_6 breaking chamber as shown in Figure 4.1 must be sealed and contamination free. Any leakage must be detected because loss of pressure will reduce the interruption capabilities. The possibility of leakage, however small, is a concern to the environmental lobby because SF_6 is a greenhouse gas and its arc byproduct is carcinogenic. In some utility service areas, outdoor SF_6 switchgear is not permitted for these environmental reasons.

Vacuum Interruption. Vacuum interrupters have been in commercial use since the late 1950s. The immediately apparent and significant advantages of switching with vacuum interrupters were extremely high switching rate, virtually no maintenance, and long life. Today, cost-effective vacuum interrupters cover the application range of 600 V to 38 kV and interrupt currents from hundreds of amps to 80 kA. Vacuum interrupters provide maximum reliability due to their extended mechanical life and minimum, even wear. Sealed contacts will not degrade the dielectric properties of the insulation medium. Figure 4.2 illustrates a schematic diagram of a typical vacuum interrupter, sometimes called a vacuum bottle. The two-interrupter contacts are immersed in a vacuum-tight, sealed

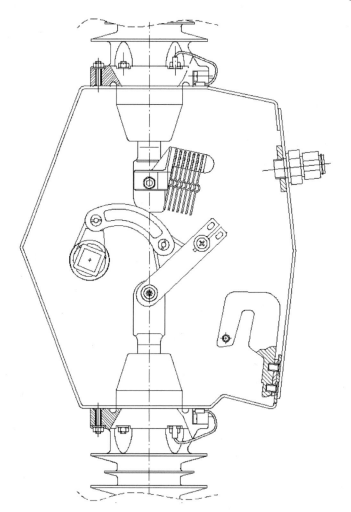

FIGURE 4.1 Cross section of an enclosed three-position SF_6 switch showing the moving contact, arc interruption chute and grounding contact (courtesy of ABB).

envelope, typically made of ceramic or glass. Depending on the voltage rating, the enclosure can be made of either one or two ceramic cylinders (6). Flexible bellows (9) provide means for mechanical movement of the contact stem (2) within the vacuum.

The contacts are surrounded by a vapor shield (4), made of a stainless steel, copper or FeNi, to protect the inside of the ceramic from arc metal vapor and preserve the dielectric integrity between the two ends of the switch. A metallic collar (7) between the two ceramic cylinders (6) serves as a seal and as support for the vapor shield (4). The contacts are typically made out of two components. The mechanical strength and means for controlling the arc are provided by copper elements (3). Special, oxygen-free, high conductivity (OFHC) copper is used for

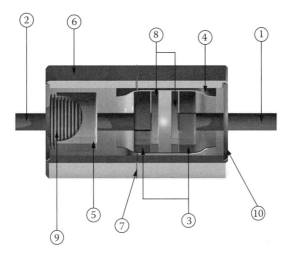

1. Fixed contact stem
2. Moving contact stem
3. Fixed and moving contact subsystems
4. Vapor (or ion) shield
5. Bellow shield
6. Ceramic enclosure (envelope) shown as two ceramic cylinders joined in the middle
7. Metal-to-ceramic sealed joints
8. Contacts
9. Metal bellows
10. Metal end plates

FIGURE 4.2 Typical vacuum interrupter.

vacuum interrupter manufacture. For higher interrupting current ratings, the contact subsystem (3) can have special geometric arrangements to generate magnetic fields to control the arcing during the arcing phase and to assist the arc interruption at current zero. The contact surfaces (8) are made from a number of specially designed materials, such as CuCr (copper chromium), CuBi (copper bismuth) or AgWC (silver tungsten carbide) to optimize the switching performance, contact life, and interrupting ratings.

In the closed position, current flows freely between moving and fixed contacts. When the moving contact is separated from the fixed contact under current, an arc is drawn. The arc vaporizes a small quantity of the metal from the surface contact. Typically, the arc voltage is independent of the flowing current and is only of the order of several volts. Therefore, the arc energy is very small (product of arc voltage, arc current and time), which allows the vacuum interrupters to be compact and have long life. When the main power frequency current approaches zero, the arc products (plasma) quickly diffuse due to the ambient vacuum. The recovery of dielectric strength between the contacts is very fast. A typical interrupter regains its full dielectric strength in a few to several microseconds. It is also significant that in most cases, the current can be interrupted in vacuum even when the contact gap is not fully open at the instant of current zero. Even a partial gap, less than a millimeter, can interrupt full current. This makes vacuum interrupters fast devices, limited in the interrupting time only by the mechanical drive.

Because the contacts are lightweight, the mechanical drive energy required by the vacuum switch is low compared to other switching technologies. The short travel distance of the vacuum interrupter contacts, low inertia of the moving parts and the small power requirements were ideal for the application of magnetic actuation operating mechanism.

In summary, the major benefits of vacuum interruption technology are minimum maintenance and long contact life in excess of 10,000 mechanical or load operations. This medium is now the preferred interruption technology for most medium-voltage applications.

4.2 PRIMARY SWITCHGEAR

Primary switchgear is deployed in primary substations (HV/MV) where subtransmission voltages are transformed to medium-voltage distribution levels, at distribution substations* (MV/LV) and along overhead lines. This section covers all switchgear types and assemblies used or associated with substations and underground cable networks.

4.2.1 Substation Circuit Breakers

A medium-voltage distribution feeder is normally connected to the source substation busbars through a circuit breaker. The main purpose of that circuit breaker is to carry load current for long periods of time and, in conjunction with a protection relay, to safely interrupt any fault that might occur on the feeder. Substation circuit breakers are specified, as a minimum, in terms of the following:

- Rated current in amps for operation on a continuous basis
- Rated current in amps for short time operation, typically 1 or 3 seconds
- Peak making capacity in kA
- Breaking capacity in kA
- Power frequency withstand voltage to earth and across open switch, in kV
- Lightning impulse withstand voltage to earth and across open switch, in kV

In Figure 4.3a, there are two incoming 115 kV overhead lines that terminate one each on the two 115/20 kV transformers. Circuit breakers are provided to connect the secondary side of each transformer to the busbar, to supply the outgoing feeders and to divide the busbar into two sections. This arrangement could be met with a switchboard similar to that shown in Figure 4.3b.

---

* The terminology used to describe the distribution network varies between European and North American systems. In North America the sequence of substations corresponding to the European is referred to as distribution substation (HV/MV), distribution transformer (MV/LV). European distribution substations are not.

Hardware for Distribution Systems 155

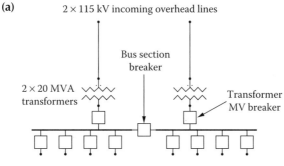

FIGURE 4.3 (a) Diagram of typical HV/MV substation and (b) typical MV switchboard.

Circuit breakers can also be used to limit the size of an outage. For example, in the diagram, a fault on an outgoing feeder would be cleared by the feeder breaker, thus limiting the outage to the faulted feeder. If the feeder circuit breakers were not present, then the transformer circuit breaker would have to clear the fault, but it would also disconnect supplies to customers on healthy feeders. Similarly, a fault on the right-hand busbar would be cleared by the right-hand transformer breaker and the bus section breaker, therefore not interrupting supplies to the left-hand busbar.

By itself, the circuit breaker cannot trip when a fault occurs. However, suitable protection relays are arranged to detect a fault condition and initiate the automatic tripping of the breaker. The protection found with underground feeder circuit breakers will generally be overcurrent and earth fault, which would be supplemented by sensitive earth fault for overhead feeders. Some authorities permit utilities to operate their MV network unearthed or earthed through a tuned coil, in which case the earth fault protection would be replaced by a directional

 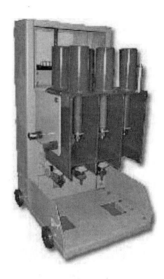

FIGURE 4.4 Retrofit vacuum circuit breaker on withdrawable truck.

earth fault protection.* Protection relays used to be discrete components with one relay per function, but nowadays one digital relay can be programmed to accommodate most if not all required functions. Circuit breakers can also be tripped by manual initiation, either from a position local to the breaker or from a remote site using SCADA.

In the past, circuit breakers, which were generally of bulk oil design, would need to be maintained after either a predetermined number of years or a predetermined number of fault clearances and so needed to be disconnected from the power system. Many designs of circuit breakers were mounted on a wheeled truck that could be racked from the service position to the disconnected position. Figure 4.4 shows a truck that can be disconnected by racking downwards and then withdrawn horizontally. The withdrawable truck also has some operational advantages, especially that a faulty breaker can easily be replaced and that earthing of the outgoing circuit can be easily achieved using a fully rated switching device.

In the 1970s, in order to make vacuum circuit breakers competitively priced and taking into account that the interrupters themselves were relatively expensive, manufacturers sought other economies in their designs of switchgear, one being to introduce a nonisolatable circuit breaker. This was supported by the fact that vacuum circuit breakers require less maintenance than their oil-filled equivalents. However, the fixed vacuum circuit breaker was unable to be used as the disconnector function associated with the withdrawable circuit breaker; hence, a separate

---

* See Chapter 5, Protection and Control.

Hardware for Distribution Systems   157

FIGURE 4.5 Modern primary substation switchgear.

disconnector was required. Also, there was some distrust of manufacturer's claims about the life of vacuum bottles. These resulted in a brief return to isolatable devices but based on non-oil switchgear.

Most manufacturers now offer nonisolatable switchgear for primary substations, and a typical example is shown in Figure 4.5.

In recent years, MV switchgear has been built to withstand internal arcing, which may be caused by failure of insulation within the switchgear. Most designs use a pressure relief system whereby the explosive overpressure resulting from the internal arc is safely discharged through some form of ventilator to the open atmosphere. Although this has been shown to work well, it generally means that the substation building has to incorporate the overpressure relief system.

For distribution substations, tests have shown that a lifting, but tethered, one-piece glass reinforced plastic (GRP) roof can be effectively used to vent internal arc products. The venting takes place at the joint between the walls and roof of the substation and so is at a high enough level to eliminate risk to persons nearby.

An alternative to controlled pressure relief is to stop the internal arc from developing beyond the initial discharge. In much the same way that a high rupturing capacity fuse operates before the first current peak, it is possible to eliminate an arcing fault between a conductor and earth by shorting that conductor to earth in a very short time period. Shorting the conductor will depress the voltage on the affected phase, hence extinguishing the arc, and also start the tripping of conventional protection. Although the protection may take approximately 100 m to operate, a small vacuum circuit breaker acting as an arc eliminator can eliminate the arc in approximately 5 m.

To achieve its fast operating speed, the arc eliminator must be controlled from an optical sensor, within the switchgear enclosure, designed to detect the light from an electric arc. If the switchgear enclosure is not light proof, then the sensor should be interlocked with a current transformer (CT) to confirm that fault current is flowing before the eliminator operates. Without this precaution, photographic flashguns can trip out complete substations.

A number of switchgear designs exist whereby an existing isolatable oil circuit breaker and its truck can be replaced by a equivalent truck consisting of a vacuum or SF_6 circuit breaker.

4.2.2 Substation Disconnectors

A disconnector is normally applied to a power system when it is required that safe working conditions are made available. Depending on the physical arrangement, they may also provide a visual indication that an item of equipment, e.g., switch, transformer or line, is disconnected from the system.

A primary substation, as in the example, may have a switching device on the incoming side of the transformer. Depending on the protection requirements, that switching device may be a circuit breaker or it may be a disconnector. Some utilities would not use a switching device but would connect the incoming cable or overhead line directly to the transformer. This type of disconnector could be a single break type, a double break type, a vertical break type or a pantograph type.

Disconnectors can be of single- or double-pole operation, operated manually or via a motor drive, and SCADA controlled disconnectors have become increasingly popular. Disconnectors are designed not to be opened when any amount of load current is flowing through them. They are normally capable of closing on to a line where the charging current is small, or to energize a small transformer.

On the outgoing (typically 20 kV) side of a primary substation, there is usually a need for some form of disconnector as already discussed for fixed circuit breakers. For full withdrawable circuit breakers, the action of physically removing the breaker is sufficient to achieve disconnection of the breaker for maintenance and, after applying circuit main grounds, for safe working on the immediately connected MV network.

4.3 GROUND-MOUNTED NETWORK SUBSTATIONS

This category covers the switchgear in all types of ground-mounted substation and switching station on MV networks. The switchgear in these substations is supplied as indoor gear for installation in building basements and small specially built buildings in the case of compact substations of weatherproof gear that is closely coupled to the transformer through an integral throat. The switchgear assemblies come in both extensible and nonextensible configurations, the simplest nonextensible configuration being the ring main unit (RMU).

Hardware for Distribution Systems

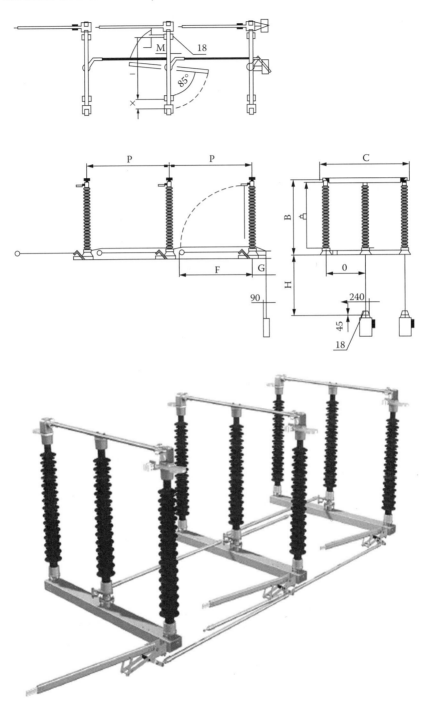

FIGURE 4.6 Rotating center post disconnector with grounding switch.

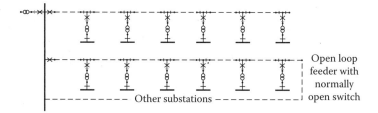

FIGURE 4.7 Underground distribution network designs.

4.3.1 Ring Main Unit

The final transformation in the electricity supply chain is normally from medium voltage to low voltage via the distribution substation and which could be ground-mounted or structure (pole)-mounted. A feeder from a primary substation could connect to a number of ground-mounted distribution substations or a number of structure-mounted distribution substations or a combination of both. The feeder may be designed as a purely radial feeder or as an open loop feeder.

In either case, the ground-mounted transformer needs to be connected to the feeder using a protective device, usually a circuit breaker or a switchfuse (see Figure 4.7). Now, if a fault occurs on the feeder, the source circuit breaker will trip, leaving the feeder and all its connected load off supply. If the location of the fault is known, then, if the feeder has a number of sectioning switches, the faulted section can be disconnected and healthy sections restored to supply. In the radial feeder, any load connected beyond the faulted section cannot be restored to supply, but in the open loop feeder, any load connected beyond the faulted section can be restored to supply by using the alternative supply beyond the normally open point. It follows, therefore, that at the distribution substation, switches for sectioning the feeder can be added to the local transformer protection to form a three-way switching system. This is often called a ring main unit because ring main is another term for an open loop feeder.

The particular ring main unit shown in Figure 4.8 is coded as CCF because, as a three-way unit, it comprises three switching devices:

- There are two cable switches, coded as C, which each have three operating positions. In the normal service position, usually known as "closed" or "on" or "1," the cable is connected to the busbar. In the next position, usually known as "open" or "off" or "0," the cable is not connected to any conductor, and if the switch includes a disconnector, then the cable is disconnected from all live conductors. In the final position, usually known as "earthed" or "earth on," the cable is connected to earth, which is normally a requirement before work on the cable is carried out. Both cable switches in Figure 4.9 are shown in the "open" position.
- There is one switchfuse, coded as F, which is used to connect the distribution transformer to the busbars and offer protection to the

Hardware for Distribution Systems

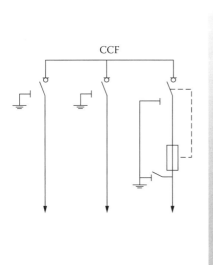
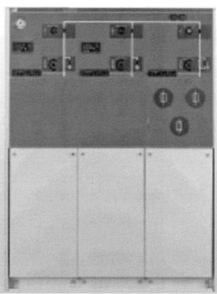

FIGURE 4.8 Ring main unit-type CCF.

transformer. In fact, protection devices are coded as T because they are generally used for transformer protection. There are subdivisions, F for a switchfuse and V for a vacuum circuit breaker. At the top of the diagram, the fuse elements are in series with a disconnector which, like the cable switch, can be earthed. Also, there is an earthing switch between the fuse and the transformer. These two switching devices enable the transformer to be disconnected when required, also for a ruptured fuse to be changed safely.

Ring main units are also available with a circuit breaker, instead of switchfuse, for the tee off connection to the transformer. Very often, this circuit breaker is rated at 200 amps, which is sufficient for transformer or direct load connection. Some network design engineers have asked for a three-way ring main unit with a 630 amp circuit breaker that can be used as part of the medium-voltage cable ring, which is not possible with a 200 amp circuit breaker. What is needed is a three-panel ring main unit-type CFV with one cable switch (for the incoming circuit), a switchfuse for local transformer protection and a 630 amp circuit breaker for the outgoing circuit.

The sectioning switches are normally live operated, fault making and load breaking switches driven by a common mechanism operating all three phases at once. However, this is not always the case; for example, many utilities employ a load break elbow in the function of a disconnector.

The majority of modern switchgear operated at medium voltage will use a standard bushing in the cable connecting area. Some users will join the MV cable

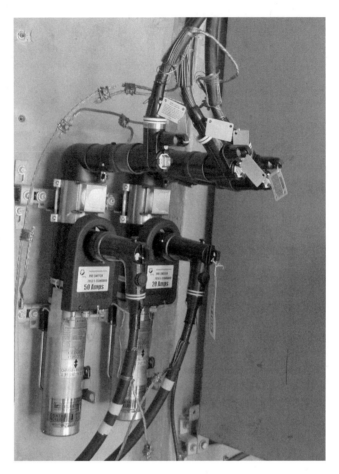

FIGURE 4.9 Disconnectable elbow.

to the bushing using a heat shrink or similar connection which, once joined, cannot be broken down without major repair. However, other users will join the MV cable onto the bushing using a preformed elbow that, under certain circumstances, can be disconnected. A typical elbow is shown in Figure 4.9.

This elbow is a load break elbow; that is, it can be operated to make and break small currents. However, many elbows are deadbreak, which means that they can only be connected to or disconnected from a bushing if the bushing and elbow are disconnected from all sources of supply.

An important point is that, if designed correctly, the disconnected elbow can provide the operator with the same function as the more conventional disconnector already described. In fact, the disconnectable elbow could be described as a form of slow operating disconnector that is manually operated, one phase at a time.

A major advantage of the standard bushing is the interchange of products that it permits. One common standard is IEEE 386-1995, to which the above

Hardware for Distribution Systems

elbow is constructed and which calls for 600 amp load rating. Another common standard is IEEE 386-1986, which covers 200 amp deadbreak elbows. IEEE 386-1995 also covers 200 amp load break elbows, which are elbows that can be used live to switch loads up to 200 amps.

4.3.2 PAD-MOUNT SWITCHGEAR

The pad-mount arrangement used in North American underground systems is a basic ring main unit constructed using disconnectable elbows, whether deadbreak or load break. The simplest ring main unit might comprise two three-phase sets of deadbreak elbows, one for the incoming cable connection and one for the outgoing cable connection. Figure 4.10 shows a typical modern unit comprising a steel cabinet, for outdoor use, which contains MV busbars and the connection to the local distribution transformer.

Pad-mount substations can be very easily upgraded to load break elbows, giving a form of switching up to 200 amps. They can also be upgraded to provide a fused connection using in line fuses but, in Figure 4.11 the ability to disconnect faulted cables is lost. They can also be upgraded to give a full switch or circuit breaker, which may be remote controlled if required by the utility.

FIGURE 4.10 Pad-mounted distribution substation (deadfront).

FIGURE 4.11 Pad-mount substation with fuses (livefront).

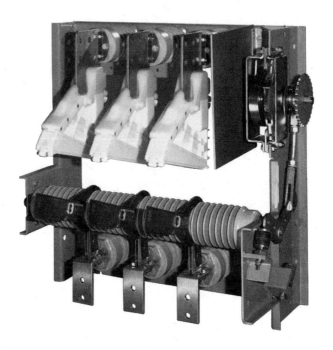

FIGURE 4.12 Fault making switch for inclusion in pad-mount switchgear.

4.4 LARGER DISTRIBUTION/COMPACT SUBSTATIONS

Ring main units are typically supplied as switch assemblies which are non-extensible; that is, that they cannot be extended once commissioned. Most new units are SF_6 where the switches are contained in one sealed enclosure, which is limited by design to a maximum number of switches (typically five). The number of switches must be specified before manufacture. Alternatively, some manufacturers' designs provide for extension bus adapters, permitting coupling of units. Again, this provision is only made during manufacture, and thus if provision for an extension is required, it must be specified in advance. These measures are necessary to keep the cost of ring main units to an absolute minimum.

Typically, larger switching assemblies are available and a code has been developed, described above, to describe all the possible configurations. Another combination might be CCFF, which would be a four-panel switchboard with two cable switches and two fused transformer protections; similarly, CCCF would be a four-panel switchboard with three cable switches and one transformer protection. Typical coded configurations are shown in Figure 4.14, where C is a cable switch, F is a switchfuse, V is a vacuum circuit breaker (200 or 630 amp rated)

Hardware for Distribution Systems 165

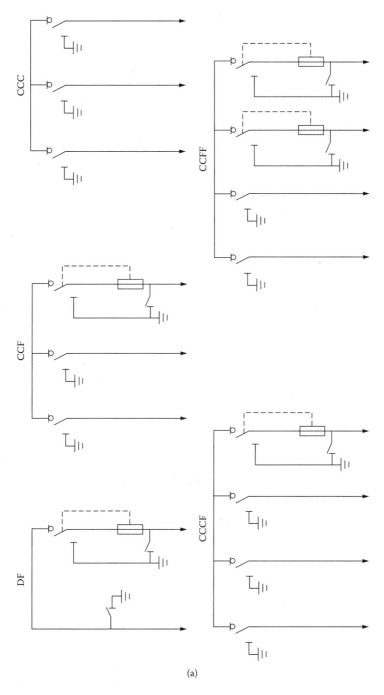

(a)

FIGURE 4.13 Typical MV switchgear combinations. *Continued.*

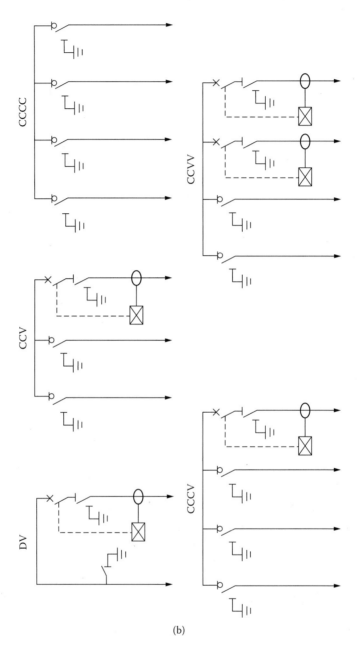

FIGURE 4.13 *Continued.*

Hardware for Distribution Systems 167

and D is a direct connection to the busbars. In these applications, D is shown as having an integral earthing switch, but this could be omitted.

4.5 POLE-MOUNTED ENCLOSED SWITCHES

Enclosed pole-mounted switches have been developed to improve the switching performance required by remote control and automation. The widening use of remote control has shown that traditional air-insulated switches may not be as reliable in adverse environmental conditions such as climates with extreme icing, see Figure 4.14 or polluted and desert-like abrasive atmospheres.

FIGURE 4.14 Overhead customer substation showing transformer and, at the pole top, an ice-encrusted air break switch.

FIGURE 4.15 Typical 24 kV SF_6 pole-mounted gas insulated switch type NXA from ABB showing mechanical operating rod for the motor actuator.

Switch moving parts are encapsulated in a stainless steel sealed housing filled with SF_6 gas as shown in Figure 4.15. Temperature compensated gas density gauges with the facility for switch lock-out at low pressures are included in switchgear designs to ensure that any leak can be detected and safe operation safeguarded. The gas is used as the insulating and arc-quenching medium. The switches are thus oil and maintenance free with a long life potential. Although operation can be achieved manually with a hook stick, most switches are installed for remote control. Operating mechanisms are manual, operator independent, quick close, quick open combined with a geared motor closing and opening mechanism. Motorized actuators are mounted at switch level with direct connection to the switch shaft or in the pole-mounted control cabinet near ground level. In the latter case, the actuator is connected to the switch with a rigid mechanical transmission rod. The motor mechanism is developed specifically to meet the requirements of automation and will operate with dead line conditions, giving opening times of approximately 0.5 seconds.

Switch position indicators are rigidly connected to the main switch shaft to fulfill the standards IEC 129 A2 (1996) and NF C-64-140 (1990). Measurements for fault location are provided by sensors integrated into the switch bushings or through CTs externally mounted around the bushings and individual pole-mounted voltage transformers (VTs).

4.6 POLE-MOUNTED RECLOSERS

Pole-mounted reclosers are produced in two basic configurations as derived from the different requirements of European (balanced three-phase) and North American (single-phase unbalanced) network design. North American systems tend to consider the option to operate individual phases separately even if operated as a three-phase bank, whereas European system design only considered operation of all three phases simultaneously. North American recloser configuration usually has three separate poles, whereas European reclosers are of the single-tank design.

Hardware for Distribution Systems

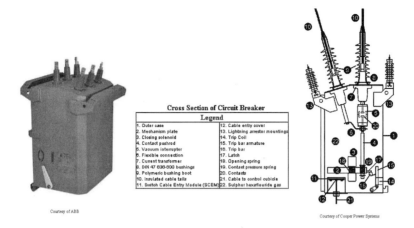

FIGURE 4.16 NULEC N series single-tank autorecloser showing main components.

There is no reason why the individual pole design cannot be used for both system types, and they are thus seeing increased acceptance in Europe.

4.6.1 SINGLE-TANK DESIGN

Such companies as Westinghouse Electric and Reyrolle introduced single-tank reclosers on distribution networks in the early 1960s. These devices mounted the three interrupters in a sealed steel container. The original reclosers relied on oil for insulation and arc interruption. The control mechanism was hydraulic with a wide tolerance on the setting making coordination difficult. The advantages of vacuum interruption and SF_6 soon obsoleted the oil recloser, and now all manufacturers supply single-tank designs with vacuum interrupted in a sealed gas-filled stainless container as shown in Figure 4.16.

The single-tank recloser is most popular in countries with European-type three-phase networks where simultaneous (ganged) operation of the three phases is required.

4.6.2 INDIVIDUAL POLE DESIGN

The individual pole design was made possible with the availability of the vacuum interrupter, which allowed encapsulation of the interrupter within the molding of insulation material (cycloaliphatic epoxy, polyurethane or silicone) of the insulator pole assembly. The three-pole assemblies shown in Figure 4.17 are mounted on a metal box that houses the magnetic actuator, visual position semaphore and hook stick operation facility. This solid pole assembly is free of gas and oil. Some recloser designs use an individual sensor mounted in a separate insulator that can be positioned on the recloser's mounting rail. This approach provides certain modularity.

A typical solid pole assembly shown in Figure 4.19 indicates the position of the vacuum interrupter and the operating mechanism enclosure on which the

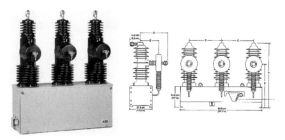

FIGURE 4.17 Three-pole solid insulation reclosers from two leading manufacturers.

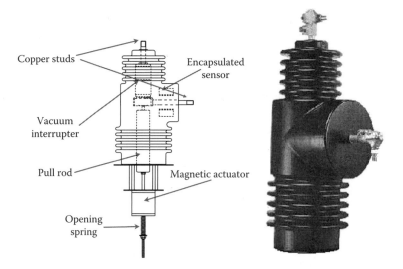

FIGURE 4.18 Example of an individual vacuum pole interrupter assembly.

poles are mounted. Most assemblies include integrally molded current sensors, and some manufacturers also include capacitive voltage dividers.

If the three individual interrupters are mechanically separate (mechanically unganged) in each phase, they can be controlled independently to provide greater protection flexibility for single-phase networks, increasing reliability. One example is the improvement in overvoltages when switching capacitors that can be made if the poles can be arranged to operate 120° apart.

All reclosers are controlled by specialized protection relays mounted in a control cabinet at the base of the pole.

4.7 POLE-MOUNTED SWITCH DISCONNECTORS AND DISCONNECTORS

The most widely used switching device on overhead distribution networks is the air break disconnector. These switches comprise two insulator posts, one sup-

Hardware for Distribution Systems

FIGURE 4.19 Typical side break switch disconnector showing operation rod and arc breaking chambers (Hubbell Power System Inc. Type AR, 15 kV 900 A continuous/interruption rating).

porting the fixed contact, the other post pivoting or rotating to move the breaking blade. Two configurations are most commonly used by the industry, a vertical break and a horizontal (side) break (see Figure 4.19). The selection of switch configuration is dependent on the line conductor spacing and geometry. The vertical break is most common because phase spacing is maintained, provided there are no lines above the switch, whereas the horizontal break will require increasing the horizontal conductor spacing around the switch location.

Traditionally, all disconnectors have been manually operated using a hook stick, a lever at the bottom of the pole (secured with a padlock) or a mechanism within a locked control cabinet. These switches can be remote controlled by adding a motor actuator and an IED.

In the simplest form, air break disconnectors interrupt small currents (e.g., 25 A at 24 kV) with a simple blade contact with breaking whip. An additional arcing chamber (see Figure 4.20) is required to increase current breaking to the nominal switch rating (e.g., 630 A at 24 kV).

4.8 OPERATING MECHANISMS AND ACTUATORS

Switchgear is operated from opened to closed or vice versa by the mechanical movement of the electrical contacts, and most switchgear can be operated by the action of a human operator. The operating mechanism falls into one of four main categories:

- Dependent manual, where the position of the contacts depends only on the position of the operating handle, and the speed of the moving contacts depends only on the speed of movement of the operating handle. This type of mechanism is common for pole-mounted switchgear and, historically, for ground-mounted switchgear. However, although inexpensive, it suffers from a major disadvantage that, if closing onto a fault and thereby experiencing high throw-off forces, successful closure rests with the skill and physical strength of the

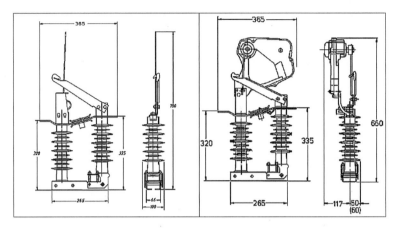

FIGURE 4.20 Air insulated pole-mounted 24 kV switch-disconnector from ABB with vertically operated blade type NPS 24 B1 with breaking whip and type NPS 24 B1-K4 with the addition of an interruption chamber for breaking load current.

operator. Failure to close the switch in a rapid and decisive manner can lead to electrical failure.
- Independent manual, where the position of the contacts depends on the energy from mechanical springs that are charged by the action of the switching operator. The operating handle charges the springs, and once there is sufficient energy stored, the electrical contacts move rapidly, overcoming the risk of the throw-off forces leading to electrical failure. This type of mechanism is fitted to the majority of ground-mounted switchgear.
- Solenoid mechanism, where a powerful electrical solenoid is used to directly operate the switchgear, normally in conjunction with the spring charge mechanism of the independent manual category. This type of mechanism is mainly used on large substation circuit breakers and is unusual on distribution switchgear.
- Motor wound spring, where an electric motor is used, replacing the manual operating handle, to charge up springs that are then released either as soon as there is sufficient stored energy or when an electrical release is operated.

4.8.1 Motorized Actuators

The most common method of operating distribution switchgear for extended control is to use the motor wound spring, which can be either fitted as part of the original switchgear or added on when extended control is needed. Although an actuator for the original switchgear will normally be provided by the original switchgear manufacturer, an add-on can be provided by any suitable manufacturer.

Figure 4.21 shows the arrangement for converting an independent manual switch into a motorized device by adding an electrically driven actuator. The

Hardware for Distribution Systems

FIGURE 4.21 Actuator for MV ring main unit. (Courtesy of W. Lucy Switchgear Ltd.)

photograph shows the two switches on an MV ring main unit, the left switch being for manual operation by the insertion of an operating handle to engage with the spring mechanism. To the left can be seen the indicator window to show the position of the switch.

The right-hand switch has been equipped for remote operation by the addition of an actuator, another of which could be fitted to the left-hand switch if required. The motor is contained in the dark colored enclosure that then operates, via a worm gear, the horizontal drive shaft, which in turn connects with a small adaptor plate onto the spring mechanism.

The motor is operated from the local DC supply and controlled by the local remote terminal unit or by local electrical pushbuttons. On the outside of the horizontal drive can be seen two auxiliary limit switches, which are used to indicate the position of the switchgear main contacts and stop the actuator when the switch has completed its operation. Although perhaps it would be ideal for the auxiliary switches to indicate the position of the switchgear contacts rather than the position of the actuator mechanism, there is a considerable cost penalty for this option because it would be necessary to open the switchgear to gain access to the main contact drive shaft. Because this would prevent the simple addition of the actuator to live switchgear, the practice of using external auxiliary contacts has become widely accepted. It can be seen that the complete actuator drive can be disconnected if needed.

A number of manufacturers use compressed gas as their energy storage instead of mechanical energy stored in a spring, and this is more common in pole-mounted switchgear. It has the advantage of possibly being a relatively cheap option but has the disadvantage that the gas bottles need to be replenished or replaced by a suitably qualified person.

4.8.2 Magnetic Actuators

In contrast to conventional stored energy mechanisms that rely on a more complex mechanical design incorporating more parts and less reliable mechanical links,

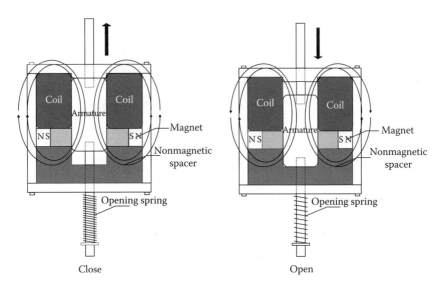

FIGURE 4.22 Magnetic actuator operating principles.

magnetic actuators are far less complex. The magnetic actuator (see Figure 4.22) is a device composed of a permanent magnet, a moving armature and an electrical coil. It is much simpler than the traditional stored energy mechanisms and solenoid-based designs. The magnetic actuator has only one moving part. This reduction in moving parts dramatically increases the unit's reliability, while drastically minimizing maintenance costs. The magnetic actuator is virtually maintenance-free, allowing thousands of operations without scheduled maintenance. A powerful neodymium iron boron (NdFeB) magnet provides the required force to hold the recloser in the closed position.

The magnetic actuator is a bistable device, meaning that it does not require energy to keep it in the open or closed position but it does need energy to enable a change of state to be made. When an open or close command is initiated, a current pulse energizes the coil for a very small period of time, enabling the required motion.

When the coil is energized with current in the proper polarity, the flux produced works together with the flux generated by the permanent magnet and drives the armature to the closed position (compressing the opening spring). Once closed, the coil is de-energized and the armature is held in position via the flux generated by the permanent magnet. In the closed position, the armature is against the top plate of the actuator, forming a low reluctance path for the magnetic flux. The static latching force of several hundred pounds is provided by the permanent magnet alone. The coil energization is not required. The magnet itself is mounted on a metal ring to prevent damage from contact with the armature as the armature assembly moves back and forth.

When the coil is momentarily energized with current in the reverse polarity, the flux produced opposes the flux generated by the permanent magnet. This temporarily offsets the holding force between the armature and the top plate. When this occurs, the opening spring moves the armature away from the top plate. As the gap increases, the holding force falls off very rapidly, and the opening spring drives the armature to the open position and holds it there without the coil being energized. The nonmagnetic spacer prevents the armature from latching to the bottom plate with the same force as exists in the closed position by inserting an air gap in the flux path. The relative simplicity of design of magnetic actuators gives the same level of maintenance-free duty as the vacuum interrupter they operate.

4.9 CURRENT AND VOLTAGE MEASURING DEVICES

Current and voltage transformers are the key components that convert measurements at high primary potential to levels acceptable to intelligent electronic protection and control devices (IEDs). Although instrument transformers with magnetic cores have traditionally been accepted as standard for providing this function, new smaller and less costly sensing devices without the saturation characteristics of iron are being introduced as an alternative.

The long-term trend in MV switchgear design has been towards smaller size. This was incompatible with the space requirements of conventional current transformers and voltage transformers, which take up a significant volume of the cubicle. The volume of the sensors is one third of conventional instrument transformers, thus allowing switchgear cubicles to be produced with a 55% reduction in volume over conventional switchgear (see Figure 4.23).

Overhead outdoor equipment is also trending towards smaller, less heavy devices with increased measurement functionality. This compactness requires a

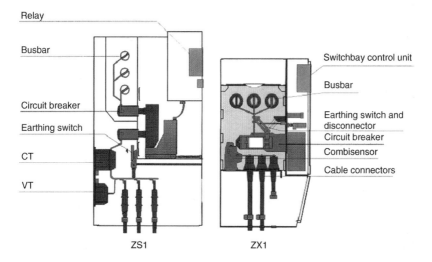

FIGURE 4.23 Comparison of switchgear cubicle size with traditional CT/VT devices (left) and new sensors (right).

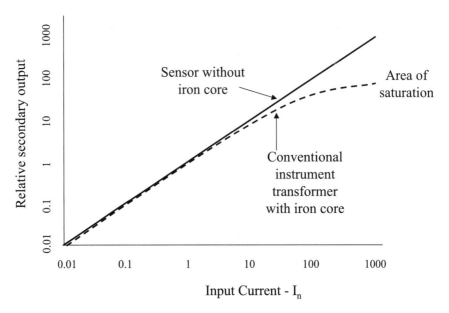

FIGURE 4.24 Comparison of conventional current transformer output with new sensor technologies.

sensor that is more easily integrated within the pole structure of the switch. However, the low power output of these new devices has still allowed traditional CTs to remain as the preferred option for outdoor pole-mounted equipment.

Selection of conventional CTs and VTs requires in advance the specification of load current and its future trend, rated voltage, secondary burden and accuracy classes if metering is required. In practice, the variety of different combinations means that instrument transformers are manufactured on demand and the effort to standardize the production of the physical cubicle that can be parameterized later is not possible. The trend for configurable and standardized switchgear with shortened delivery times is not met with conventional instrumentation, particularly for indoor gear.

New sensing technology offers the possibility to cover a very wide linear range, which for currents is typically 40–1250 A and for rated voltage 7.2–24 kV, with one device (see Figure 4.24). Table 4.2 provides a summary comparison between the conventional and new sensor technologies.

The International Electrotechnical Vocabulary defines an instrument transformer as a transformer intended to supply measuring instruments, meters, relays and other similar apparatus. Specifically:

1. A current transformer is an instrument transformer in which the secondary current, in normal conditions of use, is substantially proportional to the primary current and differs in phase from it by an angle

TABLE 4.2
Comparison of Conventional CT/VT Against Modern Sensor Technologies

Property	Conventional Devices	Sensors
Signal	1–5 A (CT) 100:√3 V (VT)	150 mV (CT) 2:√3 V (VT)
Secondary burden	1–50 VA	> 4 Mohms
Accuracy	Measurements: 0.2–1.0% Protection: 5–10%	Multipurpose 1%
Dynamic range	40 × I_n (CT) 1.9 × U_n (VT)	Unlimited
Linearity	Nonlinear	Linear
Saturation	Signal distortion	None
Ferroresonance	Destructive (VT)	None
Temperature dependence	No influence	Compensated
EMC	No influence	Shielded
Short-circuited secondary terminals	Destructive (VT)	Harmless
Open secondary terminals	Destructive (CT)	Harmless
Weight	40 kG (CT+VT)	8 kG (I+U)
Life cycle cost (LCC)	High	Low
Devices to cover operational ranges	Many	Two

which is approximately zero for the appropriate direction of the connections. CTs are defined according to IEC 60044 Part 1.

2. CTs can be subdivided into the protective CT, which is intended to supply protective relays, and the measuring CT, which is intended to supply indicating instruments, integrating meters and similar apparatus.
3. A voltage transformer is an instrument transformer in which the secondary voltage, in normal conditions of use, is substantially proportional to the primary voltage and differs in phase from it by an angle which is approximately zero for the appropriate direction of the connections. VTs are defined according to IEC 60044 Part 2.
4. Instrument transformers are given a value for burden, which is the impedance of the secondary circuit in ohms at a given power factor. However, by convention, burden is normally expressed as the apparent power (VA) at the specified power factor and the rated secondary current. It is most important to maintain the loads connected to the instrument transformer within the specified burden.

4.9.1 ELECTROMAGNETIC CURRENT TRANSFORMERS

An approximate equivalent circuit for an electromagnetic (wound) current transformer is given in Figure 4.25, where

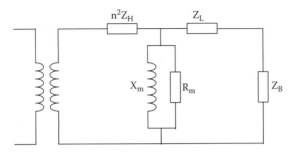

FIGURE 4.25 Equivalent circuit of electromagnetic CT.

- X_m and R_m are the core excitation and the core loss, respectively.
- Z_L is the impedance of the secondary winding.
- $n^2 Z_H$ is the primary impedance but referred to the secondary side.
- Z_B is the impedance of the burden (typically a protective relay).

Unfortunately, because of the current that circulates in the magnetizing circuit, the output of the magnetic CT is not directly proportional to the input current, and hence, some errors occur in terms of both magnitude and phase. Looking at the equivalent circuit, it can be seen that the voltage across the magnetizing circuit (X_m paralleled with R_m) is directly proportional to the secondary current. It follows that, when the primary current, and hence the secondary current, is increased, the magnetizing current increases to the point that the core saturates and the magnetizing current is large enough to produce a significant error, and this is clarified in Figure 4.26.

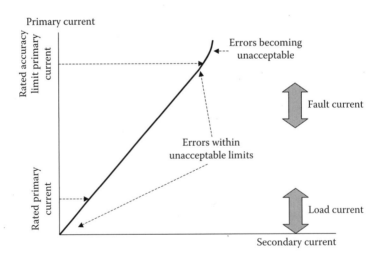

FIGURE 4.26 Nonlinearity of electromagnetic CTs.

Hardware for Distribution Systems

We can see from this diagram the major difference in user requirements between CTs that are specified for measurements and those that are specified for protection purposes.

The measurement CT needs to measure, for example, the load current in a circuit, which may be of the order of 400 amps and is near to the value for the rated primary current, typically 630 amps. In this range, the CT must be as accurate as required by the user, but outside the range of load currents, its accuracy is of less importance.

Conversely, whereas the rated primary current may still be 630 amps, the protection CT is required to operate with the fault currents that may be found on the particular system in the region of many thousands of amps. As long as the protection CT can detect the fault and cause protection to operate, it does not need to be as accurate as perhaps a measurement CT that is calculating the revenue flows for the customer. The diagram also shows that the protection CT may be asked to carry currents far in excess of its rated primary current for the short period of protection operation which is defined as the accuracy limit factor where

$$\text{accuracy limit factor} = \frac{\text{rated accuracy limit primary current}}{\text{rated primary current}}.$$

These differences are recognized in the international standards by defining two classes of CT, class M for measurement CTs and class P for protection CTs.

The most important elements of the specification of a CT can be defined in relatively few steps:

- Rated insulation level, for example, a CT that is required to operate on a 24 kV system would have a rated power frequency withstand voltage of 50 kV and a rated lightning impulse withstand voltage (peak) of either 95 or 125 kV.
- Rated primary current, selected from the preferred range of 10, 15, 20, 30, 50, and 75 amps and the decimal multiples.
- Rated secondary current, selected from the range of 1, 2, or 5 amps, but the preferred value is 5 amps.
- Rated output, selected from the standard values of 2.5, 5, 10, 15, and 30 VA with values above 30 VA being selected to suit the specific application.
- Accuracy class, which is different for protection CTs and for measuring CTs. The standard accuracy classes for protection CTs are 5P and 10P, whereas those for measuring CTs are 0.1, 0.2, 0.5, 1, 3, and 5, which is explained in more detail in Table 4.3 and Table 4.4.
- Accuracy limit factor, selected from the range of 5, 10, 15, 20, and 30 but only applicable to protection CTs.

TABLE 4.3
Limits of Error for Protective Current Transformers

Accuracy Class	Current Error at Rated Primary Current %	Phase Displacement at Rated Primary Current Minutes	Composite Error at Rated Accuracy Limit Primary Current %
5P	±1	±60	5
10P	±3	—	10

TABLE 4.4
Limits of Current Error for Measuring Current Transformers

Accuracy Class	± Percentage Current (Ratio) Error at Percentage of Rated Current Shown Below				± Phase Displacement at Percentage of Rated Current Shown Below Minutes			
	5%	20%	100%	120%	5%	20%	100%	120%
0.1	0.4	0.20	0.1	0.1	15	8	5	5
0.2	0.75	0.35	0.2	0.2	30	15	10	10
0.5	1.50	0.75	0.5	0.5	90	45	30	30
1.0	3.00	1.50	1.0	1.0	180	90	60	60
	50%		120%					
3.0	3		3		Not applicable			
5.0	5		5					

Thus, a manufacturer's nameplate on a CT may typically show "400/5 Class 5P, 15 VA," which would mean that it is protection CT, 400 amp primary, 5 amp secondary (hence ratio of 80:1) output, 15 VA and with errors within the definition of Table 4.3.

4.9.2 VOLTAGE TRANSFORMERS

The aim of the voltage transformer is to provide a voltage from the secondary winding that is at some convenient level, for example 110 volts, and that is, as near as possible, proportional to the primary voltage. In distribution systems, most VTs are inductive. From the point of view of extended control of distribution systems, the specification for a VT is relatively simple:

- Rated primary voltage, to be selected to match the system to which it will be connected but, ideally, from one of the standard IEC voltages according to IEC 60038
- Rated secondary voltage, likely to be 100 or 110 volts (Europe) or 115 or 120 V (US)
- Rated insulation level, which would be the same as for CTs

Hardware for Distribution Systems

- Rated output, selected from the preferred range of 10, 25, 50, 100, 200 and 500 VA at power factor 0.8 lagging. For three phase VTs, the rated output should be the rated output per phase

4.10 INSTRUMENT TRANSFORMERS IN EXTENDED CONTROL

The major function of instrument transformers in extended control is to provide the base signal for analog measurements of voltage and current, which can also be used for the derivation of further quantities such as active power (kW and kWh), reactive power (kVAr and kVArh), voltage unbalance and power factor.

The CTs that are required for the measurement of current depend on the accuracy of the measurement that is needed. If, for example, the customer needs to measure current to within 0.4%, then Table 4.4 shows that a measurement CT of Class 0.5 will be needed. When taking such measurements at new switchgear, it is clearly best to specify the correct CTs from the outset. However, there are cases when a customer needs only a broad indication of the current flowing, which may be the case at distribution substations. For example, the difference between perhaps 150 and 160 amps may not be material, whence the importance is to match the accuracy of the CT, and all connected equipment, to the permitted tolerance.

Some network operators may want to take current measurements using existing protection CTs, for example, fitted to circuit breakers and reclosers. If the installation of additional measurement CTs would mean the breaking down of gas-tight (SF_6) switchgear or the removal of pole-mounted switchgear, then significant savings could be made by using existing CTs. However, it is extremely important to note that the accuracy of the resulting measurement cannot be any higher than the error produced by the CT. Whereas a Class 10P protection CT may be sufficient to drive a protection relay, the 3% error may take the current measurement outside acceptable limits. Of course, trying to drive a protection relay from a measurement CT would cause the CT to saturate and probably prevent the protection from working.

Great care must be taken when using connecting cables to link CTs, whether existing or new, to a remote terminal unit under extended control. If the length of the cable is large enough, then the burden imposed by the combination of cable and detecting device may cause the rated burden of the CT to be exceeded. One possible solution is to use a transducer at the switch bay level and to communicate the output back to the IED by RS-485, RS-232, fiber cable or similar.

It should be noted that VTs, apart from providing voltage measurement, might be useable as a power supply to the IED cabinet. If this is the case, then a failure of the battery charger could be used to provide a digital indication that the supply has failed.

Depending on the type of fault passage indicator (FPI) fitted to any particular substation, it may be possible to replace the FPI sensors with the instrument transformers described in this section.

If there is a primary current in any CT whose secondary circuit is operating into an open circuit, then a higher than usual voltage will appear at the CT secondary terminals. This voltage can be high enough to damage electrical insulation and can cause a threat to life. It is, therefore, most important when working on CT secondaries to ensure that open circuits are not allowed or that the primary current be disconnected.

4.11 CURRENT AND VOLTAGE SENSORS

The nonlinearity of magnetic cores in CTs and VTs sets constraints on the measurement range and accuracy. The introduction of alternative sensing technology provides a large measurement range with high accuracy and integrated measurement and protection from one sensor.

4.11.1 CURRENT SENSOR

The principles of the Rogowski coil (RC), known since 1912, are capable of delivering the improvements sought. The RC is a uniformly wound coil that has a non magnetic core, the simplest shape being a toroidal air-cored coil. This coil has to be wound very precisely to achieve the desired accuracy band stability. The current going through the coil indices voltage e is given by the following approximate formula:

$$e = \mu_0 N A \, dI/dt = H \, dI/dt,$$

where μ_0 = permeability of free space, N = turn density (turns/m), A = single turn area (m^2), and H = coil sensitivity (Vs/A).

The design of the RC without a magnetic core eliminates the nonlinear effects of saturation and permits isolated current measurement with a megahertz bandwidth. In most respects, the RC approach is the ideal sensor for applications where measurement of DC current is not necessary. The major disadvantage is that the output is proportional to the time derivative of the current output and must be integrated. The use of digital integrators has solved the inadequacies of earlier analog integrators.

The overall accuracy of the RC approaches 0.5%, provided care is taken to minimize the major sources of inaccuracy from temperature changes, assembly tolerances and the effect of other phases (cross talk).

Temperature dependency can be lowered by use of special materials and by compensation methods. Assembly tolerances are minimized by correct mechanical installation or, in the majority of cases, integrating the sensor into the bushing assembly. Proper design of the sensor minimizes cross talk. Phase angle accuracy obtained from sensors is high, unlike iron core CTs where displacement varies with current and is worse during under- and overexcitation. Frequency response of RC sensors designed for 50 Hz operation is adequate with a range from a few Hz to 100 kHz. EMC effects on the operation of RC sensors must be considered

Hardware for Distribution Systems 183

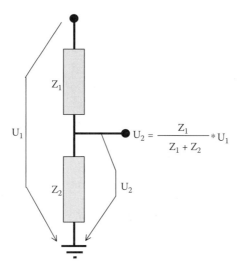

FIGURE 4.27 Principle of a voltage divider.

in the design and testing of the installation environment, because 50 Hz signals in the order of a few millivolts per ampere can affect the low signal level from the RC in difficult EMC environments.

4.11.2 Voltage Sensor

Resistive voltage dividers (Figure 4.27) are used for voltage measurement in MV switchgear providing a small lightweight device that does not contribute to nor is affected by ferroresonance. It can even be used to measure the phase to earth voltages during these resonance conditions. The construction of resistive voltage dividers must be able to withstand all normal and abnormal (fault) voltages. These conditions place very heavy demands on the divider, requiring the resistance of the divider to be very high. At these high impedance levels, the handling of stray capacitance is very important.

Sensor accuracy is dependent on the accuracy of the resistors (the division ratio). The main sources of inaccuracy occur from the resistor temperature coefficient, the resistor voltage coefficient, drift of resistors (voltage, temperature dependent), stray capacitance and effects from adjacent phases (cross talk).

Normally, an accuracy of ±0.5% is achieved and a revenue standard of 0.2% can be achieved through temperature compensation, choice of resistive material (critical for long-term accuracy) and minimization of stray capacitance. Due to the high impedance of the resistive divider, the frequency response is not as wide as the RC current sensor; however, frequencies up to a few kHz can be measured.

The same preventive measures as used for EMC interference prevention with the RC sensor should be employed for the voltage sensor because the output level is only 1 volt.

FIGURE 4.28 Traditional switchgear bushing compared with a bushing incorporating a combi sensor.

4.11.3 COMBI SENSOR AND SENSOR PACKAGING

The packaging of the sensors is critical to achieving reliability, minimum size, standardization and accuracy. Sensors are made for installation in metalclad switchgear cubicle for connection to busbars or incoming cable. This configuration is used either in new equipment or for retrofitting new technology sensors.

A second configuration is used to integrate the sensor within a standard bushing foundation (400 amp to DIN 47636-ASL-36-400), further reducing the number of physical designs because for bare conductor connection, an outer insulator boot can be forced over the bushing. This latter configuration is used for ring main units and outdoor switchgear because it allows connection by cable or the attachment of an outdoor bushing yet maintains only one configuration for both.

The ultimate packaging is to include both current and voltage sensors in one molding as a combi sensor, providing a low-cost total sensor solution for the majority of protection and monitoring needs (Figure 4.28).

Hardware for Distribution Systems

Various current sensors

Bus-bar type current and voltage comb sensors

Block-type current and voltage combi sensors

Bushing type for outdoor pole-mounted switchgear

FIGURE 4.29 Sensor family sufficient for the majority of sensing and measurement needs with distribution switchgear.

REFERENCE

1. Uwaifo, S.O., *Electric Power Distribution Planning and Development, The Nigerian Experience*, Lagos State, Nigeria: Hanon Publishers Limited, 1998.

5 Protection and Control

5.1 INTRODUCTION

The majority of electrical faults on a distribution power system provide a significant release of energy that, unless the fault is removed from the power system, can cause damage and possible injury. Such a fault is detected by the protection system, which causes a circuit breaker to trip to switch off the faulted components. Protection, therefore, is an extremely important operation on any power system. The protection must always operate when asked to do so; otherwise, the fault will persist and additional outages or damage will occur. Also, the protection must never operate when not asked to do so, otherwise unnecessary outages of healthy circuits will occur.

Faults may be between two or more phases of a multiphase network or between one or more phases and ground (earth). The value of phase fault current depends mainly on the source impedance and the impedance between the source and the point of fault. The value of earth-fault current depends on another major factor, that is, how the system is earthed, if at all. As the neutral earthing impedance increases, then the earth fault current decreases and so becomes more difficult to detect. However, system operators see other advantages in high-impedance earthing; hence, protection has been developed to be able to meet these varying demands.

Whenever a fault occurs, the owner of the system will need to know where the fault is to restore supplies to the healthy parts of the system and then make repairs to the faulted components. The fault passage indicator (FPI) is a device, several of which can be placed at strategic sites on a distribution network to provide useful information to the network owner as to whether the fault current has or has not flowed at each site. Reading these devices can help the owner to determine the location of the fault.

Protection and FPI both contribute to distribution automation and can be studied together because, although protection relays are usually more complex than FPIs, both employ similar technology.

5.2 PROTECTION USING RELAYS

The basic method of providing overcurrent relay protection on a circuit is to install current transformers (CTs) on the circuit, which then feed, into a relay, current that is proportional to the circuit current. When the current exceeds a preset value, the relay will operate at a time determined by the characteristics of the relay to initiate tripping of the associated circuit breaker.

From the basic principle of protection against excess current has evolved the graded overcurrent system, a discriminative fault protection. Correct current relay application requires a knowledge of the fault current that can flow in each part of the network and, because large-scale tests are normally impracticable, system analysis must be used. The data required for a relay setting study are as follows:

- A one-line diagram of the power system involved, showing the type, location and rating of the protective relays and their associated current transformers
- The impedances in ohms, per cent or per unit, of all feeder circuits, transformers and generators from which can be calculated the maximum and minimum values of short circuit currents that are expected to flow through each protective device
- The maximum peak load current through each protective device
- Decrement curves showing the rate of decay of the fault current
- Performance curves of the current transformers

The relay settings are first determined so as to give shortest operating times at maximum fault levels and then checked to see if operation will also be satisfactory at the minimum fault current expected. It is always advisable to plot the curves of relays and other protective devices, such as fuses, that are to operate in series, on a common scale.

The basic rules for correct relay coordination can generally be stated as follows:

- Whenever possible, use relays with the same operating characteristic in series with each other.
- Make sure that the relay farthest from the source has current settings equal to or less than the relays behind it; that is, that the primary current required to operate the relay in front is always equal to or less than the primary current required to operate the relay behind it.

Among the various possible methods used to achieve correct relay coordination are those using either time or overcurrent or a combination of both time and overcurrent. The common aim of all three methods is to give correct discrimination. That is to say, each one must select and isolate only the faulty section of the power system network, leaving the rest of the system undisturbed.

5.2.1 DISCRIMINATION BY TIME

In this method, an appropriate time interval is given by each of the relays controlling the circuit breakers in a power system to ensure that the breaker

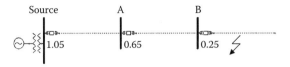

FIGURE 5.1 Discrimination by time.

nearest to the fault opens first. To illustrate the principle, a simple radial distribution system is shown in Figure 5.1.

Circuit breaker protection is provided at the source, A and B, that is, at the infeed end of each section of the power system. Each protection unit comprises a definite time delay overcurrent relay in which the operation of the current-sensitive element simply initiates the time delay element. Provided the setting of the current element is below the fault current value, this element plays no part in the achievement of discrimination.

For this reason, the relay is sometimes described as an "independent definite time delay relay" because its operating time is for practical purposes independent of the level of overcurrent. It is the time delay element, therefore, that provides the means of discrimination. The relay at B is set at the shortest time delay permissible to allow a fuse to blow for a fault on the secondary side of transformer. Typically, a time delay of 0.25 seconds is adequate. For the fault shown, the relay at B will operate in 0.25 seconds, and the subsequent operation of the circuit breaker at B will clear the fault before the relays at the source and A have time to operate. The main disadvantage of this method of discrimination is that the longest fault clearance time occurs for faults in the section closest to the power source, where the fault level, in MVA, is highest.

5.2.2 Discrimination by Current

Discrimination by current relies on the fact that the current varies with the position of the fault, because of the difference in impedance values between the source and the fault. Hence, typically, the relays controlling the various circuit breakers are set to operate at suitably tapered values such that only the relay nearest to the fault trips its breaker.

Discrimination by current has the disadvantage that there needs to be enough cable or overhead line between the two protection relays, in series, to achieve an appreciable difference in fault current between two relays.

5.2.3 Discrimination by Both Time and Current

Because of the limitations of both discrimination by time and discrimination by current, the inverse time characteristic has been developed. With this characteristic, the operating time is inversely proportional to the fault current level, and the actual characteristic is a function of both time and current.

Figure 5.2a shows the different standard inverse curves normally available in a modern relay that depend on whether IEC standards (standard inverse, extremely inverse or very inverse), ANSI standards (moderately inverse, very inverse or extremely inverse) or other standards are used. Figure 5.2b shows the front panel of a typical modern relay for overcurrent and earth fault protection.

Figure 5.3 shows three circuit breakers, each with inverse protection characteristics, at substations A, B and C. Now, as the fault current decreases as distance from the source substation increases, we might have a fault current If for a fault at the position F1. The figure shows that this current will cause relay C to operate in time t1. For this value of current, relay B would operate in time t2, which is longer than time t1; hence, provided relay C does actually operate its circuit breaker to successfully clear the fault, then relay B will reset at time t1.

If, however, relay C does not clear for some reason, for example, the circuit breaker fails to open when required, then relay B will act as a backup stage to relay C and will clear the fault at time t2.

The actual settings for relays in series must be calculated very carefully to allow for one protection operation to fail and still permit the fault to be cleared, even at a slightly later time. This later time must not be so delayed as to cause the sustained passage of fault current to permit otherwise healthy components to be damaged.

5.3 SENSITIVE EARTH FAULT AND INSTANTANEOUS PROTECTION SCHEMES

In the last section, the inverse protection characteristic was introduced, and inspection of the curve shows that, as the current increases, so the operating time decreases. These curves also show that there is a minimum current, $I_{Min\ Op}$, in Figure 5.4, that the relay will not operate. Now, the fault current that flows depends on many factors, but an important factor for earth fault is the actual ground resistance at the point of fault. As the resistance increases, the available earth fault current decreases, and it is easy to understand that there is a resistance above which the protection relay will not operate. This is likely to occur for a single-phase earth fault on an overhead network where the ground resistance is high. Such a fault could be the result of an overhead line conductor breaking and falling to the ground, which, if there are people or animals nearby, could become very serious.

Sensitive earth fault (SEF) protection is characterized by requiring a low value of earth fault current to flow but for a comparatively long period of time and is shown in the diagram. The current might be 10 amps, but this current would need to flow continuously for time t1, typically 30 seconds, before the SEF initiated trip occurs. SEF is an extremely useful protection and is often required to be indicated via a SCADA scheme.

However, in the situation where two feeders are paralleled by closing a normally open switch, the SEF on each feeder may nuisance trip if, as is likely, the parallel is left on for longer than the illustrated 30-second operating time. SEF

Protection and Control

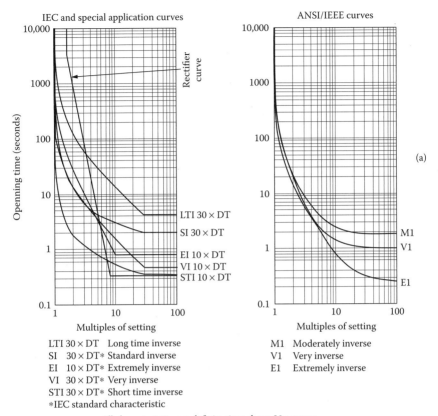

LTI 30 × DT Long time inverse
SI   30 × DT* Standard inverse
EI   10 × DT* Extremely inverse
VI   30 × DT* Very inverse
STI 30 × DT* Short time inverse
*IEC standard characteristic

All characteristics are definite time above 30x except extremely inverse and rectifier curve.

M1 Moderately inverse
V1 Very inverse
E1 Extremely inverse

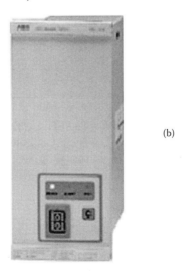

FIGURE 5.2 (a) Typical IDMT curves and (b) typical front panel.

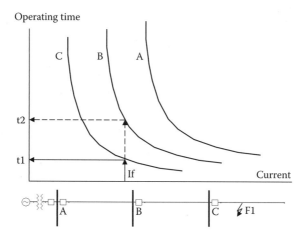

FIGURE 5.3 Grading by time.

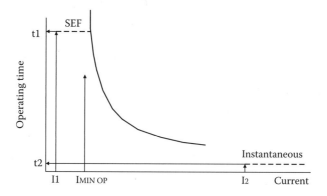

FIGURE 5.4 Sensitive earth fault protection.

should, therefore, be controllable through a SCADA scheme. Because an operation on SEF is commonly the result of a conductor on the ground, and in such a condition a person might approach the grounded conductor for a variety of reasons, then it follows that re-energizing the fault by using a recloser should be prohibited.

In a very similar but opposite way to SEF, a fault that gives a very high fault current may be regarded as being extremely serious and should, therefore, be removed in the shortest possible time period. In the diagram, this is achieved by an instantaneous (nondelayed) trip as soon as the current reaches some preset but high, value, I2. Such a scheme may also be known as HiSet.

5.4 PROTECTION USING FUSES

The fuse was the earliest protective device found in power systems. It consists of a metallic element, which will carry, continuously, the required load current

Protection and Control 193

FIGURE 5.5 MV fuses.

but which will melt, and therefore, break, at some higher current, thus preventing the passage of any further current.

Fuses for distribution systems can be classified into ground-mounted and pole-mounted. Ground-mounted fuses are normally contained within metal-clad switchgear and designed to protect a local ground-mounted transformer. They are rated up to approximately 160 amps. Pole-mounted fuses for overhead networks can protect a local transformer but may also be installed at the start of a spur line to prevent faults on the spur line affecting customers upstream of the fuses. Pole-mounted fuses are rated up to approximately 40 amps.

A typical medium-voltage fuse is shown in Figure 5.5. The exterior is a ceramic insulating tube with a metallic cap, acting as electrical contact, at each end. The electrical connection between the caps and inside the tube is a series of parallel connected silver strands, arranged so that each parallel path carries nominally the same current. These strands are fragile and so are supported on an internal structure, and the complete interior of the ceramic tube is filled with quartz sand.

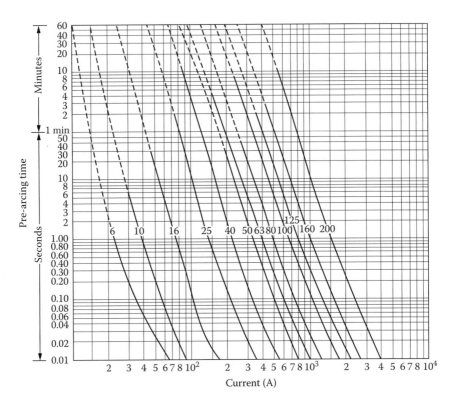

FIGURE 5.6 MV fuse characteristic.

On many distribution systems, typically in Europe, distribution transformers are three-phase and delta/star connected. The operation of a single-phase fuse on the MV system supplying the distribution transformer will give half voltage on two of the three low-voltage phases, which can damage connected customer plant and, in some countries, place the utility under financial threat. The situation is similar in the United States, where three single-phase transformers are used. If the MV star point is earthed, then the loss of one MV phase will cause zero voltage on the corresponding secondary winding. If the star point is not earthed, then the loss of one MV phase will cause zero voltage in one of the LV phases and reduced voltage in the other two phases.

One solution is to fit the MV fuse with a small, internal, explosive charge that will operate with the fuse and eject a small striker pin at one end of the fuse. This striker pin can be used to trip the switch in a combination switchfuse, thus interrupting supply to the other phases. Another application of the striker pin is to drive some form of mechanical indicator to show that the fuse has operated.

A typical time and current operating characteristic for an MV fuse is shown in Figure 5.6.

Fuses are small and relatively cheap but they suffer from the disadvantage that they need to be replaced after every operation and before load current can

Protection and Control

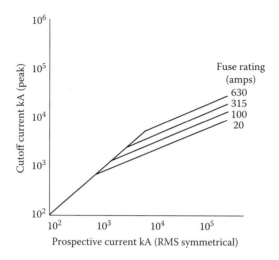

FIGURE 5.7 Current limiting action of fuse.

be passed again. By suitable design and the use of various metallic elements and filler materials, different time/current characteristics can be achieved to aid discrimination with other fuses or protective devices.

A major, and often not fully recognized, feature of the modern fuse is its ability to operate, with a high fault current, within less than one cycle. The fault current is, therefore, cut off by the operation of the fuse before the current can reach the maximum prospective value, a facility known as current limiting (Figure 5.7). For lower currents, the operating time can take some seconds to clear the fault.

For example, with 10,000 amps prospective current through a 100 amp fuse, the equivalent maximum current peak that would be reached is 1200 amps. Current limiting has been used by one manufacturer to produce a current limiting device for MV systems, the Is Fault Current Limiter. If this were to be placed in the bus section of an MV switchboard that was supplied by two transformers, one on each busbar, then the effective fault level, on each busbar, would be less than that provided by the two transformers alone because, when a fault occurs, the current limiter opens so quickly that it limits the fault contribution from each transformer before the prospective current can be reached. Some utilities and industrial customers have used this in conjunction with parallel operated generation to alleviate the need to increase the fault rating of switchgear connected to the system.

By contrast, the pole-mounted fuse, or cutout, is a less sophisticated, but still very useful, component. The fuse element is a bare element contained in an insulating tube. When the fuse needs to be replaced, the insulating tube is removed to ground level using a long insulating rod and a new element fitted to the tube. These are then returned to pole-top level by placing the bottom of the fuse assembly into the lower level electrical contact shown in Figure 5.8. Using the long insulating

FIGURE 5.8 Typical pole-mounted fuse.

rod, the fuse assembly is rotated upwards so it latches into the top-level electrical contact, thereby placing the fuse element under mechanical tension.

If the fuse blows, the mechanical tension is lost as the element ruptures and the insulating rod is, therefore, unlatched at the top and falls to the vertical position, thus showing to the utility that it has operated. This indication has earned this type of fuse the name of drop-out expulsion fuse, or simply DOEF.

The automatic sectionalizer that operates downstream of a multishot automatic recloser has been described in Chapter 4, but it is interesting to see a development of the DOEF that can be used as an alternative to the switchgear-based automatic sectionalizer.

In this development, shown in Figure 5.9, the insulating tube and fuse element have been replaced by a tubular conductor with a radially mounted CT. The CT detects the passage of fault current and is connected to a small processor contained in the tubular conductor. After the preset count of bursts of fault current, the

FIGURE 5.9 Drop-out-type autosectionalizer.

Protection and Control

processor fires a small explosive charge in the top of the conductor that makes the conductor bar drop downwards, thereby sectionalizing the fault.

The explosive charge needs to be replaced before the automatic sectionalizer can be put back into service.

In some designs of DOEF, the fuse element can be directly replaced by the automatic sectionalizing element.

When compared to the switchgear-based automatic sectionalizer, the drop-out type has the advantages of being relatively cheap, and it can be installed as a direct replacement to a drop-out expulsion fuse. Its disadvantages are that it cannot be remotely controlled or monitored, and an operator has to visit the site to replace the small explosive charge before restoring supply.

5.5 EARTH FAULT AND OVERCURRENT PROTECTION FOR SOLID/RESISTANCE EARTHED NETWORKS

To be able to measure the current in each of the three phases in a network, one CT is needed for each phase; generally, one CT is connected to each of the three-phase inputs of a relay, or, if using single-phase relays, one to each relay. This scheme will measure overcurrent, which is a separate scheme from earth fault. An earth fault scheme may also operate an overcurrent scheme if the earth fault current flowing in at least one phase exceeds the overcurrent protection setting.

Earth faults can be detected by a number of arrangements of CTs, and one common scheme is shown in Figure 5.10. Because the earth fault current is the vector sum of the three-phase currents, then the earth fault relay shown in the diagram will measure the imbalance of the three-phase CT currents and consider this as earth fault. An alternative is to remove the three overcurrent relays in the diagram and short together the input connections to each relay. This has the effect of connecting the three CTs in delta across the input to the earth fault relay.

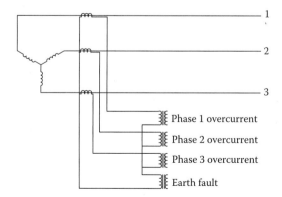

FIGURE 5.10 CT connections for overcurrent and earth fault protection schemes.

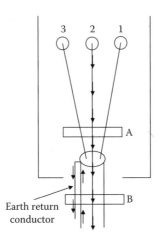

FIGURE 5.11 Zero sequence CT connections using residual CT (B).

Another alternative is to measure the earth fault current by using a single CT around the complete three-phase conductor assembly. For a three-phase overhead line, this would mean a large CT around the three conductors, insulated for the medium voltage and would be impractical. However, for underground cable networks, the installation of a single, zero sequence CT is normal and is shown in Figure 5.11.

The diagram shows a cable end box on a unit of medium-voltage switchgear. From the middle of the three cable bushings, an earth fault current is flowing to the cable sheath some distance away. A single CT around the three cores at A will detect the earth fault current and can be used to operate protection. By moving the CT to B, i.e., outside of the cable box, inspection shows that the current in the phase conductor and the return current in the cable sheath will cancel each other out, with the result that the CT would see nothing. By using an external earth return conductor and passing it through the CT, the CT will be able to detect the earth fault current and operate into the protection relay accordingly. Not using the earth return conductor has, in the past, caused maloperations of protection and fault passage indication.

5.6 EARTH FAULTS ON COMPENSATED NETWORKS

If a healthy three-phase distribution system is earthed by an inductance of value L connected between the star point and earth, the system can be tuned by variation of the inductance. Tuning, or resonance, will occur when the inductive impedance and the capacitive impedance balance, both being measured at the system frequency of 50 or 60 Hz.

The three line capacitances to earth are in parallel, and tuning will be when

$$\omega L = 1/3\omega C.$$

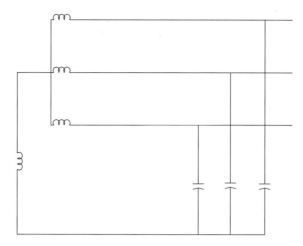

FIGURE 5.12 Capacitance and inductive components.

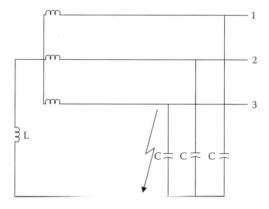

FIGURE 5.13 Earth fault on compensated system.

A single phase to earth fault on phase 3 will cause the three voltages to rotate about the earthed phase. In the example, the voltages of phases 1 and 2 will rise by root 3 and the star point will rise to the normal phase to earth voltage. This has two important consequences:

- The zero sequence voltage, measured at the star point, indicates the presence of an earth fault
- The insulation of the nonfaulted phases is stressed by a factor of root 3

This condition does not require the circuit to be tripped immediately because, under certain conditions, the circuit is stable. Those conditions are for the vector sum of the capacitive currents in phases 1 and 2 (increased because their voltage has increased by root 3) and the inductive current in the inductance L (created by the phase to earth voltage across the coil) must equal zero. From this, it can be shown that

$$\omega L = 1/3\omega C,$$

which is the same expression for tuning at healthy conditions, meaning that, if the compensation coil is tuned to the system capacitance when the system is healthy, it will also be correctly tuned for a single phase to earth fault. Provided that the inductance has a thermal rating of say, two hours, then the system can safely be left in operation for that two hours in the case of a permanent fault. Also, as the earth fault current is zero, transient faults will be self-extinguished at the first voltage zero. This system is known as either the Petersen coil system (after its inventor) or a compensated system.

The utility will be aware of the presence of a single-phase earth fault because the voltage across the coil will have risen to phase to earth voltage, and this can be measured and the alarm communicated to the control staff. However, such a condition cannot be left on forever, partially because of the fact that there could be an overhead conductor on the ground (with consequent danger to the public), the thermal rating of the coil must not be exceeded, and in some countries, local regulations require earth faults to be removed within a time period of typically up to 10 seconds.

If the source substation had, say, 10 outgoing feeders, then with a simple system such as this, the utility would have no indication as to *which* feeder contained the fault and, therefore, which feeder needed some repair work to be carried out. The original practice was to open and then close each outgoing feeder breaker in turn until one opening action caused the alarm to cancel, the fault therefore being on that feeder. This practice means that many customers would have their supply interrupted, albeit perhaps only briefly, which is becoming much less acceptable nowadays.

There are two possible courses of action that a utility could adopt in this situation:

- The first is to change the fault condition into one that the existing protection can easily detect. Suppose there was installed a resistor and an open series switch, altogether in parallel with the coil, and if that switch were closed, then the resistor provided an earth fault current of, say, 500 amps. It is a simple matter to arrange for the series switch to be closed, automatically, after the neutral point voltage alarm had been in operation for, say, 10 seconds, therefore creating an earth fault current of 500 amps. Thus, any permanent earth fault condition would be automatically disconnected after 10 seconds.
- The second is to install protection that is able to detect an earth fault and trip the faulted feeder within the appropriate timescale required by the utility.

Protection and Control

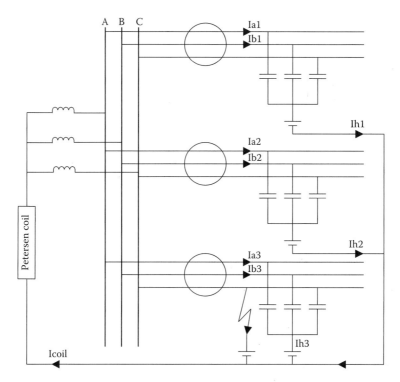

FIGURE 5.14 Earth fault on Petersen coil earthed system.

Before actually applying protective relays to provide earth fault protection on systems that are earthed with a Petersen coil, it is vital to gain an understanding of the current distributions that occur under fault conditions on such systems. With this knowledge, it is then possible to decide on the type of relay that should be applied, ensuring that it is both set and connected correctly.

In Figure 5.14, each feeder of the radial network is equipped with an earth fault relay supplied through a residual CT. Now, as the earth fault on the C phase causes the voltage on the healthy phases to rise by a factor of root 3, then the capacitive charging currents on the healthy phases increase correspondingly. Also, note that because there is no voltage on the C phase, there cannot be any C phase charging current, and the residually connected relays on the healthy feeders simply see the unbalance in the charging currents for their own feeder (i.e., vector sum of Ia1 and Ib1).

Figure 5.15 shows the vector diagram for each of the two healthy feeders when the earth fault on feeder 3 occurs. Vca and Vcb are the resultant voltages and Vo is the phase to earth voltage applied as zero sequence voltage. Ia is the "a" phase charging current, which leads the resultant voltage Vca by 90°. Ib is the "b" phase charging current, which leads the resultant voltage Vcb by 90°.

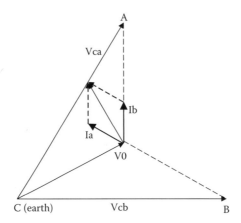

FIGURE 5.15 Vector sum of currents.

The residual CT on the healthy feeder will record Ir, the vector sum of Ia and Ib, which, from the diagram, leads the zero sequence voltage by 90°.

However, in practice, there is resistance in the coil circuit and the feeders. It can be seen that the resistance in the feeders means that the healthy phase charging currents lead the zero sequence voltage by a little less than 90°, and correspondingly, the resistance in the coil circuit means that the coil current lags the zero sequence voltage by a little less than 90°.

Now, as the coil current is the vector sum of the fault current and the resultant charging current of the two healthy phases of the three circuits, and the residual current seen by the CT in the faulted circuit is the vector sum of the fault current and resultant charging current of the two healthy phases of that circuit, then it follows that the residual current seen by the CT in the faulted circuit is the vector sum of the coil current less the resultant charging current of the two healthy phases of the two healthy circuits. The residual CT current, therefore, leads the zero sequence voltage by 90° and is of similar magnitude to the residual CT currents in the healthy phases. This type of connection, being nondirectional, would not be able to discriminate between the healthy and faulty circuit.

But the addition of a directional element would provide this discrimination because the healthy feeder residual current would appear within the restraint section of the characteristic and the residual current on the faulted feeder would lie within the operate region. In other words, the angular difference between the residual currents on healthy and faulted feeders permits the use of a directional relay whose zero torque line passes between the two currents. In practical systems, it may be found that a value of resistance is purposely inserted in parallel with the earthing coil. This serves two purposes; one is to actually increase the level of earth fault current to a more practically detectable level, and the second is to increase the angular difference between the residual signals, again to aid with the application of discriminating protection.

Protection and Control

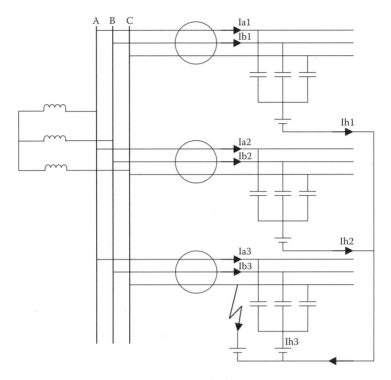

FIGURE 5.16 Earth fault on unearthed network.

5.7 EARTH FAULTS ON UNEARTHED NETWORKS

As we have already seen, before being able to apply and set a protective relay on a particular system, it is important to understand the current distributions that occur during earth fault conditions. Now, the occurrence of an earth fault on a system where the neutral is isolated from earth does not result in any earth fault current flowing, merely an imbalance in the system charging currents, a situation similar to that experienced on compensated networks. It follows, therefore, that the current distribution is also very similar to that obtained in the compensated coil earthed network.

In Figure 5.16, it can be seen that, because there is no voltage on the C phase, there cannot be any C phase charging current, and the residually connected relays on the healthy feeders simply see the unbalance in the charging currents for their own feeder (i.e., vector sum of Ia1 and Ib1). The residually connected relay on the faulted feeder, however, sees, in one direction, the unbalance in its charging current (i.e., vector sum of Ia3 and Ib3), and, in the opposite direction, the vector sum of Ih1, Ih2 and Ih3. Because the current Ih3 appears on both sides of the residually connected relay, it cancels out, meaning that the current through the CT on the faulty feeder is the charging current from the rest of the system (Ih1

FIGURE 5.17 Relay for earth fault protection (courtesy ABB).

and Ih2 in this case), but flowing 180° in opposition to the corresponding currents on the healthy feeders.

It follows, therefore, that on a healthy feeder, the residually connected CTs will see current flowing away from the source and, on a faulted feeder, towards the source which can be detected by directional relays with a characteristic angle of −90°.

5.8 AN EARTH FAULT RELAY FOR COMPENSATED AND UNEARTHED NETWORKS

A typical microprocessor-based relay for earth fault protection on compensated networks or unearthed networks is type REJ 527 as shown in Figure 5.17. Because it can use a characteristic angle of 0 or −90°, the relay can be used on compensated or unearthed networks.

Full details of the relay can be found in 1MRS750616-MUM, but this section deals only with the operation regarding earth faults on compensated or unearthed networks.

The relay uses the zero sequence voltage input for two purposes:

- As the reference voltage for directional calculations
- As a deblocking function to check that the zero sequence current seen by the relay is as a result of an earth fault rather than CT mismatch

The directional earth-fault current unit of the relay REJ 527 comprises two earth fault current stages, a low-set stage ($I_0>$) and a high-set stage ($I_0>>$) both of which can be configured as either directional or nondirectional. For the directional stages, there are two alternative operation characteristics:

- Directional earth-fault with basic angle
- Directional earth-fault with $\sin\phi$ or $\cos\phi$ characteristics

The operation of the directional earth-fault unit with basic angle is based on measuring the earth-fault current I_0 and the zero-sequence voltage U_0 and calculating the phase angle ϕ between the voltage and current. The earth-fault stage starts if the following three criteria are fulfilled at the same time:

- The earth-fault current Io exceeds the set starting level of the earth-fault stage. The earth-fault unit has two earth-fault current stages (Io>) and (Io>>).
- The zero-sequence voltage Uo exceeds the set starting level. The unit has one starting level (Uob>), serving both stages in deblocking mode.
- The phase angle ϕ between the voltage and current falls within the operation sector $\phi \pm \Delta\phi$.

The basic angle ϕ of the network is $-90°$ for isolated neutral networks and $0°$ for resonant earthed networks, earthed over an arc suppression coil (Petersen coil) with or without a parallel resistor (see Figure 5.18). The operation sector is selectable to $\Delta\phi = \pm 80°$ or $\pm 88°$, both with an optional $40°$ wide operation sector.

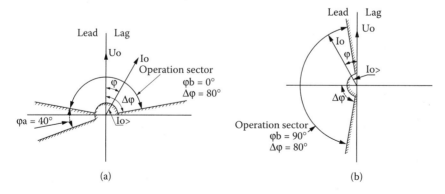

FIGURE 5.18 (a) Compensated network and (b) unearthed network.

When an earth-fault stage starts, a starting signal is generated, and simultaneously the display on the front panel indicates starting. If the above-mentioned criteria are fulfilled long enough to exceed the set operate time, the stage that started initiates a tripping signal. At the same time, the operation indicator on the front panel is lit. The red operation indicator remains lit even though the protection stage resets. The direction towards the fault location is determined by means of the angle between the voltage and current. The basic angle ϕ can be set between 90° and 0°. When the basic angle ϕ is 0°, the negative quadrant of the operation sector can be extended with ϕa. The wider operation sector ϕ is settable to 0...90°.

The operation of the directional earth-fault unit with sinϕ or cosϕ characteristics is based on measuring the earth-fault current Io and the zero-sequence voltage U_0 and calculating the phase angle ϕ between the voltage and current. The sine or cosine value for the phase angle is calculated and then multiplied by the earth fault current. This will give the directional earth-fault current Iϕ as a result. The earth-fault stage starts if the following three criteria are fulfilled at the same time:

- The directional earth-fault current Iϕ exceeds the set starting level of the earth fault stage. The earth-fault unit has two earth-fault current stages (Io>) and (Io>>).
- The zero-sequence voltage U0 exceeds the set starting level. The unit has one starting level, (Uob>), serving both stages in deblocking mode.
- The phase angle ϕ between the voltage and current falls within the operation sector of the angle correction factor ϕc = 2...7°.

When an earth-fault stage starts, a starting signal is generated, and simultaneously the display on the front panel indicates starting. If the above-mentioned criteria are fulfilled long enough to exceed the set operate time, the stage that started initiates a tripping signal. At the same time, the operation indicator on the front panel is lit. The red operation indicator remains lit even if the protection stage resets. The direction towards the fault spot is determined by means of the angle between the voltage and current. The directional earth-fault characteristic sinϕ corresponds to the earth fault protection with the basic angle –90° and cosϕ to the earth-fault protection with the basic angle 0°.

Figure 5.19 shows the characteristic for sinϕ and cosϕ operation. The directions of operation, forward or reverse, for the directional earth-fault stages can be selected independently of each other. The directional stages may also be configured separately as nondirectional protection stages.

When the earth-fault current exceeds the set starting current of the low-set stage (Io>), the earth-fault unit starts delivering a starting signal after a preset time of approximately 70 milliseconds. When either the required definite-time operation or the calculated inverse-time operation has elapsed, the earth-fault unit operates. In the same way, the high-set stage (Io>>) of the earth-fault unit delivers a starting signal after a preset time of approximately 60 milliseconds start time after the required starting current has been exceeded. When the set operate time elapses, the earth-fault unit operates.

Protection and Control

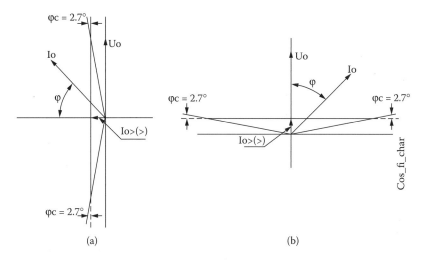

FIGURE 5.19 (a) Compensated network and (b) unearthed network.

The low-set stage of the earth-fault unit may be given a definite-time or an inverse definite minimum time (IDMT) characteristic. When the IDMT characteristic is chosen, six time/current curve groups are available. Four of the groups comply with the standards IEC 60255 and BS 142, and are called normal inverse, very inverse, extremely inverse and long time inverse. The two additional inverse time curve groups are called RI- and RD-curves. The inverse-time function of stage (Io>) can be inhibited when stage (Io>>) is started. In this case, the operating time is determined by stage (Io>>), and if not needed, the stage (Io>>) can be taken out of operation completely.

5.9 FAULT PASSAGE INDICATION

However well an electrical distribution network has been designed, constructed and operated, it is inevitable that, at some time, a fault will develop on that network. The fault could, for example, be caused by decay and deterioration due to aging and wear, by weather conditions or by a third party accidentally interfering with the network. In many network designs, a fault will mean that either customer supply is lost or the performance of the network is changed. Whatever and wherever the fault, it has to be located before it can be repaired. This section explains how the faulted section of a distribution system can be identified in all types of networks, whether that network is manually controlled or remotely controlled.

5.9.1 THE NEED FOR FPI ON DISTRIBUTION NETWORKS WITH MANUAL CONTROL

In the MV (20 kV) network shown in Figure 5.20, there is a source substation with a single transformer, and one underground feeder is shown in detail. This

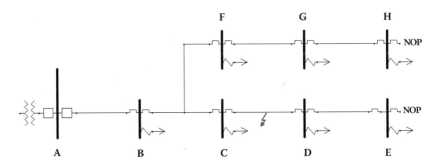

FIGURE 5.20 Principles of fault location.

feeder is controlled by the circuit breaker A and has seven substations connected, each having two cable switches and a fuse-switch to protect the distribution transformer. Two of the switches are normally open points (NOPs).

For a cable fault between substations C and D, the circuit breaker A will trip, the whole feeder will be disconnected and supply lost to all the connected customers.

Some customers will have their supply restored by switching operations, whereas others may have to wait for repairs to be made. Obviously, it is perfectly acceptable to switch back all of the healthy sections of the feeder but the question has to be asked as to which are the healthy sections and how can this be determined.

When faced with a tripped circuit breaker and no indication as to where the fault lies, a utility has a range of options as to the method by which the faulted section is identified.

The Switching Method. The utility can open one of the network switches and reclose the source circuit breaker. If the circuit breaker does not trip immediately, then the switching operation has disconnected the faulted section. The decision as to where to first open a switch depends on a combination of the network configuration and the need to restore certain supplies, for example, the commercial center of a town, as quickly as possible. Switches that could disconnect a long section of circuit or a section with a known high fault rate would be the first to operate, because it is more likely that the fault will be beyond such a switch.

In the example circuit, we could reasonably decide to open the outgoing switch at substation F and then reclose the source breaker, which of course would trip immediately. By successively opening and closing selected switches and observing whether the source circuit breaker trips, it is possible to determine where the fault is. Switches on each end of the faulted section are then opened, to disconnect the fault. In Figure 5.20, supplies to customers fed from substations A, B, C, F, G and H are restored via the source circuit breaker, whereas customers fed from substation D and E are restored by closing the normally open switch at E. However, this method has the following disadvantages:

Protection and Control

- The utility always has to re-energize the fault at least once, which may cause further damage at the point of the fault, including injury to any persons there.
- The circuit breaker may be asked to clear fault current several times yet, especially for older oil breakers, may be limited to a specific number of fault clearances before it has to be taken out of service for maintenance.
- All customers will have their supply interrupted, but some will have a second loss of supply after the circuit breaker has tripped after being closed onto the fault during fault switching. To some customers, this second interruption to supply can be as much a nuisance as the first interruption.
- Each time the circuit breaker trips on fault, there is an associated depression in the system voltage over the complete network, which can have an adverse effect on sensitive equipment.

Many utilities already do not permit deliberate re-energization of a permanent fault or are moving towards this policy; hence, the switching method is not generally acceptable nowadays.

This method for locating the fault may be based on the first opening being at the halfway point, which ensures the optimum balance of customers being restored to supply in the shortest time, or as an alternative, the first switching may be to give priority to the early restoration of a specially important customer.

The Switch and Test Method. This method is very similar to the switching method, except that once a switch has been opened, the circuit beyond that switch is tested, usually with an insulation tester at, for example, 15 kV, to determine whether it contains the fault or not. A section of the circuit is only switched back when it has been confirmed that it does not contain the fault. This method prevents the need for re-energizing the fault, but takes a considerably longer time than the switching method. During this extended time, all customer supplies are lost, but there is an advantage that no fault re-energization is needed.

5.9.2 What Is the Fault Passage Indicator, Then?

The fault passage indicator is defined as a device that can be located at some convenient point on the distribution system that will give an indication as to whether fault current has passed the point where it is located or not. It therefore has to be able to distinguish between fault current and the load current associated with the healthy feeder, and it has to have some means of displaying its operation to an operator. The simplest and most common design of fault passage indicator is the simple earth fault passage indicator. This looks for a zero sequence (earth fault) current exceeding a specified level, for example, 50 amps, and when this current flows for some specified time, for example, 50 milliseconds, it either drops a relay flag or lights a local LED to show that the fault current has gone through this location. Of course, 50 amps of load current on a balanced three-

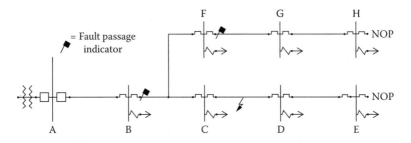

FIGURE 5.21 Fault location with fault passage indicators.

wire or four-wire MV system will not cause a zero sequence current to flow and will, therefore, not operate the fault passage indicator. The earth fault passage indicator can be driven from a split core zero sequence CT, which is easy to fit around an underground cable and is, therefore, a cheap option.

It follows that it is possible to construct a fault passage indicator to detect phase-to-phase faults, and this would need to be able to measure the phase currents via CTs. The practical place to locate these CTs is in the cable terminating area and, unless this is of air spaced construction and with sufficient physical space, then the installation of the CTs can be an expensive project. Some utilities favor combined earth fault and phase fault indication, which is technically feasible but economically unattractive because the vast majority of phase-to-phase faults on cable systems originate as an earth fault and, hence, will be detected by the cheaper earth fault passage indicator.

The fault passage indicator helps a utility to restore supplies more quickly by reducing the time that an operating crew needs to travel around the network in the search for the fault. It follows, therefore, that economic benefits in terms of crew time saved can be identified and these are discussed in Chapter 8.

Let us now suppose that two such fault passage indicators have been installed at substations B and F in Figure 5.21. For the same fault between substations C and D, the source circuit breaker will trip and fault current will pass through substations B and C, causing the indicator at B to operate.

When an operator arrives at the site, the first task is examine the two FPIs to see where fault current has passed. The FPI at B indicates that the fault does not lie in section A to B. At substation B, the operator could then open the switch towards substation C and restore substation B by closing the source circuit breaker.

The FPI at substation F tells the operator that the fault does not lie in section F to H. At substation F, the operator could open the switch towards substations B and C and restore substations F to H by closing the normal open witch at substation H.

So far, we have restored some supplies without a fault re-energization, but some supplies have yet to be switched back. With only these two FPIs, it is clear that switching or switch and test will be needed to get the remaining supplies back.

Another method of working out approximately where the fault is involves the measurement of the actual fault current that flows before fault clearance. The

Protection and Control

fault level, and hence the fault current, decreases as the distance from the source increases due to the extra impedance of the additional circuit length. If this fault level has been accurately calculated for each section of the network and is compared against the measured value, then a good indication as to the distance to the fault can be obtained. This method has three drawbacks:

- The impedance of the fault can cause significant errors, although some corrections can be applied to mitigate that effect.
- The accuracy of present-day systems is not sufficient to distinguish whether a fault is on one side or the other of a switch that could be used to disconnect a fault. For example, suppose that the measurements suggested that the fault was 1780 meters along the circuit and the inaccuracy was 3%, i.e., the fault lay between 1727 and 1833 meters and the switch that could disconnect the fault was located at 1800 meters; the operator would not be able to determine whether opening that switch would be useful or not.
- Because most circuits, especially overhead circuits, have a number of teed connections, the actual distance to the fault could be along one of several different routes, depending on which tees are considered, and this can lead to misinterpretation by the utility.

Nevertheless, the fault measurement system can provide extremely useful information to direct a repair crew to the approximate location of the fault.

Fault passage indicators can also be classified into three further categories:

- Independent devices, where the indication is locally read by a switching technician
- Communicating devices, where the indication is remotely read by a control technician, using some form of communication system, very often the communication being part of a system of extended control
- Portable devices, which are especially useful for longer overhead circuits where the distance between fixed fault passage indicators may be long and it is required to divide a section known to contain the fault into a series of smaller subsections

5.9.3 THE NEED FOR FPI ON DISTRIBUTION NETWORKS WITH EXTENDED CONTROL OR AUTOMATION

Let us now suppose that the network shown in Figure 5.22 has some automation added. This will mean the addition of a switch actuator and remote terminal unit (RTU) at substations B, F, H and E, together with power operation of the source circuit breaker. This will provide the facility to be able to remotely operate the switchgear via a SCADA outstation.

For the same fault between substations C and D, the control operator will be able to restore supplies in the same way that the manual switching operator would.

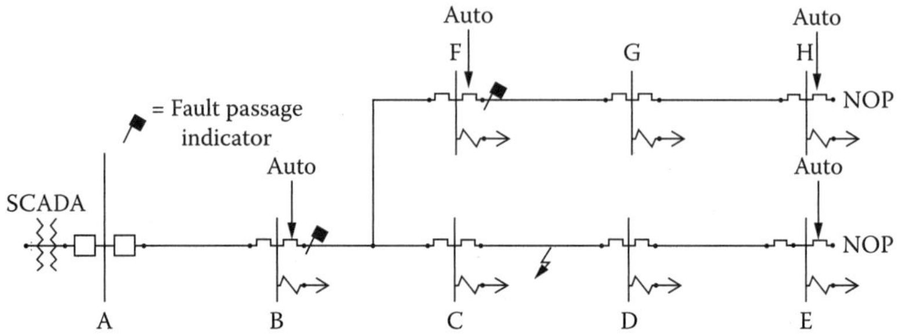

FIGURE 5.22 FPI and SCADA.

The advantage of this scheme is that switching operations can start more quickly than with the manual scheme, because there is no need for switching operators to spend time traveling to the site before they can start the switching. The control operator still needs to know the location of the fault, and this can be done by using auxiliary contacts on the FPI to connect with a digital input on the substation remote terminal unit, which will then transmit the status of the FPI to the control operator. The FPI may still give a local indication as well as the remote indication.

5.9.4 FAULT PASSAGE INDICATORS FOR USE ON CLOSED LOOP NETWORKS

The distribution networks considered above have all been radial or open loop, meaning that there is one source of supply to the network, and electrical power moves away from this one source to each load on the network. Some distribution networks have more than one source of power; that is, they are closed loop networks.

Figure 5.23 shows a closed loop network in which two sources supply the loads in a more secure manner than if the system were an open loop. If all the switching devices were load breaking only, then for the fault shown, the outgoing circuit breaker at each source substation would trip and supply to all loads would be lost.

By using circuit breakers on the network, it is possible for only the circuit breaker on each side of the fault to trip. For the fault shown, circuit breakers CBa and CBb would trip; therefore, supply is only lost to the loads at substations 2 and 3, all other supplies remaining on. The circuit breaker CBa will trip on fault energy flowing from source A, and the circuit breaker CBb will trip on fault energy from source B, and the direction of power flow from each source will be from each source towards the fault.

Generally, fault passage indicators are not directional; that is, they indicate that a fault current has passed but do not indicate in which direction the current has passed. When the fault above occurs, the network operator would find two circuit breakers tripped and the fault indicators at every substation operated. The indicators at substations 6, 4, and 3 will operate on fault energy from source B,

Protection and Control

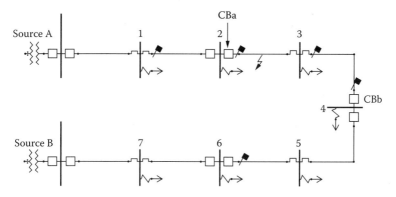

FIGURE 5.23 FPIs and closed loop networks.

and the fault indicators at substations 1 and 2 will operate from fault energy supplied from source A. The operator will know that the fault lies between substations 2 and 4 but will not know on which side of substation 3 the fault lies because, with a nondirectional indicator, the issue is which source supplied the fault energy that caused the indicator at substation 3 to operate. If, however, this indicator was to be given a directional property, then the operator would be able to see that the energy flow had been from source B, and therefore the fault must lie between substations 2 and 3. The operator would then be able to commence restoration switching knowing exactly in which section the fault lay.

Directional fault passage indicators are, therefore, needed on closed loop networks, and a limited number of suitable devices are available. They indicate whether the fault is in the upstream direction or the downstream direction when viewed from where they are located. In the same way that a directional protection relay needs a source of both current and voltage, so the directional fault passage indicator relay needs a source of both current and voltage. For an underground indicator, the current source can be earth or phase current transformers. The voltage source might be a wound voltage transformer on the switchgear or some other new alternatives, noting that a low-voltage supply from the local distribution network will not usually be sufficient.

Closed loop networks can appear on underground systems although they, presently, are rare on overhead systems. However, because utilities are now expected to provide common standards of security to all customers whether connected to underground or overhead supply networks, the use of closed loop is likely to be featured in overhead networks in the future. To this end, a number of manufacturers are developing indicators for closed loop overhead networks.

5.9.5 OTHER APPLICATIONS OF DIRECTIONAL INDICATORS

Directional fault passage indicators can also find useful application on radial (open loop) distribution networks. In Figure 5.22, we showed that for the fault considered, indicator F would not operate because there was no passage of fault

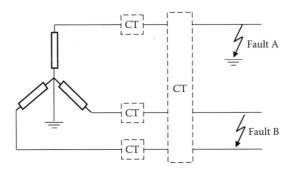

FIGURE 5.24 CT connections.

current through this indicator. If this FPI had been a directional indicator, then depending on its design and the network operating conditions, it could be arranged to give an indication of the fault as an upstream fault. This may be of use to the network operator.

A directional fault passage indicator may also find application when a second source of supply, for example, an embedded generator, is connected to an MV distribution network. This situation is very similar to the closed loop network described above, in that the generator would cause nondirectional indicators to operate and give correct but misleading information to the network operator. The way in which FPI would operate when on a system with embedded generation depends very much on how the generator is earthed.

5.10 CONNECTION OF THE FPI TO THE DISTRIBUTION SYSTEM CONDUCTOR

Unless limited in some way, the fault current that can flow on an MV distribution network may be up to 20 kA. Whole current measurement at this level is impractical; therefore, all known FPIs measure current by measuring the magnetic field associated with that current.

5.10.1 CONNECTION USING CURRENT TRANSFORMERS

Figure 5.24 shows an earthed star connected network with three separate phase CTs and one zero sequence CT that encompasses all three conductors. This is practical for new underground cable systems (retrofitting CTs into existing cable boxes can be complicated) but not practical for uninsulated overhead systems. Phase-to-phase faults can only be detected if CTs on each separate phase are supplied. Earth faults can be detected by either the zero sequence CT or the separate CTs connected in residual format.

Figure 5.25 illustrates the residual connection of three CTs to give phase-to-phase fault indication with three devices (FPI-1, FPI-2 and FPI-3) together with phase to earth fault indication (FPI-E).

Protection and Control 215

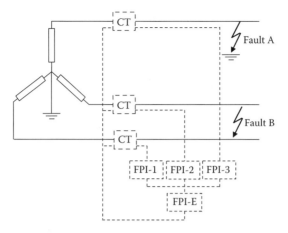

FIGURE 5.25 Residual connections.

In this connection, it is normal to short out and remove FPI-2 because it gives no benefit in return for its cost.

For phase-to-phase faults or phase-to-phase and -earth faults, it is possible to use multiple discrete detector devices (three or four above), or, in more practical terms, to use a single device with multiple inputs.

5.10.2 Connections Using CTs on Underground Systems

Figure 5.26 shows a FPI that can be used in a cable end box. Three sensors are shown which can be located on each of the three cable single cores which can then detect earth faults and phase faults.

FIGURE 5.26 Typical CT connections.

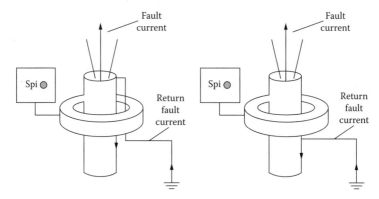

FIGURE 5.27 External CT connection.

The CTs used above are located within the cable terminating box, but this may not always be possible, for example, if the terminating box has insufficient space available or if the box is compound filled. In these cases, it is possible to use an external CT for the earth fault indication. However, care must be taken to ensure that the earth fault current returning via the cable earth, for example, the cable sheath, does not pass through the CT, as illustrated in Figure 5.27.

5.10.3 Connections Using CTs on Overhead Systems

Phase-to-phase faults and earth faults on overhead lines can be detected using conventional CTs in exactly the same way as for underground cable systems. The significant difference is that overhead systems are, for the major part, uninsulated. CTs must, therefore, either be insulated to line voltage, which adds cost, or be located in conjunction with other, already existing, insulation.

Figure 5.28 shows a typical pole-mounted switch disconnector. A CT placed on the uninsulated conductor at A would need to have primary insulation up to the line voltage. However, the potential of a thin dimensioned CT, for example, up to 50 mm, mounted around the insulator at position B would be according to the voltage gradient across the insulator and would, hence, be almost at earth potential. If this method is to be used, the initial design of the insulator must take into account the change in voltage profile caused by the CT.

From a practical point of view, the installation of CTs, whether to an underground network or an overhead network, will normally involve a shutdown for safety reasons. Fitting CTs to cable boxes is made very much more easy if the cable box is of the air-insulated type and there is sufficient physical space available to accommodate the CTs.

5.10.4 Connection without CTs on Overhead Systems (Proximity)

There are also ways of measurement of the magnetic field associated with current other than conventional CTs.

Protection and Control

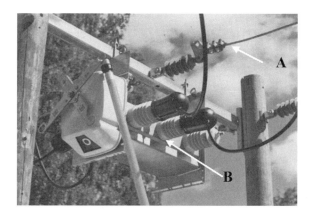

FIGURE 5.28 CT connections for pole-mounted switchgear.

In the first method, known as direct connection, the FPI is clipped directly to the bare overhead conductor using an insulated pole. Figure 5.29 shows a typical FPI, using a flashing xenon tube to indicate the passage of fault current. This method of connection is economic, but because the whole indicator is at the potential of the overhead line conductor, an auxiliary contact to give connection to an RTU will involve insulation to MV levels.

Conventional CTs detect the magnetic field associated with an electric current at a distance of, usually, a few centimeters from the conductor, and this magnetic field reduces as this distance increases. With sensitive equipment, it possible to measure this reduced field but from some more convenient location. For example, if the detector can be located beyond the safety clearance distance from a live,

FIGURE 5.29 FPI for direct connection.

FIGURE 5.30 FPI for proximity connection.

uninsulated, conductor, then installation can be made with the line live. This benefit has to be balanced against the increased cost of this technology.

In this method, known as proximity connection, the detector is mounted on the overhead line support pole, about 3 meters below the conductor. Figure 5.30 shows a typical product that has the fault detector and the local indicating device integrated into one molding. The local indication is a flashing xenon tube. Compared to the direct connection device, it is easy to provide auxiliary contacts to give remote indication via an RTU and communication system. Such auxiliary contacts would be connected to an RTU on the same pole or maybe one pole away.

Some fault passage indicators require the measurement of voltage, either as part of the fault passage detection, such as directional schemes, or as a means to know that the indicator can be reset when system voltage returns. For underground systems, the voltage on the MV network is usually taken from either a separate voltage transformer or the local distribution transformer.

For overhead systems, the voltage may, additionally, be derived using capacitance to earth. In Figure 5.31, a fault passage indicator of the proximity type is fixed to a pole that supports the live overhead conductor. There is a capacitance C1 between the overhead line conductor and the fault passage indicator, also a capacitance C2 between the fault passage indicator and earth. By potentiometric action, the voltage at the indicator can be determined and the indicator, therefore, used to decide whether or not voltage is present.

5.11 DISTRIBUTION SYSTEM EARTHING AND FAULT PASSAGE INDICATION

Because faults on a distribution network can be either between one phase and another phase, sometimes called phase faults, or between one phase and earth, sometimes called earth faults, fault passage indicators may be required to detect

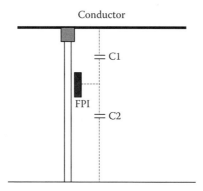

FIGURE 5.31 Voltage detection by potentiometric action.

either phase faults or earth faults or perhaps, for some users, both. Faults between more than one phase and earth can be treated as a combination of a phase fault and an earth fault. The issue is slightly complicated because fault passage indication is not quite the same as the detection of a fault for protection purposes.

For a fault passage indicator to work, there must be a change in some system condition that the fault passage indicator can detect as a fault. For some situations this is relatively easy and the associated indicator is cheap, but this is not always the case. The detection of phase-to-phase faults is usually simple, whether the indicator is connected to the power system by current transformers or proximity detectors. This is because the phase-to-phase fault current on the power system is primarily controlled by the source impedance of the power system. For example, a transformer supplying an MV distribution network might be sized at 10 MVA, which will provide sufficient phase-to-phase fault current to be easily detected.

The detection of phase to earth faults depends on how much earth fault current flows. The more the current that flows, then the simpler, and usually cheaper, the fault passage indicator can be. The earth fault current depends primarily on the method by which the MV system is earthed, and there are two basic groups of methods of system earthing. The first is called effectively earthed, when there is usually a large earth fault current available, and the second is called noneffectively earthed, where there is usually a small earth fault current available.

Effectively earthed systems have the neutral point connected to earth by solid connection, resistance connection or reactance connection. The solid connection will permit earth fault current that is limited only by the source impedance, for example, 5 kA. The resistance or reactance connection places a known impedance in the neutral circuit, which limits the earth fault current according to the value of that impedance, for example, 1 kA.

In the noneffectively earthed system, the neutral point is either insulated (not connected to earth) or connected to earth by a reactance that balances the network capacitance current, the latter being commonly known as compensated or arc suppression coil. When an earth fault occurs in the noneffectively earthed system, there is no flow of a substantial earth fault current, merely a change to the capacitive current. Detection of earth faults is difficult, although the need to disconnect the faulted component is not so high as with effectively earthed systems.

Most distribution networks in the U.K. are effectively earthed, but some overhead systems use the compensation coil. Insulated neutral systems in the U.K. are extremely rare. In other countries, compensated neutrals are common for underground systems as well as overhead systems. On some compensated systems, the reactance coil has a parallel resistor to provide additional current. This resistor may be either permanently in service or switched into service when required. If it is permanently in service, then its function will be to provide a small active current, maybe 10 amps, which can be used to operate directional earth fault protection at the substation. This 10 amps active current can be used to operate an earth fault passage indicator that is sufficiently sensitive.

The resistor may also be switched into service when required. This will normally be on a compensated system where the compensation reactance balances the network capacitance almost exactly, so that when a single phase to earth fault occurs, there is no change in condition sufficient to cause protection to disconnect the fault. This is often required in compensated systems so that customers are not disconnected for common faults, but many authorities require that a system can only operate for a certain time in this state, for example, 30 minutes. At the end of this period, the resistor can be switched into service, which will cause sufficient earth fault current to flow to cause immediate protection disconnection of the affected feeder. This earth fault current can be typically 1 kA, which will operate many types of fault passage indicator.

5.11.1 Detection of Steady-State Fault Conditions

On effectively earthed systems, the fault current rises in accordance with the frequency of the system, typically 50 or 60 hertz, and remains flowing until cleared by a circuit breaker in a timescale of, typically, between 50 and 500 milliseconds. The current in the circuit rises from the load current to the fault current, and once the fault current is flowing, it is essentially constant. If the fault current exceeds a threshold set in the indicator, then the indicator can be arranged to show that it has seen a fault. However, the load current on the system can vary as customers switch loads on or off. These variations between loads do not happen as quickly as it might rise between load and fault values, and hence, detection of the rate of rise of current may be used to improve the sensitivity of the detector. Another similar method is to seek an increase of the current in the circuit to twice its previous value and within a short time period.

A fault passage indicator that has to detect steady-state conditions, therefore, must be able to detect either a given threshold of fault current, or a given rate of rise of current, within the time period between the fault occurring and the circuit breaker clearing the fault. For earth fault detection, typical values would be a threshold of 35 to 50 amps for a period of 100 milliseconds.

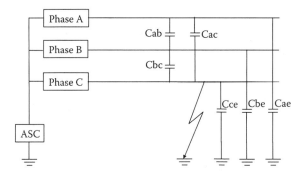

FIGURE 5.32 Detection on compensated networks.

5.11.2 Detection of Transient Fault Conditions

In noneffectively earthed systems, the earth fault current is quite different from that in effectively earthed systems. For the single phase to earth fault, if the inductive compensation current of the coil exactly balances the network capacitive current, then it may be impossible to operate conventional protection. It is also impossible to indicate fault current by conventional means. However, the initial transients of earth faults are important in unearthed or compensated networks. They provide two characteristics. First, they may confuse conventional protection relays, which may therefore need to be equipped with harmonic filters. However, they can provide a very effective method to trip the faulted line and to operate fault passage indicators.

Consider Figure 5.32, which shows a circuit connected to a single transformer that is earthed via a compensation coil. Under healthy conditions, the network voltages rotate about the center of the voltage triangle. Each phase-to-phase line capacitance is charged with the phase-to-phase voltage, and each phase to earth capacitance is charged with the phase to earth voltage.

Suppose that a fault to earth occurs on the phase C and remember that the circuit breaker will not trip for a single phase to earth fault on an unearthed or compensated system. What happens is that the network voltages rotate about the phase C instead of the center of the voltage triangle. The phase-to-phase capacitors do not see any change in voltage across themselves. However, the voltage across the Cce changes from line voltage to zero because of the fault connecting the phase C to earth. The capacitor is discharged by the fault, which causes a transient current and voltage to occur, called the *discharge transient*. At the same time, the voltage across both Cbe and Cae rise from line voltage to phase voltage, i.e., the voltage increases by a factor of 73%, which causes a transient voltage and current to occur, called the *charge transient*.

These transients can be detected and used to indicate that an earth fault has taken place on the compensated or unearthed network. In conjunction with the transient voltage changes that are detected through the capacitive divider arrangement, this can be set up to indicate the direction of the earth fault.

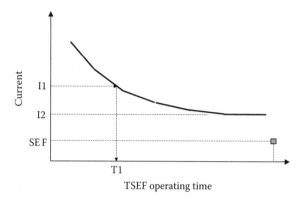

FIGURE 5.33 Indication of SEF.

5.11.3 Indication of Sensitive Earth Faults

Most protection relays for distribution networks operate with an inverse time characteristic typically shown as the curve in Figure 5.33. For example, if the current in the relay was I1, then it would operate in after a time of T1, which may typically be 200 milliseconds. If the current is below the value I2, the relay will not operate. However, on many rural overhead networks, there is a possibility that a fault will occur where the fault current is very low; this may be when an uninsulated conductor breaks and falls to the ground, where the ground resistance is high, for example, a rocky terrain.

In Section 5.2 we saw that protection engineers can detect this type of fault by using sensitive earth fault protection, which detects a small current, typically 10 amps, but which persists for a comparatively long time, for example, 20 seconds. If a fault is cleared by sensitive earth fault protection and the network operator requires fault indicators to show where the fault current has flowed, then it follows that a fault indicator with sensitive earth fault facilities is needed. Some indicators with sensitive earth fault are available.

5.12 AUTORECLOSING AND FAULT PASSAGE INDICATORS

Fault passage indicators are very similar in concept to a protection relay. The major differences are that the protection relay trips a breaker, whereas the FPI provides some form of indication only, and the protection relay is generally built to a higher, and hence more expensive, specification that includes more adjustable settings. Any circuit breaker that trips on fault, therefore, shows that fault current has passed that protection equipment, which is similar in function to the FPI. Generally, then, FPI would not be fitted to a circuit breaker. However, transient faults are usually cleared by a reclosing circuit breaker, and there may be some benefit in the operator knowing that a transient fault has occurred

Protection and Control 223

FIGURE 5.34 Fisher Pierce type 1514 FPI.

that has been successfully cleared and re-energized. If the reclosing circuit breaker does not provide an indication of the number of transient faults, then perhaps an FPI could be installed at the recloser position. Also, it may be helpful to install an FPI that indicates transient faults at strategic places on the network, for example, at the start of major spurs, to help locate sites of persistent transient faults.

5.13 THE CHOICE OF INDICATION BETWEEN PHASE FAULT AND EARTH FAULT

By using two devices, the probability of detecting all faults improves, and with three devices, all faults will be recorded. Nevertheless, a number of operators spread their indicators out so that instead of three at one location, there is one at each of three locations, claiming that this gives overall better value for money.

Because many network operators restrict the earth fault current that is allowed to flow, an earth fault passage indicator may need to be sensitive to such low currents, and typical indicators can work down to 50 amps. On the other hand, a typical indicator that looks for overcurrent faults may be set to operate at 1000 amps, because it is not usual to restrict phase-to-phase fault levels. If an operator uses one conductor-mounted single-phase device, for example, the Fisher Pierce type 1514 shown in Figure 5.34, then they would be able to detect overcurrent faults on that phase and earth faults, too, provided that the minimum operating current of the indicator is below the sometimes restricted actual earth fault current.

5.14 RESETTING THE FAULT PASSAGE INDICATOR

The correct operation of a fault passage indicator is to show that a fault current has passed that point on the network and maintain that indication until the operator has learned which indicators have operated. FPIs, therefore, must retain their indicated state for some time, after which they should reset. The reset may be after some period of time, typically three hours, and it is assumed that the operator has read them in this time. However, if the operator has read them in less time, then it would be sensible for the reset to be made as soon as possible. FPIs can usually be reset locally by an action such as pressing a reset button or moving a magnet across the device. Others, especially proximity connected overhead line indicators, can be reset by the restoration of line voltage. Indicators designed for connection to RTUs for automation usually can be reset by applying a voltage to an input terminal, that voltage being controlled by a digital output of the RTU.

5.15 GRADING OF FAULT PASSAGE INDICATORS

We have already seen that a typical earth fault passage indicator will operate for a zero sequence current of 50 amps for 50 msec. If we assume that the fault level on a network to which such an indicator is applied will always ensure that such a current will flow for all earth fault conditions,* then we will be certain that every indicator will operate to show whether or not the fault is beyond that indicator. The determination of the location of the fault depends solely on seeing which indicators have operated. The indicator farthest away from the source will show the start of the section that contains the fault, but this assumes that every indicator is working correctly, and, therefore, that an indicator that has not operated has not operated because it did not see the passage of fault current rather than the possibility that the indicator was not working correctly at the time of the fault. As with any other power system component, care must be taken that fault passage indicators are tested from time to time to ensure correct operation when required.

On every power system, the fault level decreases as distance from the source increases. Because fault passage indicators, of course, can be supplied with settings different to the above-quoted example, it follows that differently set devices can be used at any location where the fault level is sufficient to cause correct operation.

At the time of writing, at least one manufacturer can supply an indicator that, instead of operating if the threshold current is exceeded for the detection period, has a characteristic curve similar to the time lag curve fitted to protection relays. The intention is that, in the same way that the protection relays of circuit breakers in series can be graded to ensure that only the correct circuit breaker will trip for a given fault, a series of fault passage indicators on a circuit can be graded to ensure that only the one nearest to the fault will operate. The advantage of this

---

* Care must be taken to include all conceivable earth fault impedances.

Protection and Control 225

is that, when combined with an RTU as part of an extended control scheme, the RTU can report that it is at the distribution substation nearest to the fault without having to take account of the operation of upstream indicators. For extended control schemes that rely on a central logic system deciding on corrective action after taking all the available data into account, such an indicator provides little overall benefit. However, for a control scheme based solely on local logic at the RTU-equipped substation, this has the advantage of direct action being made available without checking the status of upstream indicators.

5.16 SELECTING A FAULT PASSAGE INDICATOR

Assuming that the indicator is required to have auxiliary output contacts to be connected to a digital input of the remote control RTU, then selection depends primarily on four main questions:

- Is the distribution system radial (open loop) or closed loop?
- What is the type of system neutral earthing?
- Is the indicator for an underground system or an overhead system?
- What type of indication is required — earth fault only, phase fault only or both?

There is a wide choice of fault passage indicators available for use on open loop effectively earthed systems, which provide either earth fault indication or phase fault indication or both. Some are suitable for overhead systems, whereas others are suitable for underground systems. Some detect the fault current with current transformers, some are mounted directly on the conductor and some are located a meter or two away from the conductor and rely on the changing electromagnetic field. The output of the device can be a form of visual indication, for example, a flashing xenon tube or an LED indication, whereas others have auxiliary contacts suitable for connecting to a digital input of a remote control RTU.

The range of fault passage indicators for use on closed loop underground networks is presently limited to one, although, because the improved reliability associated with closed loop systems is becoming more important for some operators, several manufacturers are developing products.

The range of fault passage indicators suitable for compensated neutral systems is presently very restricted, currently limited to two manufacturers that each make a product for underground systems and a product for overhead systems.

5.17 INTELLIGENT ELECTRONIC DEVICES

The final component in configuring a remote controlled/automated switch is the control and protection device. The term *intelligent electronic device (IED)* has been adopted by the industry as a general designation. It covers simple remote terminal units,* traditional protection relays without control or communication features, and relays with full communications interfaces.

5.17.1 REMOTE TERMINAL UNIT

The RTU is the device that allows communication of the local process to a master or central system. The size and complexity of RTUs vary drastically depending on the functionality. Large RTUs applied at transmission and large primary substations are considered part of a SCADA system. These devices typically have dimensions well in excess of inputs and outputs totaling 100. The size, complexity and cost of an RTU are directly related to

- The amount and type of data to be collected (number of input/outputs)
- The different types and number of devices to be controlled
- The amount and complexity of local data to be processed

The RTU applied at the distribution level must be of the lowest possible cost and with functionality only appropriate for the applications at this level of the network. In spite of size, all RTUs are fundamentally the same. The basic feature that has made it possible to provide an RTU function is the ability to store data. The storing of single items of data (typically measurements) enables the basic mechanism for acquiring and processing this data to be achieved. Further, these data may be transferred to the master independently of the time of collection. Data storage also provides for the transfer of the data in the opposite direction — from the master to the process. Data queuing provides additional flexibility within the RTU by making it possible to collect data in an asynchronous manner from both the process and the master. It is possible to collect a large number of changes in values over a short time period by using the buffering mechanism that is part of the queuing process. The collected data may be processed in different ways, leading to more or less advanced intelligent RTUs. The fundamental role of an RTU is as follows:

- The acquisition of various types of data from the power process
- The accumulation, packaging, and conversion of data in a form that can be communicated back to the master
- The interpretation and outputting of commands received from the master
- The performance of local filtering, calculation and processes to allow specific functions to be performed locally

Many of the collection features and processing of data performed by the RTU are basic SCADA functions that have been allocated from the master to the RTU.

---

* Although the term *RTU* is generally used throughout the telecontrol industry to denote the electronic device that interfaces measurement and status input and control outputs, the distribution industry often refers to a distribution terminal unit (DTU) to differentiate small devices from the very large RTUs used in transmission substations for energy management systems. In some parts of Europe, the distribution substation is referred to as a compact substation.

Protection and Control

This allocation of certain functions from the master to local processing improves the performance of the central master system. In distribution automation, this is an important consideration because extended control outside the substation will impose considerable burden on the central system due to the dramatic increase in number of devices to be controlled, even though the I/O points per device is small. Once more local processing is adopted it, becomes convenient, particularly for extended control where the devices are many and geographically far apart, to be able to download new process settings from the master. This ability is purely a function of the communication protocol and its implementation between the master and slave. Time-stamping of local events and the synchronization of the local time with the master is always a prerequisite for a large substation RTU. This is often an additional cost option with small RTUs. In this case, time-stamping results in burdening the master due to the necessity for regular polling of all the local locations or because provision of a separate external radio broadcast-based time-stamping has to be added.

RTUs for the control of primary switchgear are of minimum size and functionality. They are either installed in a control compartment within the switchgear cubicle/housing for indoor equipment or housed in control cabinets mounted at the base of the pole for outdoor installations. These latter installations require RTUs that can withstand severe temperature ranges (−40 to +70°C). Typically, this type of application requires integrated power supply modules for managing battery charging and also integrated modems for direct connection to the selected communication medium. RTUs for a basic remote-controlled feeder switch outside the primary substation must include sufficient I/O ports for monitoring and control of the switch and for the health checking of the switch and control cabinet. Examples of dimensions and requirements for various switch types are shown in Table 5.1.

Typically, small RTUs are configured to have, as the minimum, one board with I/O functions as 16 digital (binary) inputs and 8 digital outputs with contact ratings at 24 VAC, 0.2 amps and at 220 VAC, 0.5 amps, a battery charger board, a number of RS-232 ports for central station control, a local display control unit, and an integrated modem. This minimum configuration meets most of the low-end requirements for remotely controlled devices. The inclusion of measurements requires more careful consideration because the input level of typical analogue input boards is in the mA range and not suitable for direct connection to CTs without a transducer. Inputs from combi-sensors often require interface electronics to process the low signals from sensors such as Rogowski coils.

The programmable logic controller (PLC)* is often used as a basic IED in place of an RTU, particularly for implementing the basic logic required for auto

---

* The term *PLC* has two common meanings: (1) PLC as used above for a programmable logic controller as used predominantly by the industrial segment for a low-cost electronic device used in great quantities for local control in process industries and now being applied in utility applications, and (2) PLC as used to designate power line carrier communication used by the power industry for a communication system that sends and retrieves information sent down transmission and distribution power circuits.

TABLE 5.1
RTU I/O Counts for Three Typical Automated Device Types

RTU Dimension	Basic Airbreak Disconnector	Gas-Enclosed Switch with FPI and Earthing Switch	Gas-Enclosed Switch with FPI and Measurements	Ring Main Unit-Type CCF with FPI (Two Remote-Controlled Switches)
Digital Inputs				
Status				
Open	✓	✓	✓	✓✓
Close	✓	✓	✓	✓✓
Local-remote	✓	✓	✓	✓
Earthing Sw. position		✓		
Fault passage indicator (FPI)		✓	✓	✓
Subtotal				
Alarms				
Door open	✓	✓	✓	✓
Motor MCB	✓	✓	✓	✓
Battery failing	✓	✓	✓	✓
Low gas pressure		✓	✓	✓
Subtotal	6	9	8	10
Digital Outputs				
Control				
Open	✓	✓	✓	✓✓
Close	✓	✓	✓	✓✓
FPI reset		✓	✓	✓✓
Subtotal	2	3	3	5
Analogue Inputs				
Phase currents			✓✓✓	
Phase voltages			✓✓✓	
Subtotal	0	0	6	0
Total I/O				
Digital input	6	9	8	10
Digital output	2	3	3	5
Analogue output			6	

sectionalizing when communication is not essential. PLCs were developed for industrial process automation where the emphasis was on measuring data as well as performing logic operations on the data for control action at very low capital expense for the equipment. The basic difference between the two devices is the ability to store and pass on data to the master. However the continued evolution of microprocessor technology is eliminating the difference.

5.17.2 Protection-Based IED

Protection-based IEDs (protection relays) allow local switchgear to operate autonomously in clearing a fault. This is fundamental to clear faults with minimum damage to the power system, and they are used in primary substations to control feeder circuit breakers. Traditionally, remote control of substations was achieved by installing (hardwiring) an RTU at the substation to collect status data and control all the circuit breakers within the station. If auxiliary contacts were available in the protective relays, the operation of the particular protection element could also be communicated. The same protection philosophy was used to extend fault clearing farther down the feeder by installing line circuit breakers or reclosers, thus gaining better selectivity for fault clearance. The move to remote control required the RTU and communication features to be available at the switchgear location, hence the integration of protection and RTU functions within one device to produce the communicating protection relay. The most common communicating relay used for feeder automation is the recloser controller. This combines in one integrated assembly remote control, overcurrent protection, sensitive earth fault protection, recloser logic, battery charger and communication modem. Table 5.2 describes the features and dimensions of a typical recloser controller.

Communicating relay/controllers are typically packaged in one case (Figure 5.35) with push buttons for local operation and arrow buttons to allow selecting menu items and manipulating settings. A limited number of LED displays annunciate the state of major settings. This avoids unnecessary wiring within the control cabinet.

5.18 POWER SUPPLIES FOR EXTENDED CONTROL

Because most of the electronic control equipment for extended control is operated at DC, some form of DC power supply is clearly needed. If the control equipment was to operate only when the medium-voltage power system on which it is located was energized, then a simple AC/DC supply would be all that is needed. However, the control equipment will normally be required to operate in the postfault condition, that is, when the MV system is not energized, and in these situations a battery is needed to provide the only source of energy. Most installations for communication and control will use a sealed lead acid battery, also known as maintenance free, at the center of the DC system. The design of this system rests on four major factors:

TABLE 5.2
Dimensions and Features of a Typical Modern Recloser Controller

Function		Function	
Digital Inputs		**Digital Outputs**	
Open status	✓	Open switching device	✓
Close status	✓	Close switching device	✓
Local control selected	✓	Local control	✓
Reclose blocked	✓	Reclose block	✓
Earth fault blocked	✓	Earth fault block	✓
Protection blocked	✓	Sensitive earth fault block	✓
Alternate protection	✓	Protection block	✓
Battery test alarm	✓	System OK light	✓
Reset	✓	Battery OK light	✓
Door open alarm	✓	Disable heater	✓
Subtotal	10	Subtotal	10
Analogue Inputs		**Applications**	
Phase currents (Ia, Ib, Ic)	✓✓✓	Earth fault indication (directional)	✓
Neutral current (In)	✓	Short circuit indication	✓
Phase voltages (Va, Vb, Vc)	✓✓✓	I> nondirectional	✓
Temperature	optional	I> directional	✓
Battery voltage	✓	I_0> nondirectional	✓
Line voltages	✓✓✓	I_0> directional	✓
Zero sequence voltage	✓	I_Δ> phase unbalance	✓
Total active power	✓	RTD inputs	✓
Total reactive power	✓	Event recording	✓
Total apparent power	✓	Oscillograph	✓
Subtotal	15	Subtotal	10

- The capacity of the battery
- The nominal voltage
- The lowest voltage of the partly discharged battery that will still operate the connected equipment
- The lifespan specified

The capacity of the battery is defined in terms of the product of the load current (A) it will supply and the time (H) for which it will supply that load current, and this is normally calculated at the 20-hour discharge rate and for a minimum voltage of 1.75 volts per cell. Figure 5.36 shows the discharge characteristics of a typical sealed lead acid battery, and the graph applies to any capacity within the product range. Each curve shows the voltage that will be provided after discharging for a given time. The 1C curve applies to the nominal discharge rate for the battery and would be 24 amps for a 24 Ah battery. For

Protection and Control   231

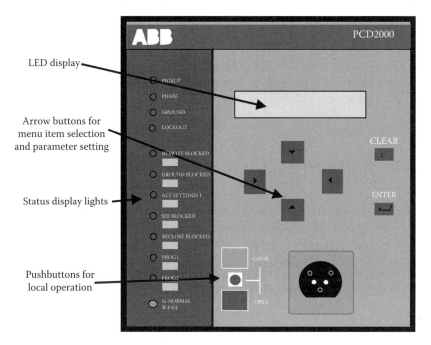

FIGURE 5.35 Facecover of a typical modern recloser controller showing the HMI and status display.

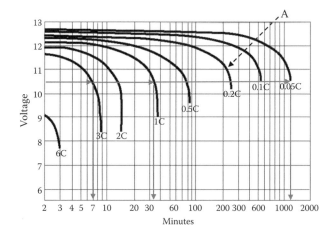

FIGURE 5.36 Discharge characteristic for typical sealed lead acid battery.

the same 24 Ah battery, the 3C curve would apply if the discharge rate was 72 amps (3 × 24) and the 0.05C curve would apply for a discharge rate of 1.2 amps (0.05 × 24).

Let us consider that this battery is of capacity 24 Ah and 12 volts. Because a 12 volt battery will have six cells in series, and let us assume that the lowest acceptable voltage is 1.75 volts per cell, then the lowest acceptable battery voltage will be 10.5 volts. We can now make a number of relevant deductions from the graph for this battery:

- When operating on the 3C curve, representing 72 amps, the lowest acceptable voltage of 10.5 volts will be reached after 7 minutes. The energy supplied will be given by 72 (A) × 7/60 (H) or 8.4 Ah.
- When operating on the 1C curve, representing 24 amps, the lowest acceptable voltage of 10.5 volts will be reached after 35 minutes. The energy supplied will be given by 24 (A) × 35/60 (H) or 14 Ah.
- When operating on the 0.05C curve, representing 1.2 amps, the lowest acceptable voltage of 10.5 volts will be reached after 1200 minutes or 20 hours. The energy supplied will be given by 1.2 (A) × 20 (H) or 24 Ah.

From this, we see that the higher the load current applied, the lower the actual energy supplied and vice versa. The manufacturer would recommend that this particular design of battery would last up to 10 years in service. A cheaper battery may last up to 5 years but the extra cost of the longer service period battery may be offset by the cost of visiting the substation after 5 years to replace the battery. The battery service life will vary with temperature, and a battery with a 10-year life at 25°C might only be expected to last for 2 years at 50°C.

In order to supply power when the charging supply is not available, the battery will need to be charged continuously when the charging supply is available. For chargers supplied by the local distribution network, utilities typically specify that the charging supply would not be available for either 6 or 24 hours and that the battery must therefore supply all load during this period. Some utilities employ a photovoltaic panel to charge the battery, and it must be recognized that the charger will only be effective during periods of high sunlight. Because of the variation of the battery-charging voltage with temperature, it is recommended that temperature-compensated charging be used.

The battery capacity is also dependent on the ambient temperature, the effect of which is shown in Figure 5.37 for a typical battery.

From this graph, we can see that, as the temperature rises, the capacity of the battery also rises, provided that the rate of discharge is substantial.

The battery charger will normally include at least two alarms, loss of charging supply (or loss of mains alarm) and battery low voltage (or lowest acceptable voltage). Although unusual, substation batteries have been known to develop high internal impedance faults such that although the open circuit voltage is correct, the application of load causes the battery output voltage to collapse. This type

Protection and Control

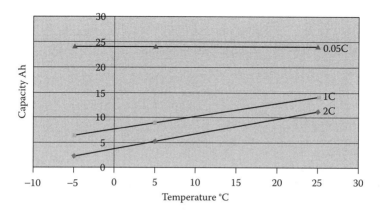

FIGURE 5.37 Temperature effect on capacity.

of fault can be detected by applying a high load current under controlled conditions and monitoring the battery voltage. In colder countries where the control cabinet has an electric heater, this may provide sufficient loading to detect this type of fault.

For loads such as emergency lighting, calculating the capacity of the battery is relatively easy. If, for example, a 15 amp lighting load is required to operate for 4 hours and the lowest acceptable voltage is 11 volts, then, from the point on Figure 5.36 marked as "A," a battery operating on the 0.2C curve will be sufficient. Now, if 0.2C corresponds to 15 amps, then 1C will be 15×5 amps, which means that a 75 Ah battery will be required.

In contrast to this, the sizing of a battery for control equipment in a substation is more complicated. The normal method is to consider a specific rating and then examine how well it will perform under several different scenarios and then balance the risk of there being insufficient capacity against the economics.

Let us consider the loads shown in Table 5.3 at a distribution substation with extended control and then consider operating on the battery alone for a period of 6 hours (360 minutes).

- For the RTU and the communications equipment operating together, the average load is 535 mA which, for a 12 Ah battery, equates to discharge of 0.05C, and from the graph, 0.05C can be sustained for 1200 minutes, well in excess of the 360-minute requirement.
- If the heater were also in circuit, then the average current would be 2.535 A, which is equivalent to discharge of 0.2C. From the curve for 0.2C, it is seen that the lowest acceptable voltage of 10.5 volts would be reached at approximately 250 minutes.
- Now, looking at the use of the switch actuator, at the current of 30 A, which is equivalent to a discharge of 2.5C, the lowest acceptable voltage would be reached after approximately 10 minutes. Now, as the switch actuator is required to operate for just 40 seconds during the

TABLE 5.3
Electrical Loads at Typical Distribution Substation

	Load Current at 24 Volts		Operating Regime	Average Current
	Quiescent State	Operating State		
RTU	NA	35 mA	Continuous	35 mA
Comms	100 mA	5 A	On for 5 seconds each minute	500 mA
Switch actuator	0	30 A	4 operations in first hour, each of 10 seconds 2 operations in last hour, each of 10 seconds	
Heater	0	2 A	Continuous	2 A

first hour, it is considered that this would be acceptable. However, the two operations during the last hour, probably for restoration switching, would probably not be acceptable as they may pull the voltage down below the 10.5 volts. Such a depression would not affect the switch actuator too severely but may affect the RTU and the comms equipment. An alternative to using a larger battery might be to switch off the control cabinet heater during the time that the battery supply is being used.

5.19 AUTOMATION READY SWITCHGEAR — FA BUILDING BLOCKS

It is clear from Chapter 1 that the marrying together of primary and secondary devices is required to produce an automation ready switchgear (ARD), and this is achieved in a number of ways:

- Existing switchgear can be converted by adding an actuator and a separate control cabinet. In the case of ring main units, it is not always possible or advisable to mount an external actuator because the physical design of the unit may never have been intended for mechanical operation. In the case of pole-mounted switchgear, the addition of a control cabinet including an actuator is feasible and cost-effective, provided the switching duty and expectations remain within the original operating specification. Complete control cabinets (see Figure 5.38) including actuators and IED are available from a number of manufacturers, particularly those that specialize in radio communications. These suppliers manufacture IEDs that have the RTU and radio communications function integrated within the same IED. Actuators are available in three forms, motor-driven rotary, motor driven linear, and an inert gas-driven piston.

Protection and Control 235

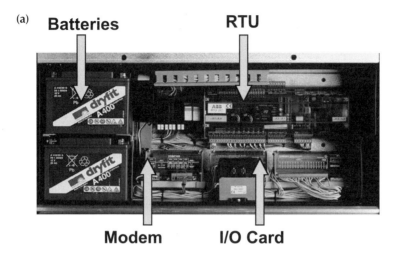

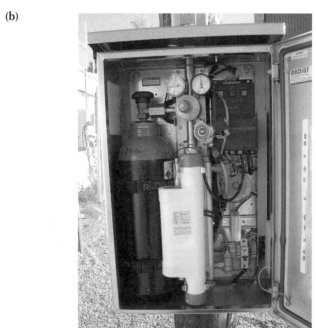

FIGURE 5.38 Typical automation ready device control enclosures and cabinets (a) enclosure within a RMU assembly (courtesy of ABB), (b) a pole-mounted cabinet showing an inert gas-powered linear actuator (courtesy of Radius Sweden AB).

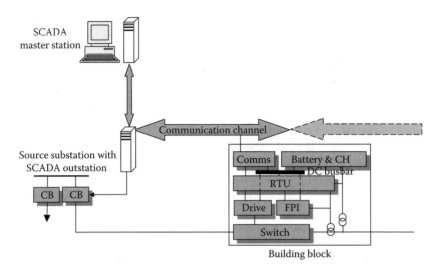

FIGURE 5.39 Structure of a typical building block.

- For new installations, it is best to consider intelligent switchgear available directly from the manufacturer, where the control facility housing has been integrated either physically within the unit, as in the case of a ring main unit, or as a separate pole-mounted cabinet. In either case, the switchgear has been provided with an integrated actuator and IED, power supply and selected communication facility (radio, fiber, PSTN, GSM, DLC etc.), all tested under the required protocol.

The concept of creating FA building blocks to perform a set of standard functions that maybe required in any control system is described in the next section, where the importance of each interface within the assembly is discussed. Experience has shown that such unified products are relatively simple to set up and commission; however, it is quite possible to set up a fully operational IED from a range of separate components supplied by different manufacturers, provided that some care is taken. For this reason, some suppliers favor a building-block approach where a package made up from subcomponents is fully tested before delivery to the end customer.

A building block can be defined as a group of components (see Figure 5.39), each of which has been tested in isolation and is known to function correctly, that have been grouped into a single assembly, which has then been tested to function correctly as the single assembly. In the same way, individual building blocks can be tested against other individual building blocks to ensure that they work together correctly. In this way, a complex scheme of extended control can be assembled from a series of pretested building blocks that are guaranteed to work together the first time.*

---

* The concept of levels of preparedness is discussed in Chapter 1.

Protection and Control

Each of the components used in the building blocks has already been described in the appropriate chapter of this book, and we will now look at how the building block is constructed. Experience shows that it is the interfaces between components that need to be tested as part of the building block; for example, it is vital that the communications channel can handle the communications protocol used by the RTU. Interfaces are examined in more detail later on.

The most common building block within extended control is the switchgear building block, whose purpose is to control a switching device. The generic diagram is given in Figure 5.39, and each component has a number of options.

5.19.1 Switch Options

The switch is the prime switching device, to be operated at MV, 11, 13.8, 24 kV etc., which will be used as the main object of the extended control scheme. The switch could, therefore, be a disconnector, a switch disconnector, a circuit breaker or a reclosing circuit breaker.

5.19.2 Drive (Actuator) Options

The drive is the method of electrically operating the switch, the most common being a direct acting motor wound spring, an electrical release for a precharged motor wound spring, a solenoid or a magnetic actuator.

5.19.3 RTU Options

The RTU is the center of the building block and may be fitted with simply digital inputs and digital outputs but may also be fitted with analog inputs for use in measuring analogue quantities. A typical RTU will have eight digital inputs (DI) and eight digital outputs (DO) and may also be fitted with up to six analog inputs.

5.19.4 CT/VT Options

There are four main reasons why CTs may be included in the building block: to operate protection relays, for measurement of load currents, for indication of fault passage and for operation of autosectionalizing logic. The CTs can be based on conventional magnetics, Rogowski coil technology and future technology, for example, optical devices. Coupling the CT to the RTU will, of course, depend very much on the technology used.

At distribution voltages, VTs are added for up to six reasons: to operate the directional element of protection relays; for the directional element in fault passage indication; to indicate loss of system voltage; to measure system voltage; for operation of autosectionalizing logic; and, depending on the rating of the VT, to provide a source of charging power for the building-block battery.

5.19.5 COMMUNICATIONS OPTIONS

One aim of the building-block approach is to achieve maximum flexibility, and the ability to interchange communication devices illustrates this principle well. For example, a utility may use different communications in different locations, and to be able to add the communications equipment to a preprepared control cabinet will give the utility the maximum benefits of scale.

5.19.6 FPI OPTIONS

The FPI is a most important component of the building block because it will indicate the location of the faulted section. FPIs are available for indication of earth faults or indication of phase faults, usually the latter including earth faults as well. As discussed elsewhere, a zero sequence CT will suffice for earth fault indication, and this can be fitted either within the cable terminating area or, if the earth return paths are suitable, over the sheath of the cable. In contrast, three separate CTs are normally needed to provide indication of phase faults (overcurrent faults).

It is not always necessary to use a separate FPI because protection relay function included in protection-based RTUs will detect fault current and can be arranged so as to provide indication of the passage of fault current to a SCADA system.

5.19.7 BATTERY OPTIONS

The battery and its charger are vital to the building block, because they provide DC power for the operation of the building block during outage time of the connected MV distribution network. The sizing of the battery has been described in Section 5.18.

5.19.8 INTERFACES WITHIN BUILDING BLOCKS

We have already mentioned that it is the interfaces between components that need to be considered in designing the building block. Not all components interface with all other components, but from Figure 5.39, we can show which interfaces between any two components are important. Table 5.4 shows the interface between each two components and, for ease of reference, numbers each interface. For example, the interface between the RTU and the CT is number 10, and we can see that, for example, the switch interfaces with the communications device, RTU, CT, VT and drive.

Now, we can look at the switchgear building block as being a combination of two sub-building blocks, which are the control building block and the item of switchgear. In Table 5.4, the lightly shaded area represents the components needed for the most basic control building block. With the addition of the area with middle shading, the most basic control building block can be converted into a control

Protection and Control

TABLE 5.4
Combinations of Interfaces in a Building Block

	Comms	Battery	RTU	FPI	CT	VT	Drive	Switch
Comms	No	1	2	No	No	No	No	3
Battery	1	No	4	5	6	7	8	No
RTU	2	4	No	9	10	11	12	13
FPI	No	5	9	No	14	15	No	No
CT	No	6	10	14	No	16	No	17
VT	No	7	11	15	16	No	No	18
Drive	No	8	12	No	No	No	No	19
Switch	3	No	13	No	17	18	19	No

building block with measurements. Taking each of these 19 interfaces in more detail, we can determine the most important factors that make a successful integration to form a building block. The darkest shading is needed for all building blocks.

5.20 EXAMPLES OF BUILDING BLOCKS

We have seen that there are a number of variable choices in the design options for building blocks, which leads to a number of individual building blocks. Table 5.6 shows the basic building blocks, but these must be supplemented by the choice in communication system and protocol.

And it is possible to identify, from Table 5.6, the five individual types of building blocks that are most commonly purchased by utilities, these being shown shaded in Table 5.4.

- Sectionalizing switch with FPI and no measurements, underground systems
- Sectionalizing switch with FPI and no measurements, overhead systems
- Sectionalizing switch with FPI and measurements, underground systems
- Sectionalizing switch with FPI and measurements, overhead systems
- Protection-based recloser for overhead systems

One common basic control building block (digital inputs and outputs only) can operate with a ground-mounted switch to form a sectionalizing switch with FPI and no measurements for underground systems and with a pole-mounted switch to form a sectionalizing switch with FPI and no measurements for overhead systems. This arrangement may, however, still require a CT (or other current sensor) to drive the fault passage indicator, even though the CT does not connect to the RTU. Similarly, a voltage-indicating device may be called for to provide a digital input to the RTU in the case of loss of supply voltage alarm. This voltage input might be derived from the VT used for the battery charger. In the same

TABLE 5.5
Coordination of Building Block Interfaces

Interface Number	Interface Between	Interface And	Coordination Needed at the Interface Location
1	Battery	Comms	Battery voltage and capacity must be suitable for comms device, preferably without the use of DC/DC converters.
2	RTU	Comms	Check RTU protocol can be carried by comms device. Check electrical connection between RTU and comms (RS-232, 485, etc.).
3	Switch	Comms	If power line carrier is being used as the comms device, there is a need to locate the coupling equipment on the electrical components of the switch, possibly within a cable end box.
4	RTU	Battery	Battery voltage and capacity must be suitable for RTU.
5	FPI	Battery	Battery voltage and capacity must be suitable for FPI if FPI needs an auxiliary supply.
6	CT	Battery	Some forms of current sensor will need auxiliary supply so check battery voltage and capacity is suitable.
7	VT	Battery	VT output must be suitable for battery charger.
8	Drive	Battery	Battery voltage and capacity must be suitable for switch drive, preferably without the use of DC/DC convertors.
9	FPI	RTU	Check electrical connections, may be serial connection or volt-free contacts on RTU. Check digital output from RTU if FPI needs a reset command.
10	CT	RTU	For current sensors, check electrical connection with RTU, for example 4–20 mA loop. For low-burden current sensors, check sufficient output to drive RTU analog input. For wound-type CT, check RTU input response during flow of fault current in CT primary.
11	VT	RTU	Check output of VT against range of input values for RTU analog input. For low-burden voltage sensors, check sufficient output to drive RTU analog input.
12	Drive	RTU	Check RTU control output contacts are rated for power input to drive mechanism. If n–1 control is used, check interposing relay requirements.
13	RTU	Switch	Check that manual operation of switch does not affect RTU control of switch and vice versa. Check that electrical parameters of auxiliary switches on main switchgear match RTU digital inputs.
14	CT	FPI	Check compatibility between CT output and FPI input. Earth fault FPI can work with zero sequence CT or three CTs in spill formation. Overcurrent FPI needs CTs on each phase under consideration.

Continued.

Protection and Control 241

TABLE 5.5 *(Continued)*
Coordination of Building Block Interfaces

Interface Number	Interface Between	And	Coordination Needed at the Interface Location
15	VT	FPI	VT may be needed with FPI as a reset mechanism when system voltage is restored. VT will be needed if FPI is of the directional type.
16	CT	VT	Applicable only if combined voltage and current sensors are proposed.
17	CT	Switch	CT has to be physically located within the electrical parts of the switch.
18	VT	Switch	VT has to be physically located within the electrical parts of the switch.
19	Switch	Drive	Drive, of whatever type, is used to operate the switch between the opened and closed states, early feasibility check is advised.

TABLE 5.6
Example of Building Blocks

Type of Switchgear	Overhead Network		Underground Network	
	Without Measurements	With Measurements	Without Measurements	With Measurements
Disconnector/sectionalizing switch	Yes	Yes	Yes	Yes
Disconnector/sectionalizing switch with local logic	Yes	Yes	No	No
Circuit breaker	Yes	Yes	Yes	Yes
Recloser	Yes	Yes	No	No

way, the control building block with measurements can be used with a ground-mounted switch to form a sectionalizing switch with FPI and measurements for underground systems and with a pole-mounted switch to form a sectionalizing switch with FPI and measurements for overhead systems.

5.21 TYPICAL INPUTS AND OUTPUTS FOR BUILDING BLOCKS

5.21.1 SECTIONALIZING SWITCH (NO MEASUREMENTS)

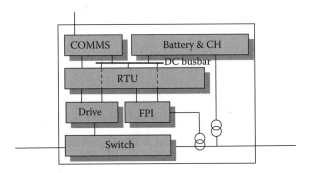

FIGURE 5.40 Typical I/O schedule for building blocks (without measurements).

Digital inputs (typical) for

1. Local control selected
2. Switch opened
3. Switch closed
4. Earth FPI operated
5. Phase FPI operated
6. Loss of charging supply
7. High temperature
8. Switchgear low gas pressure

Digital outputs (typical) for

1. Open switch
2. Close switch
3. Reset FPI

5.21.2 Sectionalizing Switch (with Measurements)

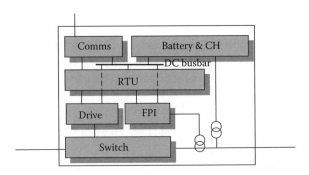

FIGURE 5.41 Typical I/O schedule for building blocks with measurements.

Protection and Control

Digital inputs (typical) for

1. Local control selected
2. Switch opened
3. Switch closed
4. Earth FPI operated
5. Phase FPI operated
6. Loss of charging supply
7. High temperature
8. Switchgear low gas pressure

Digital outputs (typical) for

1. Open switch
2. Close switch
3. Reset FPI

Analog inputs (typical) for

1. Current phase 1
2. Current phase 2
3. Current phase 3
4. Phase-to-phase voltage

5.21.3 PROTECTION-BASED RECLOSER FOR OVERHEAD SYSTEMS

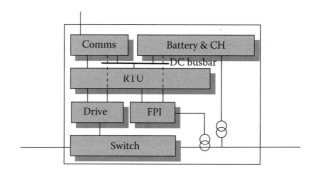

FIGURE 5.42 Typical I/O schedule for building block recloser.

Digital inputs (typical) for

1. Local control selected
2. Recloser opened
3. Recloser closed
4. Protection trip earth fault
5. Protection trip overcurrent
6. Protection trip SEF

7. Recloser locked out
8. Hot line working set
9. SEF out of service
10. Loss of charging supply
11. High temperature
12. Recloser low gas pressure

Digital outputs (typical) for

1. Open recloser
2. Close recloser
3. Set hot line working
4. Set SEF out of service
5. Reset lockout

Analog inputs (typical) for

1. Current phase 1
2. Current phase 2
3. Current phase 3
4. Phase-to-phase voltage

5.22 CONTROL BUILDING BLOCKS AND RETROFIT

We have seen how the control building block is a subcomponent of the overall distribution substation (MV/LV) or feeder line device building block. This can be very useful when extended control is to be applied to an existing distribution substation — the retrofit solution. This is because, quite often, a utility has a distribution asset where the switchgear is relatively new and extended control without replacing the switchgear can only be justified. If the switchgear either has a drive already installed (APD) or one can be supplied, probably from the original equipment manufacturer, at a reasonable cost, then the retrofit solution can be made by the addition of a control building block. Although the utility may have a wide range of distribution switches that need the addition of the control cabinet, there is clearly a need to minimize the number of variants of control building blocks to achieve the desired economies of scale. A retrofit automation strategy can be developed around the concepts described in this section.

5.23 CONTROL LOGIC

We can now consider the control logic that can be used to operate an automated system by looking at Figure 5.43.

The diagram shows two circuits, Circuit A with 1.5 automated devices (because the NOP is shared between two circuits, it counts as half a switch on

Protection and Control

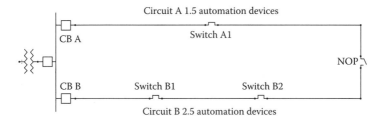

FIGURE 5.43 Circuits for automation control.

TABLE 5.7
Options Considered

Option	Automated Devices	Local FPI	Switch Control System
1	1.5 (Circuit A)	Yes	Remote
2	2.5 (Circuit B)	Yes	Remote
3	1.5 (Circuit A)	No	Remote
4	2.5 (Circuit B)	No	Remote
5	1.5 (Circuit A)	Yes	Local
6	2.5 (Circuit B)	Yes	Local
7	1.5 (Circuit A)	No	Local
8	2.5 (Circuit B)	No	Local

each circuit, also known as AIL 1.5), which is a very common arrangement, and Circuit B with 2.5 automated devices. We shall look at whether the automated devices have a local fault passage indicator and whether the devices are under remote control or local control. This gives eight options (Table 5.7) to examine, but we will also consider, as a special case, the combination of automatic multishot reclosing and automatic sectionalizing.

The automation control scheme can be described as being permissive or active. The permissive scheme will decide that a fault has occurred and recommend a course of action to the control engineer, who would be invited to accept the recommendations before any switching action took place. In contrast to this, the active scheme would carry out its recommended switching actions and then inform the utility control engineer that the routine had been completed.

5.23.1 OPTION 1, CIRCUIT A WITH 1.5 SWITCH AUTOMATION, FPI AND REMOTE CONTROL OF SWITCHES

If CB A opened, the required actions would be

- Check that CB A had tripped on the operation of protection.
- Note that for overhead systems fitted with automatic reclosing, it is assumed that the automatic reclosing has locked out, thereby confirming a permanent fault.

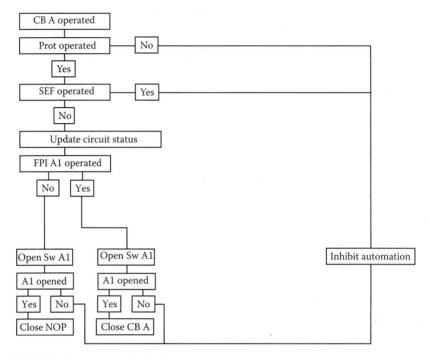

FIGURE 5.44 Logic for Option 1.

- Check that the protection was not sensitive earth fault because this may indicate damaged switchgear.
- Check the fault passage indicator at Sw A1.
- If the FPI has not operated, then the fault is between CB A and Sw A1, so we would open Sw A1 and close the NOP.
- If the FPI has operated, then the fault is between Sw A1 and the NOP, so we would open Sw A1 and close CB A.

These switching actions can be drawn as the logic chart shown in Figure 5.44.

5.23.2 OPTION 2, CIRCUIT B WITH 2.5 SWITCH AUTOMATION, FPI AND REMOTE CONTROL OF SWITCHES

If CB B opened, the required actions would be

- Check that CB B had tripped on the operation of protection.
- Note that for overhead systems fitted with automatic reclosing, it is assumed that the automatic reclosing has locked out, thereby confirming a permanent fault.
- Check that the protection was not sensitive earth fault because this may indicate damaged switchgear.

Protection and Control

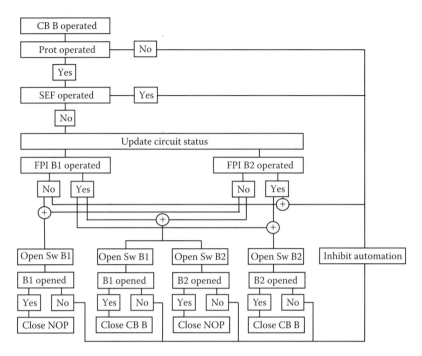

FIGURE 5.45 Logic for Option 2.

- Check the fault passage indicator at Sw B1 and at Sw B2.
- If the FPIs at Sw B1 and at Sw B2 have both not operated, then the fault is between CB B and Sw B1, so we would open Sw B1 and close the NOP.
- If the FPI at Sw B1 has operated but the FPI at Sw B2 has not operated, then the fault is between Sw B1 and Sw B2, so we would open Sw B1, close CB B, open Sw B2 and close the NOP.
- If the FPIs and Sw B1 and at Sw B2 have both operated, then the fault is between Sw B2 and the NOP, so we would open Sw B2 and close CB B.
- If the FPI at Sw B1 has not operated and the FPI at Sw B2 has operated, then we would indicate this as a possible malfunction and inhibit the automation logic.

These switching actions can be drawn as the logic chart shown in Figure 5.45.

5.23.3 Options 3 and 4, No Fault Passage Indicators

The logic for the switching routines can easily be derived from Figures 5.42 and 5.43 but with one major exception. If there is no FPI, then we can only determine where the fault is by some other form of testing, which will take time. This testing could be re-energizing the circuit, after opening one or more switches in the hope of being able to disconnect the faulted section, by closing the circuit breaker, but this will always incur an additional re-energization of the fault. Because this is not a good operating practice, it is stressed that switch control with FPI is always recommended.

5.23.4 Options 5 and 7, Local Control Only

For these two options, Sw A1 is not controlled from the scheme automation controller via a communication link but only from logic built into its own local controller. This local logic is commonly based on the detection of voltage on the incoming side, in which case the required actions would be

- Check that CB A has tripped on the operation of protection.
- Note that for overhead systems fitted with automatic reclosing, it is assumed that the automatic reclosing has locked out, thereby confirming a permanent fault.
- Because of the sudden loss of voltage at its incoming side, the local logic at Sw A1 opens Sw A1.
- If the protection tripping of CB A was not sensitive earth fault, then CB A closes after a time delay of 1 minute, during which time it is assumed that Sw A1 has opened correctly.
- The return of voltage at the Sw A1 local controller suggests that the fault might be between Sw A1 and the NOP but this indication is not firm.
- Sw A1 therefore closes.
- If Sw A1 has closed onto a fault, then CB A will immediately trip, causing a second loss of voltage at the Sw A1 local controller, which instructs Sw A1 to open, but now to lock in the open position.
- We have therefore re-energized the fault once.
- CB A closes after a time period of 1 minute.
- To cater for a fault between CB A and Sw A1, the NOP is fitted with a voltage detector in the direction of Sw A1.
- Following the loss of voltage caused by the first tripping of CB A, the local logic at the NOP will close the NOP after a time period of 3 minutes.
- If the fault is between CB A and Sw A1, and Sw A1 has correctly opened, then the closure of the NOP will reconnect supplies between the NOP and Sw A1.
- If the fault is between Sw A1 and the NOP, then the closure of the NOP will re-energize the fault a second time and cause the CB B to trip, thereby interrupting additional load on Circuit B.
- The second loss of voltage at the NOP will cause the NOP to open and lock open.
- The CB B will close after a time period of 1 minute.

Protection and Control

This method of local control logic has the advantage that a communication channel is not needed for the scheme to operate and, therefore, has some economic attractions. But because it always involves at least one re-energization of the fault, for the same reasons as given for Options 3 and 4, it is always recommended that remote control with communications is applied.

5.23.5 OPTIONS 6 AND 8, LOCAL CONTROL ONLY

The application of noncommunicating local controllers to the 2.5 device Circuit B follows a similar but more complex operating regime.

5.23.6 SPECIAL CASE OF MULTISHOT RECLOSING AND AUTOMATIC SECTIONALIZING

One major attribute of overhead systems is that they experience transient faults, that is, a fault, for example, a bird on the line, which causes a circuit breaker to trip out. However, once the fault arc has been cleared, then it is safe to re-energize the line; hence, the circuit breaker is automatically reclosed after a brief period.

Figure 5.46 shows the control regime of the automatic recloser. When the fault occurs, the recloser opens and waits for the dead time of, say, 5 seconds and then recloses. If the fault was transient, then the circuit is re-energized and no further action is taken. But if the fault is still present, then the recloser will trip a second time, stay open for a second dead time and reclose. If, after this second reclosure the fault is still present, the recloser will trip and lock out.

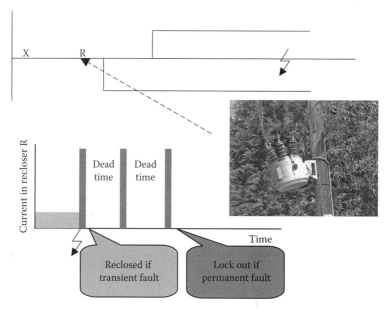

FIGURE 5.46 Transient faults and the automatic recloser.

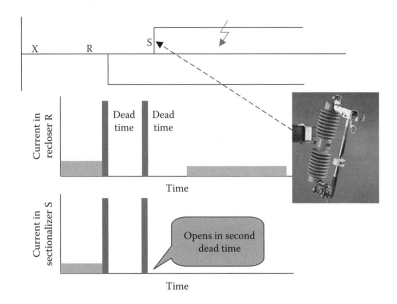

FIGURE 5.47 Application of the multishot mecloser and autosectionalizer.

Because the recloser can operate a number of times in quick succession, it is commonly known as a multishot recloser.

Because of the complexity of some overhead systems, the utility may want to improve system performance by adding further reclosers to the network. This can introduce protection discrimination problems, and so the automatic sectionalizer has seen increased application recently. The automatic sectionalizer, or autosectionalizer, is fitted with local control logic to detect the bursts of fault current caused by a multishot recloser tripping through to lock out on a permanent fault, and its operating principle is shown in Figure 5.47.

For the permanent fault shown, the autosectionalizer, S, will see the fault current that is cleared by the first tripping of the multishot recloser and register this first short burst of fault current. When the recloser makes its first reclosure, a second burst of fault current will flow through the autosectionalizer, which will be registered by the control logic. As soon as this second burst of fault current has stopped, signifying that the recloser is open and it is safe for a switching device without fault clearing capability to open, the control logic will cause the autosectionalizer to open. The recloser will then make its second reclosure, thereby re-energizing healthy network sections, the autosectionalizer having disconnected the fault during the second dead time of the recloser.

The autosectionalizer shown in Figure 5.47 is of the explosive charge dropout type with a CT fitted round its body to detect the fault current. In the alternative, an autosectionalizer could be based on any switching device that can be fitted with a power actuator, for example, a pole-mounted switch disconnector.

6 Performance of Distribution Systems

6.1 FAULTS ON DISTRIBUTION NETWORKS

6.1.1 Types of Faults

In order to understand how protection, fault passage indicators and control systems apply to distribution networks, it is first necessary to examine in more detail the mechanism of faults. In the most fundamental terms, the majority of faults on a power system are caused by a partial or complete failure of electrical insulation, which leads to an increase in the current. The most common types of fault are shown in Figure 6.1. A fault can be between one phase conductor and earth (ground), in which case it is an earth (ground) fault, or it can be between two phases, in which case it is a phase-to-phase fault. Additional combinations exist; for example, a fault on a three-phase cable could be two phases to earth and one phase left completely healthy. The value of the insulation after the fault has occurred may range from 0 to several hundred ohms, which can be taxing on both protection system design and field procedures for localizing the fault. Even when the utility has identified which section of the circuit contains the fault, the field staff will need to pinpoint the exact location of the fault before repairs can begin, and high-resistance faults are not always easy to locate. The flashing fault is of keen interest here because the voltage that causes failure to occur is below the normal system voltage but above the voltage of test equipment. When the circuit is energized, the insulation fails and the circuit trips, but when the test equipment voltages are applied, there is no fault.

However, a fault can also be defined as when a component is unable to carry the load current for which it was designed, in which case a series open circuit would be included as a fault.

Some examples of an open circuit fault would be

- A ferrule in a cable joint in which the cable had been mechanically pulled out, but without any arcing taking place
- Fault energy release, which had burned away a section of phase conductor
- An overhead line conductor that failed mechanically without causing electrical failure, typically a broken jumper

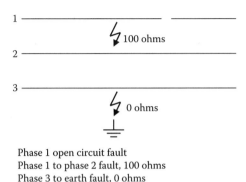

Phase 1 open circuit fault
Phase 1 to phase 2 fault, 100 ohms
Phase 3 to earth fault, 0 ohms

FIGURE 6.1 Examples of fault conditions.

The type of fault can also be categorized into three supplementary categories, self-clearing, transient and permanent:

- The self-clearing fault, further explained in Table 6.1, is characterized by there being no successful operation of any protective devices, and in this category there are some typical examples:
  a. The open circuit fault; for example, the failed overhead line jumper.
  b. The developing fault, where there is a short-term discharge that has the effect of locally heating the point of fault, drying out moisture and hence removing the developing fault. This type of fault is more common in underground cables but can also occur on overhead line insulators. The reason why there is no protection operation is because the discharge stops before the protection relay has operated. The discharge may well start again, sometimes within a few moments of the previous discharge, sometimes many months later.
  c. The self-extinguishing fault, where an arc does develop but is of such a small value that it can be extinguished in free air. The power arc will extinguish at the first current zero, and if the recovery voltage is less than the voltage strength of the now de-ionized path, the arc will not restrike. The main factors that determine whether self-extinction will occur are the magnitude of the current and the recovery voltage. Self-extinguishing faults do not occur in networks with solid or resistance earths, but in compensated networks and unearthed networks, self-extinction up to 30 or 40 amps may be possible.
- The transient fault (or nondamage fault) is characterized by there being a successful operation of a protective device but which can be restored to service without there being a need for permanent repairs, and in this category there are some typical examples:
  a. Because the insulation of an overhead line is based on air, the circuit can be restored to service once its circuit breaker, or other protective device, has been tripped and the air has de-ionized. There is no need

Performance of Distribution Systems

to carry out repairs to the circuit; hence, the restoration can be made after a brief period of time, typically 1 second up to 30 seconds. Typical examples would be a tree branch touching an uninsulated overhead line conductor, a squirrel sitting on the earthed case of a pole-mounted transformer while touching the incoming MV uninsulated conductors or a power followup arc resulting from a lightning strike. The tripping and reclosing can be made on the circuit breaker, which protects a complete overhead line fed from a source substation, in which case all the customers fed from that circuit will see an interruption to their supply. However, many utilities use a recloser, which is a circuit breaker equipped with its own, locally based, protection and automatic reclosing sequence. Because the recloser is small and relatively cheap, utilities will often use several on a feeder, together with the source circuit breaker. This means that only those customers downstream of the tripped recloser will see an interruption to their supply.
b. The developing fault, where the discharge is long enough for protection systems to operate but the passage of fault current dries out a water-caused fault sufficiently for the circuit to be restored.
- The permanent fault (or damage) is characterized by there being a successful operation of a protective device but which can be restored to service only after permanent repairs have been made at the point of the fault.

In the medium-voltage overhead networks of utilities across the world, approximately 80% of faults are transient and 80% of faults [1] involve one phase to earth only, from which the following rules of thumb can be proposed:

- 64% of faults are transient and involve one phase to earth.
- 16% of faults are transient and involve more than one phase to earth.
- 16% of faults are permanent and involve one phase to earth.
- 4% of faults are permanent and involve more than one phase to earth.

One major factor determining whether such incidents do require repair is the speed that the power arc can be interrupted by the protection and circuit breaker. For example, an insulator flashover that is left with a power arc for 3 seconds can sustain such severe burning damage that it would probably need to be replaced, whereas it could be reused if the arc had been cleared more quickly. Had the fast interruption taken place, then the interrupting circuit breaker could perhaps be reclosed after a short period, say, 5 seconds, so that the loads could be restored after such a brief interruption. This is termed reclosing and is widely used by some utilities on their overhead networks.

These devices are variously known as reclosers or automatic circuit reclosers, but in fact they do not have an official nomenclature within international standardization authorities. They are covered in fuller detail in Chapter 4, and a

FIGURE 6.2 Typical MV reclosing circuit breaker. (Courtesy S&C Electric [2].)

typical modern device is illustrated in Figure 6.2, seen on a wooden pole with an unearthed crossarm.

6.1.2 THE EFFECTS OF FAULTS

The way in which faults affect a distribution system depends on the type of fault and the protection that is available, which in turn depends on the type of distribution system. Table 6.1 summarizes the protection operations for each type of fault while Table 6.2 summarizes the corresponding effects.

6.1.3 TRANSIENT FAULTS, RECLOSERS, AND COMPENSATED NETWORKS

It has already been shown, from a technical point of view, that unless required by local regulations, a single phase to earth fault in the compensated network does not need to operate protection. Adding a single recloser would mean that customers downstream of the recloser would see some brief interruptions (during the operating sequence of the recloser) for transient phase-to-phase faults. These account for typically 16% of faults, and the benefit of the recloser on this type of network is often not justified. However, adding a second recloser has two effects:

- A transient phase-to-phase fault beyond the second recloser would not cause brief interruptions to customers connected upstream of this recloser; hence, these customers see an improvement in supply.

TABLE 6.1
Protection Operation for Different Types of Fault

Type of Fault			Type of Neutral Earth		
			Solid/ Resistance	Compensated[a]	Unearthed
Self-clearing	Open circuit	Phase only	Overcurrent and earth fault protection inoperative but could detect by SCADA if loss of volts is catered for		
	Developing	Phase fault	No protection will operate to trip circuit because the fault clears before protection can operate		
		Earth fault			
	Self-extinguishing	Phase fault	Cannot occur	Up to 40 amps may self-clear, otherwise protection operates	
		Earth fault			
Transient	Air insulation	Phase fault	Recloser may be used		
		Earth fault	Recloser may be used	Recloser with directional EF protection could operate but no urgent need	
	Developing	Phase fault		Recloser may be used	
		Earth fault		Recloser with directional EF protection could operate, but possibly no urgent need	
Permanent		Phase fault	Overcurrent protection must operate		
		Earth fault	Earth fault protection must operate	Directional earth fault must operate, but possibly no urgent need	

[a] Also known as arc suppression coil earthed, Petersen coil earthed or resonant earthed.

- Because a recloser can clear permanent faults, any permanent phase-to-phase fault downstream of the second recloser will trip that recloser, and customers upstream of that recloser will not see the fault other than as a depression in voltage while the fault is being cleared.

The recloser can provide a significant increase in benefits if it is used on a solid (and resistance) earthed network or where local regulations require that tripping is made for earth faults on compensated or unearthed networks. However, the protection included with many types of recloser does not include the directional earth fault scheme usually required to operate on compensated or unearthed systems. Consider Figure 6.3.

Table 6.3 shows the customer effects of transient or permanent faults in each situation.

The benefit of adding the second recloser is that it operates for faults downstream of itself without causing interruptions to the customers at load 1; it also improves the reliability of supply to load 1, although load 1 customers still see a voltage depression, each time the recloser sees fault current. The benefits are shown in Table 6.3 in italics. An interesting rule can be derived from these

TABLE 6.2
Effects of Different Types of Fault

Type of Fault			Type of Neutral Earth		
			Solid/ Resistance	Compensated	Unearthed
Self-clearing	Open circuit	Phase only	Some LV customers will have half voltage[a]		
	Developing	Phase fault	No interruption to supply, maybe voltage depression		
		Earth fault			
	Self-extinguishing	Phase fault	No interruption to supply, maybe voltage depression		
		Earth fault			
Transient	Air insulation	Phase fault	Recloser ensures interruption is as brief as possible		
		Earth fault	Recloser ensures interruption is as brief as possible	Outage can be avoided	
	Developing	Phase fault		Recloser ensures interruption is as brief as possible	
		Earth fault		Outage can be avoided	
Permanent		Phase fault	Outage for all customers		
		Earth fault	Outage for all customers	Outage can be avoided	

[a] For the loss of one incoming phase to a three-phase distribution transformer, some customers connected to the low voltage side will experience irregular voltage. For delta star-connected transformers (a very common connection), two of the LV phases will see reduced voltage while one phase will have normal voltage. For a star-connected transformer, two of the LV phases will see reduced voltage while one will see zero voltage.

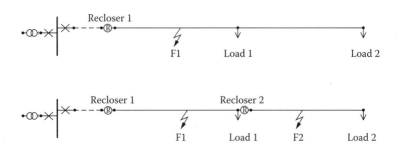

FIGURE 6.3 MV network with reclosers.

Performance of Distribution Systems

TABLE 6.3
Customer Effects of Faults

Fault Condition		Customers at Load 1	Customers at Load 2
Fault at position 1 with one recloser	Earth fault, temporary	Interruptions during auto sequence only	Interruptions during auto sequence only
	Earth fault, permanent	Sustained outage	Sustained outage
	Phase fault, temporary	Interruptions during auto sequence only	Interruptions during auto sequence only
	Phase fault, permanent	Sustained outage	Sustained outage
Fault at position 1 with two reclosers	Earth fault, temporary	Interruptions during auto sequence only	Interruptions during auto sequence only
	Earth fault, permanent	Sustained outage	Sustained outage
	Phase fault, temporary	Interruptions during auto sequence only	Interruptions during auto sequence only
	Phase fault, permanent	Sustained outage	Sustained outage
Fault at position 2 with two reclosers	Earth fault, temporary	*Voltage depression only*	Interruptions during auto sequence only
	Earth fault, permanent	*Voltage depression only*	Sustained outage
	Phase fault, temporary	*Voltage depression only*	Interruptions during auto sequence only
	Phase fault, permanent	*Voltage depression only*	Sustained outage

benefits, which will be expanded and developed in Chapter 3. A recloser protects the quality of supply of customers upstream of the recloser from faults that occur downstream of the recloser. In summary, then, we have seen that many utilities can choose between a compensated/unearthed scheme and a solid scheme with reclosers, the prime differences being:

- The compensated/unearthed scheme, where the single phase to earth fault is permitted to remain on the system for some time, permits continuity of supply for approximately 64% of all faults. A single recloser has benefits for transient phase-to-phase faults only, and multiple reclosers can sectionalize a network and provide smaller zones of outage, provided suitable directional protection is available.
- The solid or resistance earthed scheme can, if used with a single recloser, see transient phase or earth faults as a number of brief interruptions during the operating cycle of the recloser. Multiple reclosers can sectionalize a network and provide smaller zones of outage.

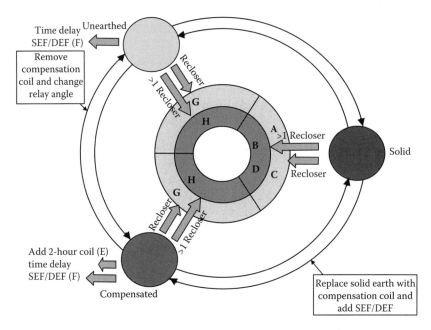

FIGURE 6.4 Relationship between reclosers and earthing.

Figure 6.4 summarizes the differences between the alternatives and gives an indication of the changes needed if moving from one scheme to another, where

- Adding a recloser to a solidly earthed system (A) reduces the effects of transient faults.
- Adding more than one recloser to a solidly earthed system (B) reduces the number of customers affected by transient faults.
- Adding one recloser to a solidly earthed system (C) reduces the effect of permanent faults.
- Adding more than one recloser to a solidly earthed system (D) reduces the number of customers affected by permanent faults.
- Adding a (typically) 2-hour rating (E) to the existing compensation coil eliminates the effect of permanent single phase to earth faults, hence eliminating momentary interruptions so caused.
- Adding a time delay (F), say up to 0.6 seconds, to the directional earth fault protection, either by delayed inverse time trip or sensitive earth fault trip, eliminates the effect of transient single phase to earth faults associated with either a compensated or unearthed network, hence eliminating momentary interruptions so caused.
- Adding one recloser (G) to a compensated network reduces the effects of transient phase to phase faults.
- Adding more than one recloser (H) to a compensated network reduces the number of customers affected by permanent phase-to-phase faults.

6.2 PERFORMANCE AND BASIC RELIABILITY CALCULATIONS

Quantifying the reliability of distribution networks has now evolved into a set of defined reliability indices that are recognized throughout the industry. This is not to say that utilities have not developed their own measures in order to set specific business goals within their organization. We will only concentrate on the commonly accepted indices as defined by the IEEE and discuss how they are applied to assess and compare the performance of different distribution networks and automation strategies.* These indices indicate the annual average performance of the network in terms of interruption frequency and duration. They are weighted by the number of customers or energy supplied and are either presented on a systemwide or customer basis. The index can apply to the entire system or areas as long as the data are consistent with the area for which the index is being computed.

6.2.1 SYSTEM INDICES

System average interruption duration index (SAIDI)† is the average duration of all interruptions per utility customer during the period of analysis (usually annually). For each stage of each interruption, the product of the number of customers interrupted and the corresponding duration is calculated and is known as customer-minutes. For the total number of faults in the period under review, the total customer-minutes interrupted are added up and divided by the total number of customers served for the system or area of the system under evaluation.

$$\text{SAIDI} = \frac{\text{sum of duration}\ddagger \text{ of all customer interruption}}{\text{total customers in system}} \text{ during the period}$$

System average interruption frequency index (SAIFI) is the average number of interruptions (sustained) per utility customer during the period of analysis. Simply, this is the number of customer interruptions per year, divided by the total customers on the system.

$$\text{SAIFI} = \frac{\text{number of customer interruptions}}{\text{total customers in system}} \text{ during the period}$$

---

* The theory of reliability is treated in more detail in other texts dedicated to the subject, which should be referred to for a more thorough treatment.
† This is the same value as customer minutes lost (CML), a term that has been used extensively in the industry at the outset of network performance assessment.
‡ An interruption for which the duration is counted is for a sustained outage and must be greater than the designated duration defining a momentary from a sustained interruption.

Momentary average interruption index (MAIFI) is the average number of momentary interruptions per utility per number of customers in the system. Typically, momentary interruptions are below a specified duration and are calculated separately. In regulated environments, the duration threshold is set by the regulator and is a key measure for counting duration-based interruptions, resulting in the application of penalties.

$$\text{MAIFI} = \frac{\text{number of customer momentary interruptions}}{\text{total customers in the system}} \text{ during the period}$$

Customer total interruption duration index (CAIDI) is the average total duration of interruptions (sustained) among customers experiencing at least one outage in the period.

$$\text{CAIDI} = \frac{\text{sum of the durations of all customer interruptions}}{\text{number of customers experiencing one or more interruptions}} \text{ during the period}$$

These are the indices we will use to calculate network performance improvements through automation and determine the associated economic benefits.

6.2.2 Calculating the Reliability Performance of Networks

The mathematics of calculating the reliability performance of a distribution network is relatively simple but the quantity of data for any network other than extremely small ones means that a software solution is the only practical way. Nevertheless, it is vitally important that the principles of the calculation be understood by the utility because it is always wise to be able to check that the software calculations agree broadly with the experience of engineers and their quick calculations. The major advantage of using software for the calculation is that it is simple to study the effect of changes in the network, for example, the effect of applying extended control to a selection of network switches.

To illustrate how the manual calculations work, a model network was devised according to Figure 6.5. It will be seen that the loads on each circuit are the same, so comparison of the results for each circuit will show the effects of adding source reclosing, a switched alternative supply and midpoint reclosing:

- Each circuit comprised of four sections of overhead line, each section of length 0.5 miles with a permanent fault rate of 0.2 faults per mile per year and a transient fault rate of 0.6 faults per mile per year. The total circuit length on the network was, therefore, 12 miles.

Performance of Distribution Systems

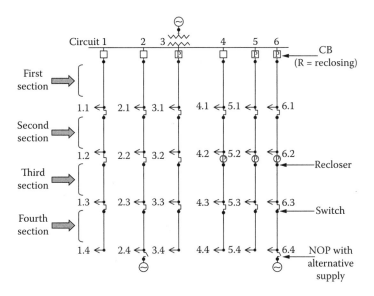

FIGURE 6.5 Circuit diagrams for model network.

- Each circuit comprised four load points, each with 100 customers, and the identity of each load point is shown in the diagram.
- Switchgear consisted of line disconnectors that, for some circuits, were replaced by reclosers. Some circuits were fitted with source reclosing circuit breakers and some with switched alternative supplies.
- Switching time was 1 hour for all switching devices, except for protection-operated circuit breakers.
- With the exception of the overhead lines, all other equipment had zero fault rate. The repair time for overhead line permanent faults was 5 hours.

The switching algorithm assumes that downstream restoration (via a normally open switch to an alternative supply), where possible, would be done first, giving a downstream restoration time of 1 hour. After that, upstream restoration would be made, giving an upstream restoration time of 2 hours. The repair time of 5 hours included all switching actions as well as physical repairs. The principles of the calculations made by hand are shown in Table 6.4, which applies to the circuit number 6.

6.2.3 Calculation of Sustained Interruptions (SAIDI)

For each load point, the spreadsheet calculates the annual restoration time for a fault in each section of the network. For example, a fault in the first section will put load 6.1 off supply for the repair time of 5 hours, as shown in cell C3. Because the first section has a sustained fault rate of 0.2 faults per mile per year and it is

TABLE 6.4
Table of Calculations

	A	B	C	D	E	F	G	H	I	J	K	L
1			\multicolumn{9}{c}{Calculations for SAIDI for Permanent Fault Rate of 0.1}	
2	Load Point and Customers		Restoration Time Hours for Faults in Section				Annual Outage Hours Due to Faults in Section				SAIDI	Customer Mult by SAIDI
			1st	2nd	3rd	4th	1st	2nd	3rd	4th		
3	6.1	100	5	1	0	0	0.5	0.1	0	0	0.6	60
4	6.2	100	2	5	0	0	0.2	0.5	0	0	0.7	70
5	6.3	100	2	2	5	1	0.2	0.2	0.5	0.1	1.0	100
6	6.4	100	2	2	2	5	0.2	0.2	0.2	0.5	1.1	110
7	Tot	400										340
8			\multicolumn{9}{c}{From which the average SAIDI is 340/400 or 0.85}	
9			\multicolumn{9}{c}{Calculations for SAIFI for Permanent Fault Rate of 0.1}	
10	Load Point and Customers		Restoration Time Hours for Faults in Section				Annual Sustained Interruptions for Faults in Section				SAIFI	Customer Mult by SAIFI
			1st	2nd	3rd	4th	1st	2nd	3rd	4th		
11	6.1	100					10	10	0	0	0.2	20
12	6.2	100					10	10	0	0	0.2	20
13	6.3	100					10	10	10	10	0.4	40
14	6.4	100					10	10	10	10	0.4	40
15	Tot	400										120
16			\multicolumn{9}{c}{From which the average SAIFI is 120/400 or 0.3}	
17			\multicolumn{9}{c}{Calculations for MAIFI for Transient Fault Rate of 0.3}	
18	Load Point and Customers		Restoration Time Hours for Faults in Section				Annual Momentary Interruptions for Faults in Section				MAIFI	Customer Mult by MAIFI
			1st	2nd	3rd	4th	1st	2nd	3rd	4th		
19	6.1	100					30	30	0	0	0.6	60
20	6.2	100					30	30	0	0	0.6	60
21	6.3	100					30	30	30	30	1.2	120
22	6.4	100					30	30	30	30	1.2	120
23	Tot	400										360
24			\multicolumn{9}{c}{From which the average MAIFI is 360/400 or 0.9}	

Performance of Distribution Systems 263

of length 0.5 miles, it will experience 0.1 faults per year (Row 1). The annual outage hours for load 6.1 arising from faults in the first section is therefore 0.5, calculated in cell G3 as the product of C3 and the 0.1 in Row 1.

The restoration time in cells C3, D4, E5 and F6 are set to the 5-hour repair time. Cells C4, C5, C6, D5, D6 and E6 are set to the switched restoration time for upstream customer load points of 2 hours. Downstream restoration is set to 1 hour, but, because of the protection-operated circuit breaker at load 34, load 6.1 does not experience any outages for faults in the third and fourth sections. The restoration times in cells E3, E4, F3 and F4 are, therefore, set to zero (because there is no restoration time for loads 6.1 and 6.2 for faults in these two sections). Cells H3, I3 and J3 are calculated in the same way.

The total annual outage hours for load 6.1 is, therefore, the sum of cells G3 to J3, which is entered into cell K3.

Annual outage hours for loads 6.2, 6.3 and 6.3 are then calculated in the same manner.

To calculate the average annual outage time for the four loads, i.e., SAIDI, we need to weight the individual load outages by the customers at each load point. This is done in cell L3, which is the product of cells K3 and B3, the units of which are customer hours (CHR). The total customer hours is the sum of cells L3 to L5, which is entered in cell L7. This is then divided by the total number of customers on the feeder from cell B7 to arrive at the SAIDI for the feeder in Row 8.

6.2.4 Calculation of Sustained Interruption Frequency (SAIFI)

For each load point, the spreadsheet calculates the annual number of sustained outages for a fault in each section of the network. For example, a fault in the first section will give 10 sustained customer outages a year for customers at load point 6.1. This is because the first section will experience 0.1 sustained faults per year (see above for reasoning), and this applies to the 100 customers at the load point. This number is shown in cell G11.

Cells H11, I11 and J11 are calculated in the same way but, because of the circuit breaker at load 6.2, load 6.1 does not experience any outages for faults in the third and fourth sections. Cells I11 and J11 are, therefore, set to zero.

The total sustained interruptions for load 6.1 is, therefore, the sum of cells G11 to J11. However, this is for each of the 100 customers at load point 6.1, so to arrive at the sustained interruptions per customer, we must divide the sum of cells G11 to J11 by the number of customers, which is entered into cell K11.

To calculate the average sustained interruptions for the four loads, i.e., SAIFI, we need to weight the individual sustained interruptions by the customers at each load point. This is done in cell L11, which is the product of cells K11 and B11. The total sustained interruptions is the sum of cells L11 to H11, which is entered in cell L15. This is then divided by the total number of customers on the feeder from cell B15 to arrive at the SAIFI for the feeder in Row 16.

6.2.5 Calculation of Momentary Interruption Frequency (MAIFI)

For each load point, the spreadsheet calculates the annual number of momentary outages for a fault in each section of the network. For example, a fault in the first section will give 30 momentary customer outages a year for customers at load point 6.1. This is because the first section will experience 0.3 sustained faults per year (see above for reasoning), and this applies to the 100 customers at the load point. This number is shown in cell G19.

Cells H19, I19 and J19 are calculated in the same way but, because of the reclosing circuit breaker at load 6.2, load 6.1 does not experience any outages for faults in the third and fourth sections. Cells I19 and J19 are, therefore, set to zero.

The total momentary interruptions for load 6.1 is, therefore, the sum of cells G19 to J19. However, this is for each of the 100 customers at load point 6.1, so to arrive at the momentary interruptions per customer, we must divide the sum of cells G19 to J19 by the number of customers, which is entered into cell K19.

To calculate the average momentary interruptions for the four loads, i.e., MAIFI, we need to weight the individual momentary interruptions by the customers at each load point. This is done in cell L19, which is the product of cells K19 and B19. The total momentary interruptions is the sum of cells L19 to L22, which is entered in cell L23. This is then divided by the total number of customers on the feeder from cell B23 to arrive at the MAIFI for the feeder in Row 24.

6.2.6 Summary of Calculated Results

Using the procedure in the spreadsheet, we can calculate the reliability performance for each load point and for each circuit as a whole, the results being shown in Table 6.5.

Customers on the basic circuit, 1, will see a SAIDI of 5.6 hours per year, although there is a variation between the best served, at the start of the feeder (3.2 hours), and the worst served, at the end of the feeder (8 hours). The value for SAIFI does not vary along the feeder, and because there is no reclosing function, the value of MAIFI is zero. It can be seen that for any customers seeking a certain level of supply reliability, their location on the feeder can be important.

By moving to circuit 2, we add a switched alternative supply, and it can be seen that it decreases the value of SAIDI the nearer the customer is connected to the open point. There is still a variation along the circuit, but the worst served customer SAIDI is now reduced to 4.4 hours.

By moving to circuit 3, we add reclosing at the source circuit breaker. The effect of this is that transient faults are cleared and reclosed. Because the nonreclosing circuit breaker in feeder 1 will trip and stay open for both transient and sustained faults, the reclosing function will permit the transient faults to become momentary interruptions, seen by MAIFI becoming positive, while the sustained fault interruptions decrease. Note that because the total number of faults remains constant, the sum of MAIFI and SAIFI for feeder 3 equals the SAIFI for feeder 1.

TABLE 6.5
Summary of Calculated Results

			Reliability Performance		
Circuit	Load Point	Customers	MAIFI	SAIFI	SAIDI
1	1.1	100	0	1.6	3.2
1	1.2	100	0	1.6	4.8
1	1.3	100	0	1.6	6.4
1	1.4	100	0	1.6	8.0
Total for Circuit	NA	**400**	**0**	**1.6**	**5.6**
2	2.1	100	0	1.6	3.2
2	2.2	100	0	1.6	3.6
2	2.3	100	0	1.6	4.0
2	2.4	100	0	1.6	4.4
Total for Circuit	NA	**400**	**0**	**1.6**	**3.8**
3	3.1	100	1.2	0.4	0.8
3	3.2	100	1.2	0.4	1.2
3	3.3	100	1.2	0.4	1.6
3	3.4	100	1.2	0.4	2.0
Total for Circuit	NA	**400**	**1.2**	**0.4**	**1.4**
4	4.1	100	0	0.8	2.4
4	4.2	100	0	0.8	4.0
4	4.3	100	0.6	1.0	4.6
4	4.4	100	0.6	1.0	5.0
Total for Circuit	NA	**400**	**0.3**	**0.9**	**4.0**
5	5.1	100	0.6	0.2	0.6
5	5.2	100	0.6	0.2	1.0
5	5.3	100	1.2	0.4	1.6
5	5.4	100	1.2	0.4	2.0
Total for Circuit	NA	**400**	**0.9**	**0.3**	**1.3**
6	6.1	100	0.6	0.2	0.6
6	6.2	100	0.6	0.2	0.7
6	6.3	100	1.2	0.4	1.1
6	6.4	100	1.2	0.4	1.0
Total for Circuit	NA	**400**	**0.9**	**0.3**	**0.85**

We can examine the effects of a midpoint recloser by comparing the results for circuit 1 and circuit 4, where we see that the SAIFI for customers upstream of the recloser (4.1 and 4.2) is decreased. This is because the recloser clears all faults beyond it, thereby not subjecting these two substations to the effect of those faults. For customers beyond the recloser, note that the sum of MAIFI and SAIFI remains at 1.6.

Circuit 5 combines the additional plant of feeder 3 and feeder 4. Note that customers upstream of the midpoint recloser now have momentary interruptions but that customers beyond the recloser have more momentary, and less sustained, interruptions than with feeder 4. This is because transient faults upstream of the midpoint recloser are prevented from becoming sustained faults by the action of the source recloser.

Circuit 6 combines the additional plant of feeder 2 and feeder 5 to provide the best performance that can be obtained with this amount of hardware. There is still a variation in the SAIDI between the feeder customer load points, although the range is reduced when compared to the other arrangements. It is important to note that if we reverse the assumed switching routine to carry out upstream restoration before downstream restoration, then the disposition of SAIDI between customer load points will alter. It therefore follows that customers seeking supply at a particular reliability must discuss the restoration switching routine with the utility.

6.2.7 CALCULATING THE EFFECTS OF EXTENDED CONTROL

We have already seen that adding extended control to a switch on a distribution network means that the switching time is reduced from the time taken for an operator to arrive at the site, or a group of sites, and perform the switching tasks, 1 or 2 hours in the example, to the time for a dispatch engineer to instruct the switch to operate, which may be, say, 10 minutes. Apart from the dispatch engineer, the switchgear could be controlled by some form of preprogrammed control scheme, possibly controlling the complete network without any immediate human intervention, in which case the switching time would be determined by that logic.

Let us re-examine circuit 6 in the example. Cell D3 shows a restoration time of 1 hour, which is assumed to be the time taken to travel to the site, open the switch at substation 6.1 and close the normally open point. If this operation were to be completed by extended control in a time of 10 minutes, then the value in cell D3 would be replaced by 10 minutes, but expressed in hours, or 0.167 hours. In the same way, if the dispatch engineer could use the extended control to carry out the upstream restoration in 15 minutes (0.25 hours), then we would replace the value of 2 hours in cells C4, C5, C6, D5, D6 and E6 by 0.25 hours. The spreadsheet would then calculate the effect of extended control, and it follows that the annual outage hours, and hence SAIDI, will be improved.

Looking at the spreadsheet, the restoration time does not affect the frequency of sustained interruptions or momentary interruptions. In this context, it is wise to recall that momentary interruptions are those interruptions caused by temporary faults and cleared by the action of reclosing circuit breakers. The actual interruption times will depend on the reclosing sequences but usually range from 1 second for high-speed autoreclosers to 1 minute for substation breakers. Now, utilities and their regulators define momentary interruptions as any interruption lasting for less than a certain period, for example, 5 minutes. It is very important to see that if the extended control can carry out its actions in a period less than the definition of the momentary interruption, then extended control will add interruptions in the momentary category, but of course, the addition to the momentary category is balanced by an exactly corresponding reduction in the sustained category.

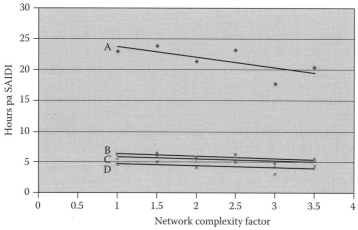

FIGURE 6.6 Performance with respect to network complexity and level of automation.

When using software to calculate network performance, it is therefore of paramount importance to co-ordinate the split between momentary and sustained interruptions in the software with that required by the utility.

6.2.8 Performance as a Function of Network Complexity Factor

In Chapter 3, we defined the network complexity factor (NCF), and we can now use appropriate software to calculate the reliability performance of each feeder and plot the performance against NCF. Figure 6.6 shows the annual SAIDI, the first line showing the situation where there is no extended control, and the lower three lines showing, respectively:

- The line "Source CB" is where extended control has been added to the circuit breaker at the start of the circuit, providing the facility to close the circuit breaker from a control room, and thereby removing the need for the crew to travel to the source substation to operate the circuit breaker.
- The line "Source CB & NOP" is where, in addition to the source circuit breaker, extended control has been added to each of the normally open points.
- The line "All switches" is where, in addition to the source circuit breaker and the normally open point, extended control has been added to each of the line switches on the circuit.

We can see that there is an improvement in SAIDI as the network becomes more complex. This is because the more complex networks have more alternative supplies available, and hence more sections can be restored by postfault switching than with the less complex feeders.

It also shows the very great improvement in SAIDI that can be achieved by adding reclosing to the source circuit breaker, caused by the removal of transient faults from the calculation for SAIDI and SAIFI. However, the values of MAIFI will increase. The improvements gained by adding extended control appear small from the diagram, but these differences can be very significant to customers on the feeder.

The graph also demonstrates an alternative concept for extended control whereby a SAIDI of 5 can be achieved for a complex circuit with source autoreclose or a more simple circuit with extended control of all the line switches including the normally open switch. This concept is extremely important as it gives the utility an additional dimension in network design that became available only following the introduction of extended control.

6.2.9 Improving Performance without Automation

The model that we have discussed in Section 6.2.2 uses relatively short lengths of overhead line. In practice, overhead lines can be much longer than this, commonly up to 35 km, and if we adjusted the model to allow for 35 km circuits, we would find that the average annual outage duration increased from 0.85 hours to maybe 20 hours, and the average annual number of interruptions increased from 0.3 to maybe 10. This is simply because the increase in line length will proportionately increase the number of faults.

It has become a generally accepted fact that overhead lines have more fault outages than underground cables. This is because the bare conductors are more susceptible to direct contact and weather-induced faults than underground cables, and is sometimes linked to the fact that overhead lines are the cheaper to install. The trade-off is in terms of reliability, but few utilities or customers would argue that quality of 20 hours and 10 interruptions per year is acceptable. For this reason, most utilities have set targets for the performance of their distribution networks, and these are influenced by one or more of the following:

- The operational cost to the utility of faults and repairs
- For deregulated locations, the requirements of the regulator and the penalties that are imposed for failure. These are dealt with in more detail in Chapter 8 but can be summarized as
  - A penalty applied for each kWh lost due to outages
  - A penalty applied if the average annual outage duration exceeds a predefined value
  - A penalty applied if the average annual interruption count (sustained or momentary) exceeds a predefined value
- The cost to the customer of the outage(s)
- The loss of customer revenue associated with fault outages

Performance of Distribution Systems

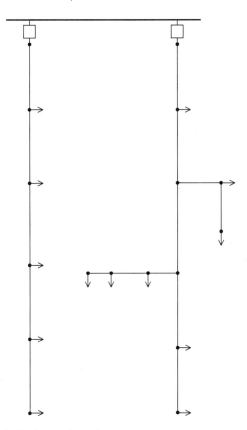

FIGURE 6.7 Sample feeder configurations.

And they have led to investment decisions by the utility on behalf of the customer, although it must be stressed that these factors are now rapidly changing. We can now investigate how these investment decisions can be affected by network design.

The most fundamental requirement of the distribution network is to earn revenue for the utility operator, and for that, an MV network need only consist of cables (or overhead lines) to transmit the electrical power and transformers to connect to the LV network. There would only be one form of circuit protection, that is, a circuit breaker at the source substation. The network could be a single line, as in the left-hand circuit in Figure 6.7, but because networks have developed in a more complex way, the right-hand circuit is perhaps more representative and will, therefore, be used as our model circuit for the time being.

Although this model circuit would gather revenue for the utility, it would be somewhat impractical for a number of reasons:

- Customer outages following a fault are longer than necessary. For example, because a fault anywhere on the left-hand spur cannot be

separated from the main line, all customers would be off supply until repairs had been made, which for a cable fault could be 12 hours.
- Because disconnection from live circuits can only be made at the source circuit breaker, construction and maintenance work would necessitate disconnection of all customer supplies unless live line working was possible.
- Location of a fault on such a multiended network can be very complex and time consuming, and would extend customer outage times.
- Depending on the winding arrangements, it is likely that faults on the secondary winding of the distribution transformer would not be cleared by the source circuit breaker, thereby creating a dangerous situation. Some form of local protection is needed.
- Transient faults, as found on overhead line networks, would cause the source circuit breaker to trip, interrupting supplies to all customers until the breaker could be reclosed by an operator attending the substation. As we will see in Chapter 7 on distribution automation logic, transient faults are removed by de-energization, and if automatic reclosing is added to the breaker, then supplies can be fully restored after a brief period of time.

What we have seen is that we have a cheap to install (but expensive to operate) network but it does not perform very well — the customers would receive more and longer interruptions than are necessary, and because of this, most utilities do not design and operate as simple networks as this example. They have taken account of the sum of the capital costs and the revenue costs and decided, on behalf of the customer, that a higher performance is both possible and needed, thus additional investment has been made in the network to achieve these aims.

The term *reliability* reflects all the outages seen by the customer, which may be caused by some or all of the following:

- Maintenance and construction outages (usually prearranged with affected customers)
- Failures of generation
- Failures of transmission network
- Failures of subtransmission network
- Failures of medium- or low-voltage (the distribution) networks

The performance of a network depends primarily on how it has been designed, in terms of the circuit lengths, the switchgear type and content, the protection policies,* the provision of alternative supplies and the control policies. These all affect how the network performs when a fault occurs, but we can also influence the number and type of faults that occur.

---

* Chapter 5, Protection and Control.

TABLE 6.6
Description of Design Methods for Overhead and Underground Networks

Design Method Options Available	Design Options Applicable to		
	Overhead	Underground	Mixed
Add manually operated sectionalizing switches	Yes	Yes	Yes
Add manually operated switched alternative supply	Yes	Yes	Yes
Add automatic in line protection	Yes	Yes	Yes
Add continuous alternative supply	Yes	Yes	Yes
Add source auto reclose with local automatic control	Yes	No	For overhead sections
Add in line auto reclose with local automatic control	Yes	No	For overhead sections
Add autosectionalizer with automatic local control[a]	Yes	No	For overhead sections
Add extended control to switchgear	Yes	Yes	Yes

[a] The automatic sectionalizer may be based on a switch disconnector or the modifications to drop-out expulsion fusegear, both with local automatic control. They are described in Chapter 4.

For example, we could use underground cables instead of overhead lines because the former generally has a lower fault rate. In addition, we might reduce the fault rate of overhead lines by preventative maintenance such as tree trimming work. We might strengthen the supports to overhead lines or we might replace bare conductors with insulated conductors. A high cable fault rate may indicate that the cable is at the end of its useful service life and it might be replaced. Switchgear and transformers could be subjected to diagnostic measurements and substantially refurbished as the need arose; for example, the moving portion of an isolatable circuit breaker could be renewed without replacing the busbars.

The type of neutral earthing on the system can play a very important part in preventing an outage when a fault develops. Isolated neutral systems are not common, but a single phase to earth fault will not usually cause protection to operate. Similarly, systems that are earthed with a Petersen coil can be arranged not to cause an outage following a single phase to earth fault. We will now examine ways to improve the reliability level by simply adding different types of line switching devices in different locations. These devices will be treated as design tools, and all except the last tool are manually or protection operated devices. The last tools consider adding the remote control capability to the device. This will be in preparation for development later of specific stages of automation complexity that will be used as standard solutions.* The design tools and their applicability to either underground or overhead networks are given in Table 6.6.

---

* Chapter 5, Automation Logic.

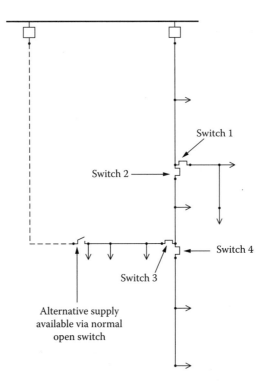

FIGURE 6.8 Model circuit with sectionalizing switches.

6.3 IMPROVING THE RELIABILITY OF UNDERGROUND NETWORKS

6.3.1 Design Method 1 — Addition of Manually Operated Sectionalizing Switches

To improve the performance of our model circuit (Figure 6.8), we could add some switches that can be opened to cut the network into sections (to sectionalize). If a fault occurs on the spur beyond switch 1, the source circuit breaker will trip, disconnecting all supplies. But if switch 1 can be opened, say, within 1 hour, then although customers on this spur will still be off supply for the repair time, all the other customers can be restored because opening this switch has disconnected the fault from the rest of the network.

We can say that the addition of this switch has, for faults downstream of the switch, improved the performance of the network for customers upstream of the switch. For those customers, we have changed the outage time from repair time of, say, 12 hours, to a switching time of, say, 1 hour. A very similar argument applies to the application of switch 4, and we can consider switch 2 in the same way because for a fault beyond this switch, customers upstream can be restored after the switch has been opened.

6.3.2 Design Method 2 — Addition of Manually Switched Alternative Supply

Careful inspection of the diagram shows that switch 2 can do a lot more to improve performance. Suppose the fault is on the main line and between the source circuit breaker and switch 2. We could open switch 2 to disconnect the fault but could not close the source circuit breaker because the fault is still present. But the diagram shows that, in dotted lines, an alternative supply is available via a normally open switch. Once switch 2 has been opened, then the normally open switch can be closed, thereby bringing supply back to customers between the normally open switch and switch 2.

We can say that, as for switch 1, the addition of this switch has, for faults downstream of the switch, improved the performance of the network for customers upstream of the switch. When used in conjunction with a normally open switch, the addition of this switch has, for faults upstream of the switch, improved the performance of the network for customers downstream of the switch. For these latter customers, we have changed the outage time from repair time of, say, 12 hours, to a switching time of, say, 1 hour.

By adding a normally open switch and an alternative supply, the designation of network changes from radial to open loop. The switched alternative supply cannot always be used to restore supplies to healthy sections of the network; for example, we have already seen that for a fault immediately adjacent to the source, we cannot restore supplies to the spur beyond switch 1 by using the alternative supply because the fault is still connected. If a switch was to be installed on the main line, immediately upstream of switch 1, then the faulted section could be disconnected and the spur restored to supply.

From this reasoning, we can say that, to disconnect a faulted section, we need a point of disconnection at each end. Those customers within the faulted section stay off supply for repair time, but other customers, on healthy sections, can be restored in switching time. The more points of disconnection there are, the fewer customers there are in each section, and so fewer customers are subject to repair time outages.

Underground (cable) networks normally use either a switchfuse or a circuit breaker adjacent to the distribution transformer to give protection in the event of a transformer fault. Such transformer faults are rare, but without the protection the fault would either trip the source breaker, disconnecting all supplies, or as mentioned already, fail to trip depending on the fault condition. These networks normally contain local switchgear in the ring. The network discussed already would appear as a cable network shown in Figure 6.9.

The local switchgear shown here is a ring main unit, comprising two switches on the ring and local protection for the transformer. It can be seen that a fault in any position, except one, can be disconnected and supplies restored to all customers by switching, using either the normal supply or one of the switched alternative supplies. The exception is the tee substation. If the fault was to occur between the ring main units either side of this substation, supply to this substation would, normally, need repairs to be made.

274 Control and Automation of Electric Power Distribution Systems

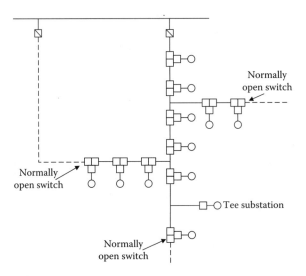

FIGURE 6.9 Model circuit with additional tie points.

By adding switches to the network, we have seen how the outage duration for customers can be reduced from repair time to switching time, and we will investigate later how extended control of the switchgear can improve outage duration even more. We must also note that the switching time must include three major components:

- The time taken to identify where the fault actually lies. This will depend of fault localization techniques including fault passage indication, the latter being covered in more detail in Chapter 5. We will see how extended control can reduce the effect of fault localization time.
- The time taken to travel to the switchgear that will be used to disconnect the faulted section. On longer networks and those where traffic congestion means slow travel speeds, this time may be several hours. We will see how extended control can reduce the effect of travel time.
- The time taken to actually operate the switch. This time is usually only a minute or two for live operated and fully rated switchgear. However, for dead operated switchgear or switchgear without a full rating, then the actual switching time may typically be 1 hour.

6.3.3 Design Method 3 — Add Automatic in Line Protection

Using this design method, we could replace one of the ring main switches at the third substation with a circuit breaker as shown in Figure 6.10. It is important to note that, when a circuit breaker is used to protect the local transformer, it only needs a load rating to match the transformer, many such breakers being rated at 100 or 200 amps. But in order to be on the ring, the rating needs to match other switchgear on the ring, typically 630 amps.

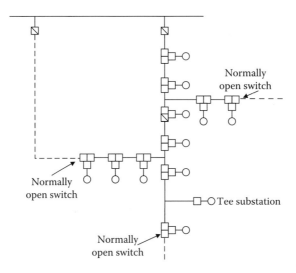

FIGURE 6.10 Model with additional automatic in line protection.

If a fault were to occur downstream of this additional circuit breaker, and assuming that the protection gave the correct degree of discrimination, then the fault would be cleared by this additional (in line) circuit breaker. Customers between the source breaker and the in line breaker would not be affected, so it can be said that the addition of the in line circuit breaker will improve the performance for customers upstream for faults downstream.

This is a similar improvement to adding a sectionalizing switch but there is a major difference. For the sectionalizing switch, customers between the switch and the source will lose their supply when the source breaker trips, and the supply will be restored only when the switch has been opened and the source breaker closed. This depends on the switching time, which may be 1 hour. But with the in line breaker, these same customers do not lose supply at all; hence, the in line breaker removes completely a switching time interruption for customers upstream. To be absolutely accurate, these customers do see a reduction in voltage during the fault clearance time, this reduction in voltage arising from the passage of fault current through the impedance of the system, but all the customers fed from the source busbar will see the same voltage reduction. The addition of one or more in line circuit breakers protects customers upstream from faults downstream but cannot protect customers downstream from faults upstream, although this latter improvement could be gained if the continuous alternative supply was used.

6.3.4 DESIGN METHOD 4 — ADD CONTINUOUS ALTERNATIVE SUPPLY

For the fault shown in the Figure 6.11, circuit breaker B will trip to clear the fault, and all customers downstream of B will lose supply. Customers on feeder A will not be affected. However, if we were to operate the network with the

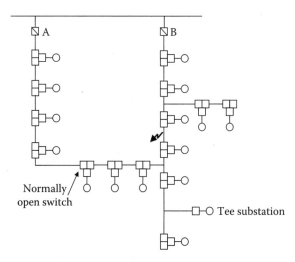

FIGURE 6.11 Model circuit with additional continuous alternative supply.

normally open switch running closed (that is, the alternative supply becomes a continuous one) then for the same fault, both source circuit breakers would trip.

Closing the open point, therefore, degrades the supply to customers on circuit A because with the open point left open, they would not be affected by faults on circuit B (the model circuit). Operating as a closed loop, therefore, worsens the situation unless we add some in line circuit breakers with suitable protection.

The simplest continuous alternative supply is shown in Figure 6.12. There are two circuits arranged to supply a single load, and the four circuit breakers are all closed. This type of network is, therefore, commonly known as a closed loop network.

The great advantage is that, if there is a fault on either circuit, the protection at each end of that circuit trips and the supply to the load is continuously met from the second circuit. Both circuits must be capable of taking the full load of the customer, and normally, unit protection is required on the two circuits.

The disadvantages of this arrangement are the cost of the second circuit and the cost and complexity of the unit protection, although for such a simple scheme, it would be possible to use time graded and directional protection to save the cost of the communication channel between the circuit ends if unit protection is used. However, this type of supply is common for important loads, for example, industrial customers and hospitals, where the additional cost to the utility is balanced by the additional benefit to the customer.

We can now consider a practical distribution network using closed loops and circuit breakers. From Figure 6.13, we can see that a fault at F1 will be cleared by the operation of source breaker A and line breaker CB1. No other customers are involved. With similar logic, we can see that faults in the other three locations will be cleared by the protection at each end of the protected zone. But if we consider F1 in more detail, we can see that, if we did not use unit protection

Performance of Distribution Systems

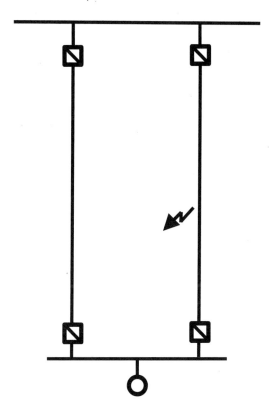

FIGURE 6.12 Simplest continuous alternative supply.

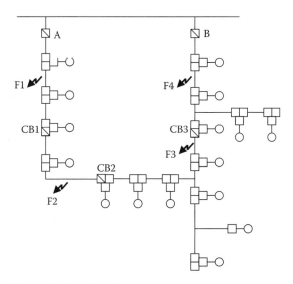

FIGURE 6.13 Model circuit with unit protection.

schemes, then it would be necessary to grade CB2, CB3 and source CB B to get discrimination.

This level of discrimination is very difficult to achieve with conventional protection unless significantly long operating times for protection are acceptable. In general, closed loop schemes only work with unit protection. The communication channel is normally a pilot wire (private or rented) but other channels can be considered.

The closed loop network has some considerable advantages over other schemes, albeit at increased capital and revenue costs. In summary, the faulted section is disconnected immediately, and customers on healthy sections do not lose supply. Inspection of the diagram shows that, because there are only a limited number of in line circuit breakers, the automatic disconnection of the fault, for example, F1, will leave the circuit between source breaker A and in line breaker CB1 disconnected. Now, this fault is between the first and second distribution substations, and it is perfectly acceptable to

- Open the ring main unit switches at both ends of the faulted cable
- Re-energize the first distribution substation by closing the source breaker
- Re-energize the second and third distribution substations by closing the in line breaker CB1

Fault passage indicators can be used to identify exactly which section contains the fault. However, most FPIs operate on a nondirectional basis, the theory being that fault current will only flow, at least on a radial or open loop network, from one point only. With the closed loop network, fault current will flow from both ends; hence, the FPI will need to have the capability of showing in which direction the fault current has flowed.

It is worth mentioning that as distributed generation becomes a reality, utilities may be forced into operating this type of network because active bidirectional operation rather than passive networks will be required.

6.4 IMPROVING THE RELIABILITY OF OVERHEAD NETWORKS (DESIGN METHODS 5, 6, AND 7)

There are two major types of MV overhead networks, the urban overhead and the rural overhead. The urban design is typically applied to small to medium towns, sometimes even in cities, and has served the United States especially well. Load density is high, and pole-mounted transformers may be located every 50 or 100 meters. By contrast, the rural network is found in villages and the more open countryside, where load density is low and pole-mounted transformers may be located every 2 km.

There is a critical difference between the performance of overhead line networks and that of the underground networks that we have considered so far. That

Performance of Distribution Systems   279

difference is that the overhead networks can suffer temporary faults (transient faults), that is, faults that, once the fault is cleared by the operation of protection, the fault arc is removed and, after a brief period to allow ionized gases at the fault position to disperse, the circuit can be safely re-energized without any repairs. Because approximately 80% of faults on overhead lines are transient, this makes a major problem for network operators because, after the operation of protection, the circuit would not be restored until after a switching time, typically 1 hour.

However, if the transient fault is cleared by a circuit breaker, and that circuit breaker is fitted with an autoclose mechanism capable of automatically reclosing the breaker after a brief period, then effects of transient faults is reduced from a 1-hour outage to a brief outage. The reclosing circuit breaker can be either the source breaker fitted with a reclosing relay and power operation mechanism or a discrete, integrated recloser that is fitted at some strategic point on the network. If the same fault is cleared by the operation of a fuse, which can only be reclosed after replacement of the ruptured fuse element, typically after 1 hour, then the customer interruptions are sustained. The operation of the reclosing breakers, automatic sectionalizers and fuses is covered in more detail in Chapter 5.

Hence, customers on an overhead network can experience momentary outages from two very different causes:

- Automatic reclosing following a transient fault, which only applies to the overhead network
- Extended control restoration switching of healthy network sections following a permanent fault that has been disconnected from the healthy sections, which can apply to both underground networks and overhead networks

Because of similarities between overhead systems and underground systems, we can improve the performance of overhead systems with the same tools (methods 1 to 4) as for underground systems. However, method 5 (addition of source autoreclosing), 6 (addition of in line autoreclosing) and 7 (addition of autosectionalizer) are extra tools that apply to overhead networks. It should be noted that the incidence of transient faults on underground systems is extremely unusual; hence, underground systems do not use automatic reclosing. We can illustrate the effects of these three tools with a worked example. Consider the network shown in Figure 6.14, which shows six different circuits fed from a substation.

Each circuit consists of 10 kM overhead line and four substations, each with 100 connected customers and:

- Circuit 1 has no in line switchgear.
- Circuit 2 has one in line switch at the midpoint.
- Circuit 3 has one pole-mounted recloser at the midpoint.
- Circuit 4 has the source circuit breaker fitted with autoreclose together with one in line switch at the midpoint.

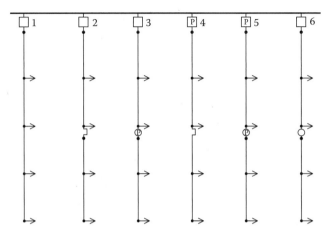

FIGURE 6.14 Improvements to overhead circuits.

TABLE 6.7
Improving Performance of Overhead Circuits

	Circuit 1	Circuit 2	Circuit 3	Circuit 4	Circuit 5	Circuit 6
SAIDI	18	10.5	4.5	9.5	9.00	9.00
SAIFI	4	4.0	2.25	1.0	0.75	0.75
MAIFI	0	0	0.75	3.0	2.25	3.25

- Circuit 5 has the source circuit breaker fitted with autoreclose together with one pole-mounted recloser at the midpoint.
- Circuit 6 has the source circuit breaker fitted with autoreclose together with one automatic sectionalizer at the midpoint.

We shall assume that the fault rate for the overhead lines is 0.3 faults per km per year for transient faults and 0.1 faults per km per year for permanent faults. The switching time is 2 hours and the repair time is 12 hours. We shall also assume that the operating time of the automatic reclosing function and the auto-sectionalizer is less than the reporting difference between momentary and sustained interruptions.

Analysis shows the performance statistics in Table 6.7, from which we can see that

- Adding the sectionalizing switch to move from circuit 1 to circuit 2 reduces SAIDI because for faults downstream of the switch, customers upstream can be restored in switching time of 2 hours after the switch has been opened to disconnect the fault, instead of the 12 hours to repair the fault if it could not be disconnected from the healthy sections.

Performance of Distribution Systems

- By comparing the results for circuit 5 and circuit 6, we can see that the autosectionalizer increases the average number of momentary interruptions when compared to a recloser. This is because the recloser clears all faults beyond itself, whereas the autosectionalizer requires that customers between itself and the reclosing source circuit breaker suffer a momentary interruption during the counting cycle of the autosectionalizer. The autosectionalizer can be in two basic forms, either based on an automation ready switch (it may even be a cutdown autorecloser) or based on dropout expulsion fuses. Both styles will permit the same generic function, but the latter cannot be automatically reset and closed after its operation. However, the latter type is normally considerably cheaper in capital cost than the former.

6.5 IMPROVING PERFORMANCE WITH AUTOMATION

We have seen that one of the factors that controls the performance of the network is the switching time, or the time taken for a person to visit a substation and open or close an item of switchgear. We have used a time of 1 hour in the illustrations, which is the total of traveling time and switch operating time. For many, but not all, types of switchgear, the operating time is maybe 45 seconds, that is the time for an operator to open an access door or gate, remove a padlock and operate the switch by pulling a lever or pressing a button. The actual time for the switch contacts to move is, of course, in the order of 300 milliseconds.

But this type of switchgear is not always the cheapest, and therefore, many utilities use the cheaper, dead operating switchgear where the actual operating time is much longer that 45 seconds. Before this type of switchgear can be operated, the operator must prove that it is dead and, therefore, safe to work on. If the switchgear is of the disconnectable elbow type, then it may take 15 minutes to disconnect the elbows and place them in the parking position. The switching is also more complex because the switchgear needs to be made dead from some other location, and that location must, of course, be one in which live break switchgear is installed.

Most manufacturers nowadays provide a motor drive (or power actuator) as a standard option on their MV switchgear. When this is then fitted with a remote terminal unit and a communication channel, then it is said to be capable of extended control (otherwise known as remote control); that is, it can be controlled from a distant site.

Figure 6.15 also shows how the fault passage indicator can be integrated with the remote terminal unit to give remote indication of the passage of fault current.

The effect of extended control on this type of switch is to make a dramatic decrease in the switching time. The actuator may take a few seconds to run up to speed, and the switch contacts will still operate in about 300 milliseconds but the major saving of extended control is that a switch operator does not need to

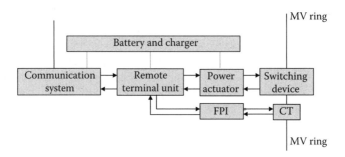

FIGURE 6.15 Components of a remote-controlled switch.

visit the substation to operate the switchgear, and hence, the travel time component disappears to zero. The effect of extended control on switching times is, therefore, to reduce it from, say, 1 hour to 1 or 2 minutes, depending in practice on the workload of the dispatch controller and the time required for the communication channel to transmit the control commands. Customers who were interrupted for a 1-hour switching time can, therefore, now be interrupted for a 2-minute switching time, which represents a very significant improvement. Now, when a fault occurs on a circuit, all the customers downstream of the protective device that has operated lose supply. Those who have a 2-minute restoration time still lose supply but their outage time is reduced. Most utilities define the breakpoint between a momentary interruption and a sustained interruption as 3 or 5 minutes, and it follows that, if the extended control switching time is less than this period, an outage that was classified as a sustained outage becomes a momentary outage when extended control is added. For circuits that do not have transient faults, that is, underground circuits, we can say that the sum of the momentary outages and the sustained outages after extended control is applied is the same as the number of sustained outages before extended control is added.

6.6 IMPROVEMENTS BY COMBINING DESIGN METHODS 1, 2, 3, 4, AND 8 ON UNDERGROUND CIRCUITS

We can now look at the calculated effects of using the design methods 1, 2, 3, 4, and 8 which are given in Table 6.8.

- The column "Basic" shows the results for the most basic network with no switchgear other than the source circuit breaker. Because extended control can only be applied to sectionalizing switches that are on the circuit (and there are no sectionalizing switches on this network), there can be no improvement in performance due to the addition of extended control.

Performance of Distribution Systems

TABLE 6.8
Combinations of Methods of Circuit Performance Improvement

Control Level	Performance Parameter	Basic	Sectrs	Underground Network Type Open Loop	Open Loop and CB	Closed Loop
Nil	Momentary interruptions	0.00	0.00	0.00	0.00	0.00
	Sustained interruptions	0.73	0.93	0.94	0.71	0.52
	Total interruptions	0.73	0.93	0.94	0.71	0.52
	Duration	10.51	7.39	5.25	4.96	4.39
Partial	Momentary interruptions	0.00	0.22	0.22	0.00	0.00
	Sustained interruptions	0.73	0.71	0.71	0.71	0.52
	Total interruptions	0.73	0.93	0.94	0.71	0.52
	Duration	10.51	7.18	4.65	4.53	4.39
Full	Momentary interruptions	0.00	0.43	0.44	0.22	0.30
	Sustained interruptions	0.73	0.50	0.50	0.49	0.21
	Total interruptions	0.73	0.93	0.94	0.71	0.52
	Duration	10.51	6.97	4.31	4.24	4.10

- The column "Sectrs" (sectionalizers) shows the results of adding ring main-type switchgear at each substation; that is, two switches on the ring for each load point.
- The column "Open Loop" shows the results of adding a normally open supply to the network with sectionalizers.
- The column "Open Loop + CB" shows the results of adding one line circuit breaker to the open loop network.
- The column "Closed Loop" shows the results of adding four line circuit breakers to the open loop network.
- The four rows for "Control Level Partial" apply to the addition of extended control to some of the sectionalizing switches plus the source circuit breaker.
- The four rows for "Control Level Full" apply to the addition of extended control to all of the sectionalizing switches plus the source circuit breaker.

The results of this calculation are summarized graphically in Figure 6.16, which shows a number of interesting themes:

- The average annual outage duration improves rapidly as sectionalizers and the switched alternative supply are added to the basic network.
- For a given required average annual outage duration; the addition of extended control adds an additional dimension, for the performance shown as a dotted line, we could either use a sectionalized network

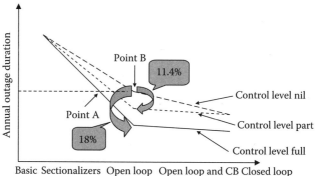

FIGURE 6.16 Variation in outage durations with possible improvements to the basic circuit.

with automation (point A) or we could use an open loop structure without automation (point B).
- The performance of an existing network can be improved by adding extended control. For example, the open loop network without extended control could be improved, on average, by 11.4% by adding partial extended control and by 18% by adding full extended control. This range level of improvement, between 10 and 20%, is very often the target range that regulators will demand that utilities make, and this improvement could be made very quickly, only subject to the delivery and installation time of the additional control equipment.

However, it must be noted that the improvements shown in the graph relate to the average situation, whereas in reality, there will be a distribution of the actual outage durations about the average. For the example we have considered so far, the histograms of outage duration at each load point are shown in Figure 6.17. We can see that

- With control level nil, two load points were off supply more than 7 hours. Control level part has removed this category.
- With control level nil, the largest group of outages was 4 to 5 hours. Control level part has reduced this to 3 to 4 hours.
- If a penalty was payable for customers whose average annual outage exceeded 5 hours, this penalty would be reduced by adding control level part.

In the same way that we have examined the changes in the outage durations caused by combining design methods 1, 2, 3, 4, and 8 on an underground circuit, we can now look in more detail at how the interruptions (sustained and momen-

Performance of Distribution Systems

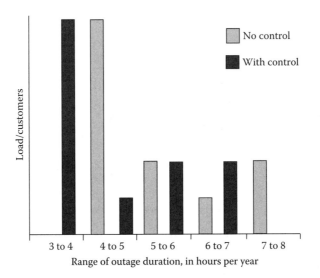

FIGURE 6.17 Distribution of outage duration.

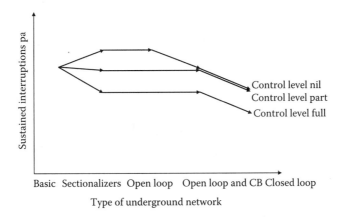

FIGURE 6.18 Variation in sustained interruptions with possible improvements to the basic circuit.

tary) vary with the application of the tools considered so far. Figure 6.18 shows the variation in sustained interruptions.

Referring to Figure 6.18 and starting with the basic network, there are a calculated number of sustained interruptions per year, and we can develop the basic network by moving to sectionalizers, for the moment only examining the case with control level nil, where we see that adding a number of sectionalizers increases the sustained faults. This is because the sectionalizers themselves will fail from time to time, and it is important that equipment added to a network, to improve its performance, must have a significantly positive net benefit. Similarly,

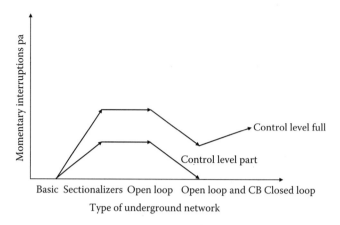

FIGURE 6.19 Variation in momentary interruptions with possible improvements to the basic circuit.

the addition of a normally open switch will, per se, cause a small increase in interruptions, although it will significantly reduce the annual average outage duration. Adding the in line circuit breaker starts to reduce sustained interruptions because faults downstream of the circuit breaker no longer interrupt customers upstream of the breaker. For the same reason, more customer interruptions are avoided by the addition of more circuit breakers and the continuous alternative supply found in the closed loop option.

We shall now examine the case with control level part, where we see that, by adding extended control, some of the interruptions that would have been subject to a manual switching time of 1 hour (and therefore sustained) become subject to an extended control switching time of 2 minutes (and therefore momentary). The figure shows a decrease in sustained interruptions but Figure 6.19 shows that the momentary interruptions increase from zero. If the sectionalizers had a zero fault rate, the improvement in sustained interruptions would have been larger.

It is left to the reader to follow through the above two diagrams, but it is interesting to note that Figure 6.19 shows an increase in momentary interruptions as full control is added to the closed loop network. Referring back to Figure 6.13, let us consider the cable fault F4, which is cleared by the correct operation of source circuit breaker B and network circuit breaker CB3. The faulted section can be disconnected by the opening of the two switches defining the edge of this faulted section, and supplies can then be restored to all customers affected by closing the two circuit breakers that had previously tripped. With no extended control, the restoration will take the switching time of 1 hour, and therefore, all customers who lose supply will be classed as receiving a sustained interruption. By adding extended control to the intermediate switches between these two circuit breakers, the supply can be restored in 2 minutes; therefore, all customers who lose supply will be classed as receiving a momentary interruption.

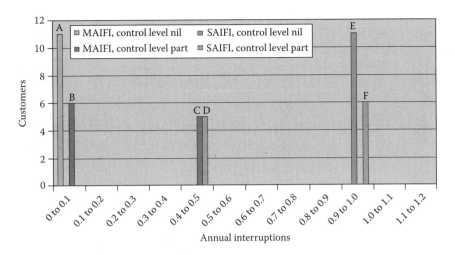

FIGURE 6.20 Distribution of interruptions.

In the same way as for outage durations, the calculations regarding interruptions can be examined in terms of each load point, rather than simply the average situation. Figure 6.20 shows the momentary and sustained interruptions for the example that we have considered so far, using control levels nil and part, and it shows that with control level nil, all 11 customer load points have zero momentary interruptions (column A) and between 0.9 and 1.0 sustained interruptions (column E). By moving to control level part, five of these customer load points reduce their sustained interruptions from between 0.9 and 1 to between 0.4 and 0.5 (column D) but at the expense of their momentary interruptions increasing from zero to between 0.4 and 0.5 (column C). Because a momentary interruption is less annoyance than a sustained interruption, these five customer load points have benefited by the inclusion of extended control. The remaining six customer load points do not change their momentary interruptions (column B) and their sustained interruptions (column F).

REFERENCES

1. U.K. National Fault and Interruption Reporting Scheme.
2. S&C Electric Company Inc., Web site, http://www.sandc.com.

7 Communication Systems for Control and Automation

7.1 INTRODUCTION

The communication link is a vital component of distribution automation. Although local automation schemes may only utilize loss of voltage or other criteria to initiate switching operations, most large-scale deployments require communications to initiate an action or report the action to a central control center. Communication systems have been used for decades by the electric utility industry in a wide variety of applications. In simplest terms, a communication system provides the connecting link between the sending end (transmitter) and the receiver. Many different media are used to transmit the signals ranging from copper circuits, radio, microwave, optical fibers and even satellite. DA can rarely rely on a "greenfield" situation because an existing infrastructure may be present for both communications and control, where media and protocols have already been established. Often, DA communication facilities must extend, replace, supplement or include existing media and embed them into general communication architecture.

A survey conducted in the United States during the mid-1990s on 26 early feeder automation projects identified the communication media and type being applied in the projects. As shown in Figure 7.1, the predominant type was radio.

The survey further concluded that when considering the average age of installation for each communications media type, there were distinct phases and that the industry has moved to phase III, where radio is predominating (Figure 7.2).

The trends of this early survey have continued as confirmed by a survey conducted 5 years later with utilities outside the United States. This showed two-way radio (Figure 7.3) as the most used form of communication but with new types of communication like fiber optics, cellular and even satellite emerging as candidates.

7.2 COMMUNICATIONS AND DISTRIBUTION AUTOMATION

Communications is a very complex topic in its own right. The components of a communication system are most often referenced in terms of the ISO Open

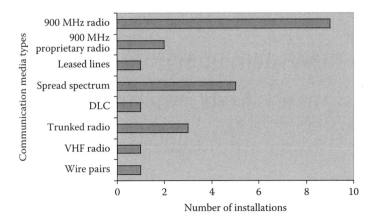

FIGURE 7.1 Primary communication systems in use for DA (Chartwell 1996).

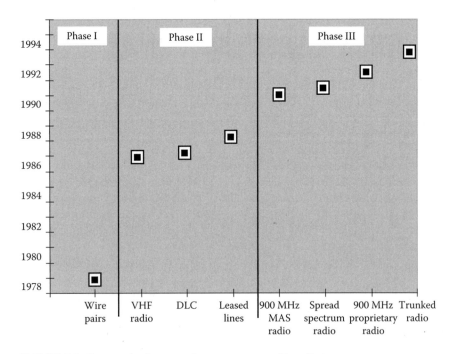

FIGURE 7.2 Communication types by average year of installation.

Communication Systems for Control and Automation

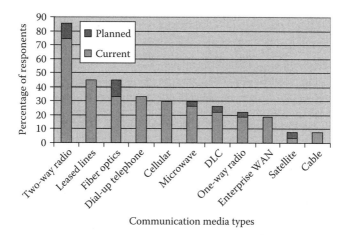

FIGURE 7.3 Types of communication currently in use or planned for DA by responding utilities.

System Interconnection (OSI) Seven-Layer Reference Model (ISO Standard 7498). The layers are Application Layer, Presentation Layer, Session Layer, Transport Layer, Network Layer, Data Link Layer, and Physical Layer. All of these layers may or may not be applicable in the systems we discuss. Because the goal of this chapter is to highlight DA communications, only basic concepts are listed below and covered in the following sections.

The physical link, described in more detail in Section 7.3, provides the communication medium between the feeder equipment transmitting and receiving units. It is the physical medium such as copper wires that the communication signals are transmitted on. For feeder automation, fiber optic lines, wire, or wireless physical links are often used. In a simple case, an RS-232 cable from one device to another can be the physical link.

The communication protocol, described in more detail in Section 7.6, may specify the address of the device the message is being sent, the address of the device that sent the message, information about the type of data in the message (e.g., a control command), the data itself, as well as error detection or other information. Some protocols have become industry standards but most are just widely accepted in a particular field. Whether a standard exists or not, most protocols leave some areas for interpretation that inevitably must be dealt with in a real implementation.

Different advantages and drawbacks are encountered among the different communication options available, and the appropriate selection of a communication technology depends on many factors. These are described in more detail in Section 7.6. Different utility requirements and objectives, physical electric network configuration, existing communication systems, and other factors may affect how the communication system is to be deployed. The goal of DA is to improve system performance, which implies high-quality services and improved communication facilities. These services must provide more individual control and automation possibilities, to obtain an improved overview of the network through measurement, to increase the supply quality and to move towards automatic meter reading and the provision of other services. To be successful and cost-effective, new communication architectures and protocols may be necessary. Compared with transmission system communication, many more points must be integrated, but with smaller volumes of data per point. The data vary in importance and must be handled according to priority. Apart from understanding the purely communication aspects, detailed knowledge of power distribution operation and practices is essential in order to devise suitable communications for power distribution systems. Considering communication system architectures, there is an advantage in being able to support hybrid communications through a system concept that allows for autonomous subregions to be managed according to the data, topology, and type of communication. Such a concept involves the structuring of communication facilities linked together via intelligent node controllers or gateways that can handle communication interfaces, data and protocol transformation, and independent downloaded control algorithms for automation and demand side management (DSM).

7.3 DA COMMUNICATION PHYSICAL LINK OPTIONS

Figure 7.4 shows the different communication technologies that can be used for distribution automation applications.

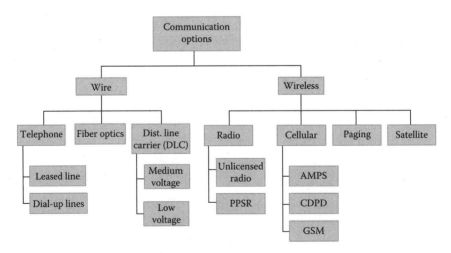

FIGURE 7.4 Distribution automation communication technology options.

7.4 WIRELESS COMMUNICATION

Wireless communication technologies refer to those that do not require a physical link between the transmitter and the receiver. Recent advancements in communications technology have spawned new interest in the transmission of data for use by the electric utility industry. In the past, the standard medium used for this process has been dedicated leased telephone lines. However, transmitting data via cellular, satellite, and other wireless communications has become more feasible due to the technological advancements in recent years.

7.4.1 UNLICENSED SPREAD SPECTRUM RADIO

This communication media includes analog or digital technology to communicate between master and slave packet radios. Spread spectrum RF packet radio uses a network of fixed packet radio nodes with low emission power and usually occupies the 902–928 MHz band. Relaying packets of information from node to node is the way to transmit data, and each node is assigned to a specific address. Interference and collisions are minimized by programming the radios to cycle continuously through hundreds of channels, usually spaced at 0.1 kHz intervals.

7.4.2 VHF, UHF NARROW BANDWIDTH PACKAGED DATA RADIO (LICENSED/UNLICENSED)

Data radio, if correctly utilized, is a robust way of communication. The investment and service cost is fairly low. Especially at the VHF band, huge coverage can be achieved; 100-km jumps (no line of sight) is quite possible. The available channel spacing is normally 12.5/25 kHz. The possibility to use frequencies that are licensed in the VHF and UHF bands increases the reliability because no other user is licensed to use the same frequency. Lately, forward error correction (FEC) with interleaving has found its way into these products as well, increasing coverage and security even further. Other important features are collision avoidance, peer-to-peer functionality and radio network test functions. The speed is typically 9.6–19.2 kbps depending on channel spacing, and the typical range is 10–100 km (system dependent).

Both analogue and digital systems are available, but digital systems are preferred in DA communications.

7.4.3 RADIO NETWORK THEORY

In both spread spectrum and narrow bandwidth data radio systems, the distribution network operator will have to operate the communication network. This is normally performed by its own telecom department, outsourced to a separate telecom company or the communication network equipment supplier. To facilitate this, some knowledge about radio network theory is required.

FIGURE 7.5 Typical narrow band VHF/UHF data radio

Normally, the communication is point-to-point or point-to-multipoint, as shown in Figure 7.6 and Figure 7.7.

Antennas. Selecting the correct antennas can be crucial for the communication links. Designs of antennas and antenna systems include quite complicated physics, mathematics and circuit theory. However, below are some basics that can be helpful in understanding the principles of antennas in a radio network.

The antenna gain is a measure of how well the antenna will send or pick up a radio signal. The gain of an antenna is measured in decibels-isotropic (dBi) or in decibel-dipole (dBd). The decibel is a unit of comparison to a reference. The letter following the "dB" indicates the reference used. The dBi is a unit measuring how much better the antenna is compared to an isotropic radiator.

An isotropic radiator is an antenna transmitting signals equally in all directions, including up and down (vertically). An antenna of this type has 0 dBi gain. (An isotropic antenna is only a theoretical model that has no real design.)

The higher the decibel number is, the higher the gain will be of the antenna. For instance, a 6 dBi gain antenna will receive a signal at a higher level than a

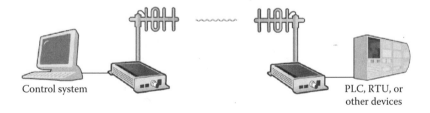

FIGURE 7.6 Point-to-point communication.

Communication Systems for Control and Automation

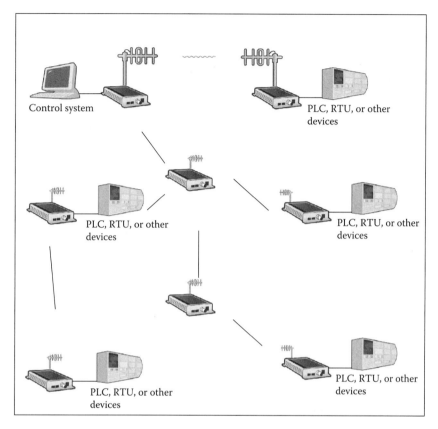

FIGURE 7.7 Point-to-multipoint communication.

3 dBi antenna. A dBd unit is a measurement of how much better an antenna performs against a dipole antenna. As a result a dipole antenna has a 0 dBd gain. However, a dipole antenna typically has a 2.4 dBi gain, as dipole antennas have more usable gain than isotropic radiators. Any dBi measurement may be converted to dBd by adding 2.4.

The only way to increase antenna gain is to concentrate the antenna signal radiation/reception pattern — the electromagnetic field — in a smaller area than the omnidirectional pattern of an isotropic antenna. This can be compared to using a pair of binoculars. You will see the object better, but you will see a smaller area.

Concentrating and focusing the EM field creates gain that is then achieved by the physical design of the antenna. There are basically two categories of antennas, directional and omnidirectional. A directional antenna radiates in one direction only, whereas an omnidirectional antenna radiates in all directions. An omnidirectional antenna should not be confused with an isotropic radiator. Although an isotropic radiator will radiate in all three dimensional directions, an omnidirectional antenna may not radiate vertically (up or down).

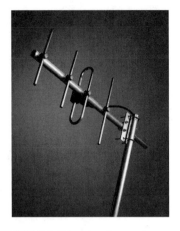 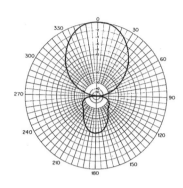

FIGURE 7.8 Directional antenna.

 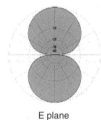 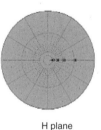

E plane   H plane

FIGURE 7.9 Omni-directional antenna.

Figure 7.8 shows an example of a 14 dBi gain directional antenna. The field strength diagrams show the direction of maximum radiation of the antenna. The E plane represents the electric field, and the H plane represents the magnetic field. The E plane and the H plane are orthogonal to each other.

Figure 7.9 shows an example of an omnidirectional antenna. In this case, it is a ground plane antenna.

An important factor is the balance between antenna gain and radio output power, and it should be noted that increasing the receiver sensitivity by 3 dBm is equal to doubling the output power. The receiver sensitivity is, except from the radio design itself, dependent on the antenna and the antenna installation. Using a high gain antenna is in many cases much more effective than increasing the transmitter power. For example, using a 6 dBm gain antenna, instead of a 0 dBm gain, on a 2 W radio transmitter is equivalent to increasing the transmitter power from 2 to 8 W.

Fade Margin. Fade margin is an expression for how much margin — in dB — there is between the received signal strength level and the receiver sensitivity of the radio and is illustrated in Figure 7.10. Site A is transmitting with 33 dBm (2 W) power and, after the distance to site B, the signal level has dropped to 100 dBm. This gives a margin of −10 dBm because the receiver sensitivity of the radio at site B is 110 dBm.

Communication Systems for Control and Automation   297

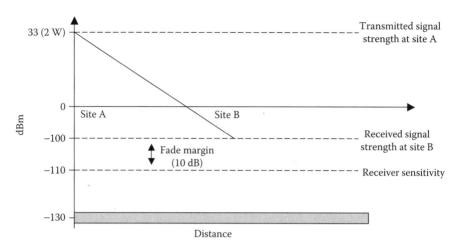

FIGURE 7.10 Fade margin.

In very noisy environments, the level of the noise floor can be higher than the receiver sensitivity (for example, greater than 110 dBm in the above example). In this case, it does not help to increase the receiver sensitivity or use a higher gain-receiving antenna. The only solution if the source of the noise cannot be eliminated is then to increase the power of the transmission so that signal strength at the receiving radio is higher than the noise. However, in some cases, moving the receiving antenna can reduce the noise impact.

Radio Link Calculations. This section contains some useful formulas for radio "link budget" calculations for the cases of both line-of-sight and non-line-of-sight. Note that the calculations are based on assumptions of certain circumstances and should only be used as a guideline for radio network design. In reality, many factors can have an impact on the radio link.

Planning of a radio network is usually done by first simulating the network in a desktop propagation study using a professional software package. The results are then verified by performing a field survey. However, the formulas below can be used to calculate if the radio link has an acceptable fade margin or, if not, how much antenna gain needs to be added or if repeaters have to be used.

The known factors are often:

- The distance between two sites
- The (possible) height of the antennas
- The transmit power of the radio
- The receiver sensitivity of the radio
- The antenna gain

Calculations for Non-line-of-Sight Propagation. The first calculation to be made is the propagation loss. This value tells how much the signal strength is decreased due to the distance between the transmitter and the receiver. For this we use the Egli model. The Egli model is a simplified model that assumes "gently rolling terrain with average hill heights of approximately 50 feet (15 meters)" [Singer, E.N., *Land Mobile Radio Systems*, PTR Prentice Hall, 1994, p. 196]. Because of this assumption, no terrain elevation data between the transmitter and receiver facilities are needed. Instead, the free-space propagation loss is adjusted for the height of the transmitter and receiver antennas aboveground. As with many other propagation models, Egli is based on measured propagation paths and then reduced to a mathematical model. In the case of Egli, the model consists of a single equation for the propagation loss:

$$A = 117 + 40 \times \log \mathbf{D} + 20 \times \log \mathbf{F} - 20 \times \log (\mathbf{Ht} \times \mathbf{Hr})$$

where $\mathbf{A}$ = Attenuation (dB), $\mathbf{D}$ = Distance between the antennas (miles), $\mathbf{F}$ = Frequency (MHz), $\mathbf{Ht}$ = Height of transmitting antenna (feet), $\mathbf{Hr}$ = Height of receiving antenna (feet), 1 mile = 1610 meters = 1.61 kilometers, and 1 foot = 0.305 meter.

Metric system users can use the formula:

$$A = 117 + 40 \times \log (\mathbf{D} \times 1.61) + 20 \times \log \mathbf{F} - 20 \times \log ((\mathbf{Ht} \times 0.305) \times (\mathbf{Hr} \times 0.305)),$$

where $\mathbf{D}$ = Distance between the antennas (kilometers), $\mathbf{Ht}$ = Height of transmitting antenna (meters), and $\mathbf{Hr}$ = Height of receiving antenna (meters).

EXAMPLE 1

$\mathbf{D}$ = 12.5 miles (20 kilometers), $\mathbf{F}$ = 142 MHz, $\mathbf{Ht}$ = 65 feet (20 meters), and $\mathbf{Hr}$ = 16 feet (5 meters).

$$A = 117 + 40 \times \log \mathbf{12.5} + 20 \times \log \mathbf{142} - 20 \times \log(\mathbf{65} \times \mathbf{16})$$

$$A = 117 + 43.8764 + 43.0457 - 60.3406 = 143.6 \text{dB}$$

This example shows that a 142 MHz RF signal will be attenuated 143.6 dB over a distance of 12.5 miles (20 kilometers).

EXAMPLE 2

Increasing the antenna heights × 2, $\mathbf{D}$ = 12.5 miles (20 kilometers), $\mathbf{F}$ = 142 MHz, $\mathbf{Ht}$ = 130 feet (40 meters), and $\mathbf{Hr}$ = 32 feet (10 meters)

$$A = 117 + 40 \times \log \mathbf{12.5} + 20 \times \log \mathbf{142} - 20 \times \log(\mathbf{130} \times \mathbf{32})$$

$$117 + 43.8764 + 43.0457 - 60.3406 = 131.6 \text{ dB}$$

Communication Systems for Control and Automation

Increasing the antenna heights by a factor 2 gained 12 dB, which means that the signal is attenuated 12 dB less over a distance of 12.5 miles (20 kilometers).

The calculations can now be extended using the formula below, which can be used to calculate any of the factors within:

$$FM = Srx + Ptx + Gtx + A + Grx - Cl,$$

where FM = Fade margin, Srx = Sensitivity of the receiver (dBm) (using +dBm instead of –dBm), Ptx = Transmitter RF output power (dBm), Gtx = TX antenna gain (dB), A = Over-air attenuation (dB) (see above), Grx = Receiver (RX) Antenna Gain (dB), and Cl = Cable/Connector Loss (dB).*

EXAMPLE 1

Is the radio link theoretically possible? Calculation of fade margin FM:

Distance = 3 miles (5 km)
Antenna height 1 = 65 feet (20 meters)
Antenna height 2 = 16 feet (5 meters)
Radio Tx power = 33 dBm (2 W)
Radio Rx sensitivity = 110 dBm
Frequency = 456 MHz
Antenna gain 1 = 3 dBd
Antenna gain 2 = 6 dBd
Cable/connector losses = 4 dB total
Fade margin = To be calculated

$$A = 117 + 40 \times \log 12.5 + 20 \times \log 456 - 20 \times \log(65 \times 16) = 129 \text{ dB}$$

$$FM = Srx + Ptx + Gtx + A + Grx - Cl$$

$$FM = 110 \text{ dBm} + 33 \text{ dBm} + 3 \text{ dBd } 129 \text{ dB} + 6 \text{ dBd} - 4 \text{ dB} = 19 \text{ dB}$$

The fade margin is 19 dB, which is an acceptable level. The radio link should be possible.

EXAMPLE 2

How high should the master antenna be mounted?

Distance = 12.5 miles (20 km)
Antenna height 1 = To be calculated
Antenna height 2 = 64 feet (20 meters)
Radio Tx power = 33 dBm (2 W)

---

* To simplify, we use an average cable/connector loss value, including both the Tx and Rx site, of 4 dB. This is, in some cases, a high value, but still assumes a correct antenna cable installation.

Radio Rx sensitivity = 110 dBm
Frequency = 142 MHz
Antenna gain 1 = 3 dBd
Antenna gain 2 = 6 dBd
Cable/connector losses = 4 dB total
Fade margin = 20 dB

Calculating highest allowed over-air attenuation:

$$FM = Srx + Ptx + Gtx + A + Grx - Cl$$

$$A = Srx + Ptx + Gtx + Grx - Cl - FM$$

$$A = 110 \text{ dBm} + 33 \text{ dBm} + 3 \text{ dBd} + 6 \text{ dBd} - 4 \text{ dB} - 20 \text{ dB} = 128 \text{ dB}$$

Calculating antenna height using A from above:

$$A = 128 = 117 + 40 \times \log 12.5 + 20 \times \log 142 - 20 \times \log(X \times 64)$$

$$20 \times \log (X \times 64) = 117 + 40 \times \log \mathbf{12.5} + 20 \times \log 142 - 128$$

$$20 \times \log (X \times 64) = 75.9$$

$$X = (10^{75.9/20})/16 = 97.5 \text{ feet} = 30 \text{ meters}$$

Calculations for Line-of-Sight Propagation. The formulas below are to be used when there is no terrain or obstacles that can interfere with the radio signal. The first calculation to be made is the free space loss (FSL). The FSL value tells how much the signal strength is decreased due to the distance between the transmitter and the receiver.

FSL = free space loss

$$\mathbf{FSL} \text{ (dB)} = 20 \times \log(\lambda/4\pi \times \mathbf{R}),$$

where **R** = Distance between Rx and Tx antenna (line of sight) in meters.

$$\lambda = C/f = 300/f$$

EXAMPLE 1

$$f = 460 \text{ MHz}$$

$$\text{Distance} = 10{,}000 \text{ meters (10 km) (6.2 miles)}$$

$$\lambda = C/f = 300/460 = 0.65 \text{ meters}$$

Communication Systems for Control and Automation 301

$$FSL = 20 \times \log(0.65/4\pi \times 10{,}000) = \mathbf{-105.7 \text{ dB}}$$

EXAMPLE 2

$$f = 460 \text{ MHz}$$

$$\text{Distance} = 20{,}000 \text{ meters } (20 \text{ km}) (12.4 \text{ miles})$$

$$\lambda = C/f = 300/460 = 0.65 \text{ meters}$$

$$FSL = 20 \times \log(0.65/4\pi \times 20{,}000) = \mathbf{-111.7 \text{ dB}}$$

EXAMPLE 3

$$f = 142 \text{ MHz}$$

$$\text{Distance} = 10{,}000 \text{ meters } (10 \text{ km}) (6.2 \text{ miles})$$

$$\lambda = C/f = 300/142 = 2.11 \text{ meters}$$

$$FSL = 20 \times \log(2.11/4\pi \times 10000) = \mathbf{-95.5 \text{ dB}}$$

EXAMPLE 4

$$f = 142 \text{ MHz}$$

$$\text{Distance} = 40{,}000 \text{ meters } (40 \text{ km}) (24.8 \text{ miles})$$

$$\lambda = C/f = 300/142 = 2.11 \text{ meters}$$

$$FSL = 20 \times \log(2.11/4\pi \times 40{,}000) = \mathbf{-107.5 dB}$$

From which we can conclude that, in Example 1, if the transmission power is 33 dBm (2 W), the signal strength at the receiver antenna will be 33 db – 105.7 db = –72.7 db (not taking cable losses at the transmitter in consideration). If the receiver sensitivity is 110 dBm, this will work correctly. However, there are more factors for the link feasibility calculation. Those are calculated by using the same formula as for no-line-of-sight links but by replacing the over-air attenuation A with free space loss.

The calculations can now be extended using the formula below, which can be used to calculate any of the factors within:

$$FM = Srx + Ptx + Gtx + FSL + Grx - Cl$$

can be used to calculate any of the factors within.

FM = Fade margin
Srx = Sensitivity of the receiver (dBm) (using +dBm instead of –dBm)
Ptx = Transmitter RF output power (dBm)
Gtx = TX antenna gain (dB)
FSL = Free space loss (dB) (see above)
Grx = Receiver (RX) antenna gain (dB)
Cl = Cable/connector loss (dB)*

EXAMPLE 5: RECEIVER SENSITIVITY NEEDED

Using Ex 2 above and assuming 3 dB antennas at each site:

$$FM = Srx + Ptx + Gtx + FSL + Grx - Cl$$

$$Srx = Ptx + Gtx + FSL + Grx - Cl - FM$$

$$Srx = 33 \text{ dBm} + 3 \text{ dBd} - 111.7 \text{ dB} + 3 \text{ dBd} - 4 \text{ dB} - 20 \text{ dB} = -96.7 \text{ dBm}$$

EXAMPLE 6: TOTAL ANTENNA GAIN NEEDED

Using Example 4 above, 40 km (24.8 miles) line of sight:

$$FM = Srx + Ptx + Gtx + FSL + Grx - Cl$$

$$Gtx + Grx = GA \text{ (Total gain)}$$

$$FM = Srx + Ptx + FSL + GA - Cl$$

$$GA = FM + Srx - Ptx - FSL + Cl$$

$$GA = 20 \text{ dB} + 110 \text{ dBm} - 33 \text{ dBm} - 107.5 \text{ dB} + 4 \text{ dB} = -6.5 \text{ dB total}$$

Conclusion: 3 dBd antenna gain is needed at each site to get approximately 20 dB fade margin.

7.4.5 TRUNKED SYSTEMS (PUBLIC PACKET-SWITCHED RADIO)

Public packet-switched radio (PPSR) transmits data packets to a radio base station, which in turn transmits over a public, commercial network. PPSR networks use dedicated, multichannel radio frequencies outside the cellular bands to provide two-way communications between intelligent devices. This communication technology operates in the 810 and 855 MHz bands and uses proprietary

---

* To simplify, we use a general cable/connector loss, including both the Tx and Rx site, of 4 dB. This is a high value for most installations, but still assumes a correct antenna cable installation.

packet data protocols. This technology is available nearly nationwide, and the throughput is typically 19.2 kbps.

7.4.6 Cellular

Circuit-Switched Data over cellular (AMPS). Circuit-switched data over cellular Advanced Mobile Phone Service via a cellular modem uses the analog cellular network for data transmission. Data streams are transmitted over the cellular AMPS network to a cellular base station, which in turn transmits over a public network to the destination. A dedicated cellular, modem-to-modem circuit is established and remains open for the entire data transmission session. This communication technology is best suited to lengthy data-intensive transmission.

Cellular Digital Packet Data (CDPD). CDPD is a packet-switched data over cellular technology. CDPD was designed and developed by a group of major cellular carriers and is compatible with the existing AMPS cellular infrastructure. CDPD transmits data packets to a cellular base station, which in turn transmits over the public network. CDPD breaks the data stream into packets and sends them in bursts, "hopping" along idle channels as needed.

Global System for Mobile Communications (GSM). GSM is a digital cellular communications system that has rapidly gained acceptance worldwide and was initially developed in Europe to handle data transmissions. This technology offers various data services with user bit rates up to 9600 bps. A unique service of GSM, the Short Message Service (SMS), allows users to send and receive point-to-point alphanumeric messages up to a few tens of bytes. GSM was designed to have interoperability with Integrated Services Digital Network (ISDN). GSM has the potential to provide a global solution, and its digital transmissions provide good clarity, low static, and a high level of security.

7.4.7 Paging Technology

Paging is a wireless technology, typically satellite-based, that provides one-way communication transmission of alphanumeric and full-text data. It can be an alternative for cellular technology when the data transmission requirements are in only one direction.

7.4.8 Satellite Communications — Low Earth Orbit

Satellite communications is an intriguing technology, mostly because it has the capability to provide a global automation solution for utilities. Satellite communications via commercial orbiters are used for high speed, point-to-point applications. Emerging technologies consist of sets or "constellations" of low earth orbiting satellites (LEOs). LEOs are positioned in non-stationary constellations about 400–700 miles above the Earth and use bands below 1 GHz ("small" LEOs) and in the 1–3 GHz range ("big" LEOs). These satellites offer a wide coverage due to the large number of satellites. Many LEOs are already in place or will be in place within the next several years.

7.5 WIRE COMMUNICATIONS

Wire communication technologies include a physical link between the transmitter and the receiver, of which there are a number of options described below.

7.5.1 TELEPHONE LINE

Dedicated connections using a leased line between utility monitoring points and operations center are one of the most popular media for real-time data communications in utility applications. They are found in many parts of the communications backbone. This technology is widely available in most geographic areas, particularly urban/suburban areas. It is a mature technology, and many methods exist to transmit information. A drawback is the operation- and maintenance-associated costs.

Dial-up telephone lines are used to transmit data over the voice public switched telephone network (PSTN). Data is transferred after establishing a modem path between the devices, usually the master terminal unit and the remote terminal unit (RTU). This technology is not suited for real-time applications due to the cost of the telephone service (if applicable) and may be limited by the relatively low speed of data transmission and lower capacity.

7.5.2 FIBER OPTICS

Fiber optic cables are becoming very popular among electric utilities. Fiber optic cables transport bursts of light and have virtually unlimited channel capacity. Fiber optic cable is commonly used in underground distribution automation applications; however, the fibers can also be employed in overhead distribution feeders. A drawback is the cost to lay the fiber.

7.5.3 DISTRIBUTION LINE CARRIER

Distribution line carrier (DLC) uses the distribution power line as the conductor for the signal, thus can be cost-effective particularly in urban areas where most substations are underground and served by cable. DLC can also be used on mixed underground and overhead networks. Modern DLC technology allows within the CENELEC frequency range a gross data transfer rate of 36 kps, thus permitting fast transmission of control commands, alarms and measurements.

Communication over power lines, although reported as a recent phenomena in IT news circles, is in fact an old standard among electricity providers and the first patent was issued on this topic over 100 years ago. Classical DLC via high voltage lines for protection and communication purposes started in the early 1920s, and one-way communication systems for load scheduling and tariff switching have been an integral part of life in many European countries for several decades. The use of PLC-based baby monitoring phones has been a standard application for over two decades.

Up to now, commercial PLC was either used as a high-end expensive technology only affordable for utilities or it was regarded as a junk product that probably did not work very well. Nowadays, the goal is to use this technology for commercial applications ranging from narrowband utility operations to delivering broadband Internet to the general public. The goal is to provide commercial and residential PLC as a plug-and-play technology with little or no requirement for technical or engineering preconfigurations with maximum reliability for the price per unit of a baby phone.

The enormous development over last the 15 years in PLC has been mainly driven by three factors:

1. The price reduction for development and production of smart digital signal processing-based communication PLC modems
2. The deregulation of the worldwide power market in conjunction with the layoff of hundreds of thousands of employees and the parallel increasing demand for remote control to keep costs at bay
3. An increasing demand for computer and broadband networking within flats, buildings and communities; the current emerging technologies in this area can be divided into in-house ("indoor") technologies and access ("outdoor") technologies

PLC is made complicated because of the wide range of power systems conductors, and their impedances, over which it is required to operate, together with the legal environment surrounding the communications industry. The utility power distribution network is designed to efficiently distribute power at 50 or 60 Hz and was not originally designed to work in conjunction with other communication frequencies. The challenges that arise are

- Signal attenuation mainly in cable networks
- Signal interference based on all possible types of noise
- Line conditions that can vary greatly over time
- Noise injected by connected devices or injected by radio stations and other EM sources
- There are still many issues undefined by the government regulators as well as inconsistencies between the dedicated assigned bands for utilities and the international ITUT radio frequency assignments; an example of this is the normal time transceivers working in conflict with the upper range of the CENELEC-A band
- The technology has to compete with already well-developed and deployed systems like radio, GSM or PSTN; even though PLC brings a huge advantage in terms of infrastructure cost, the business and working model requires proper planning and maintenance

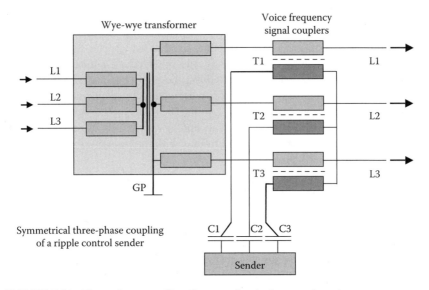

FIGURE 7.11 Three-phase coupling diagram of a ripple control sender.

Ripple Control Systems. Ripple control systems are mainly used for load scheduling and tariff switching. The signals are injected with very high power at frequencies below 3 kHz, typically below 1.5 kHz. Such signals have the advantage that they are still close to the power frequency and so are able to propagate through the transformers and over the grid with very low attenuation. The data rate is very low, and only one-way communication is possible. Ripple control transmitters (Figure 7.11) are very large and expensive, and even though the current technology is somewhat obsolete it is still the most used and available commercial power line-based communication technology. However, harmonics of the power frequency can sometimes interfere with the ripple control system.

Ripple control systems can be compared with radio broadcast stations, and like a broadcast station the ripple control system struggles to obtain 100% coverage. The typical output power of a ripple control system is between 10 and 100 kW; under special conditions it can be as high as 1 MW.

Classical PLC. High-voltage PLC has been a common technology since the 1920s and is used as a utility proprietary communication network via the high-voltage lines. The application is mainly network control, grid protection and supervision. Historically, PLC was a domain for analogue communication systems, but nowadays more and more of it is replaced by digital communication systems allowing data rates of up to several hundred kbps. Typical systems work currently with up to 36 kbps per channel.

The usable frequency range for PLC systems is between 15 and 500 kHz (typically above 30 kHz). The lower frequency is provided due to increasing costs for coupling equipment. High-voltage PLC uses a channel spacing of 4 kHz. The lower the frequency, the higher will be the possible distance to bridge without a

Communication Systems for Control and Automation

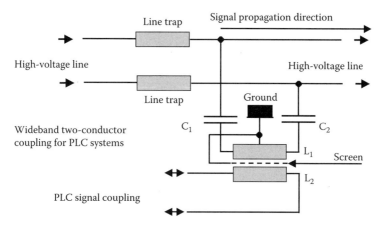

FIGURE 7.12 Wideband two-conductor coupling for high-voltage PLC systems.

repeater. There are examples in service of links up to 1000 km without one intermediate repeater.

The maximum output power of a PLC system is limited to 10 W. The relatively low output power in combination with the guided characteristic of the waves via the lines causes very low RF interference.

A high-voltage PLC system is in general a point-to-point communication system with much defined coupling impedances and attenuation characteristics.

Line traps as depicted in Figure 7.12 are huge coils working as tuned resonators for the signal frequency and define the signal propagation direction as well as the coupling impedance; such devices can be very bulky.

Narrowband PLC. All communication technologies working in the so-called CENELEC bands according to EN 50065 are commonly summarized as narrowband technologies (Figure 7.13).

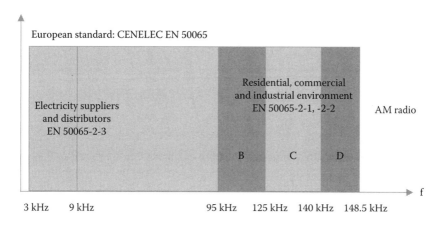

FIGURE 7.13 Frequency assignment according to EN 50065.

There are different rules in the United States and Japan. These countries have an upper frequency limit for PLC systems of about 500 kHz because they do not use long-wave radio systems. Most of the high-speed PLC systems, working in the CENELEC bands, with data rates up to 1 Mbps, are designed to work in the U.S. and Japanese market. The EN 50065-2-1/2/3 frequency band assignment makes sense for grids directly connected to LV customers. For communication systems working via medium-voltage power lines (1 to 36 kV), this assignment makes no sense because there are no residential systems connected. Medium-voltage PLC systems are allowed to work in all bands, according to EN 50065.

Broadband PLC. Broadband PLC systems work with frequencies up to 30 MHz. These technologies are mainly used for IT-related communication systems. There are already mature systems on the market with a data rate up to 30 Mbps.

The broadband PLC technologies can be divided in three major classes.

1. In-house (indoor) PLC systems are systems replacing wire- or radio-based LANs. The advantages in comparison to radio are the low emission of electromagnetic waves and the high data rate. The advantage in comparison to wires is the small or nonexistent requirement for infrastructure installation. The main technology in this area is based on a chipset from Intellon working together with numerous product integrators.

2. Access (outdoor) PLC systems are systems transferring high data rates between buildings and intermediate backbone networks. The average distance such a system can work without a repeater on low-voltage power lines is about 300 m. There is, in general, a data concentrator located in an MV/LV substation or in the basement of a so-called friendly customer. From this point, they send the data either via leased lines or via medium-voltage lines to the next backbone connection.

The major companies producing equipment and working as system integrators are MainNet, DS2, Xeline and ASCOM. DS2 has a lot of licensees producing modems based on its chipset. From the communication point of view, DS2 would appear to have the most mature technology; from the system approach, MainNet would appear to have the leading position in cooperation with PPC, its associated German integrator. ASCOM would appear to be somehow in the middle between DS2 and MainNet. The ASCOM coupling network is shown in Figure 7.15.

3. Broadband MV PLC systems are PLC-based high-speed communication systems working via medium-voltage power lines. There is no currently available standard for these systems. Besides advanced coupling schemes, the technologies utilized are similar to what is currently being used on low-voltage lines. Capacitive coupling equipment for signal coupling to MV lines can be obtained from PPC, Eichhof and Effen.

Carrier Communication Technologies. This section deals with all major parts required to set up a functional communication network via power lines. The main focuses are communication systems over medium-voltage power lines used for automation, monitoring and control. The major differences between low-voltage and medium-voltage PLC systems are the required reliability and the couplings utilized.

Communication Systems for Control and Automation 309

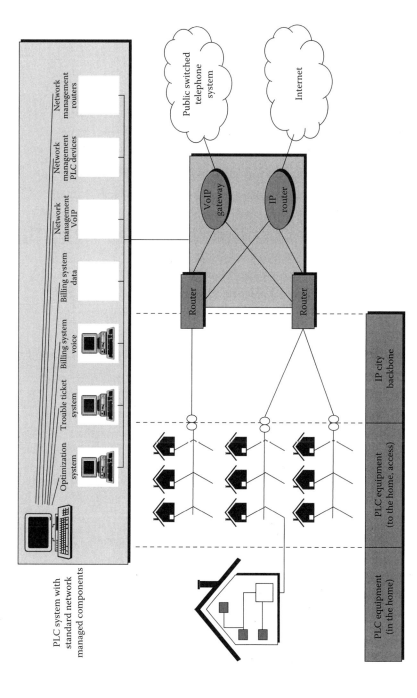

FIGURE 7.14 PLC system with standard network management components.

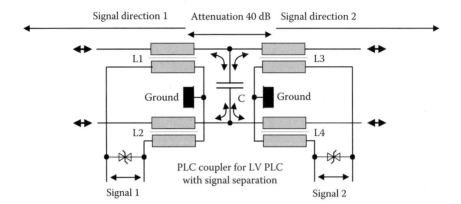

FIGURE 7.15 Coupling network proposed by ASCOM to isolate PLC systems.

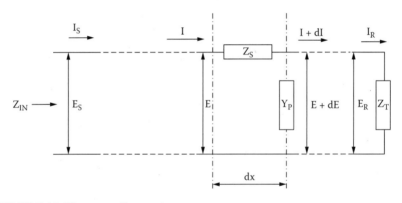

FIGURE 7.16 Elementary line section.

Lines and Cables. Wave propagation via power lines is complicated because we must deal with multiconductor power lines. Many calculations require demanding mathematics, and even for the advanced practitioners, the results are sometime mysterious. Some basic concepts and some terminology are explained in the following paragraphs. The transmission parameters of a line are derived from its primary parameters. For readers interested in the derivation of the equations used in this chapter, dedicated literature dealing with such topics is required. All values are referenced according to Figure 7.16.

Where the general solution is given by the equations with γ as the so-called propagation coefficient,

$$\frac{d^2 I}{dx^2} = -\frac{dE}{dx}(G + j \times \omega \times C) \times I = (R + j \times \varpi \times L)(G + j \times \varpi \times C) = \gamma^2 \times I$$

$$\gamma = \sqrt{((R + j \times \varpi \times L)(G + j \times \varpi \times C))}$$

Communication Systems for Control and Automation

the characteristic impedance is given by

$$Z_0 = \frac{E_S}{I_S} = \sqrt{\frac{(R + j \cdot \omega \cdot L)}{(G + j \cdot \omega \cdot C)}}$$

and the variations in typical characteristic impedances for overhead lines are shown in Figure 7.17.

The variations in typical characteristic impedances for underground cables are shown in Figure 7.18.

The line equations are given by

$$Z_{IN} = Z_0 \cdot \frac{Z_T + Z_0 \tanh \gamma \cdot l}{Z_0 + Z_T \tanh \gamma \cdot l} \qquad \text{Input impedance}$$

$$\alpha = |\gamma| \cdot \cos(rad(\gamma)) \cdot 8.686 \qquad \text{Attenuation in dB/km}$$

$$\beta = |\gamma| \cdot \sin(rad(\gamma)) \qquad \text{Radians/km}$$

When a line is not terminated with its characteristic impedance, a signal traveling down the line is reflected partially (or completely if the line is open or short-circuited), and the reflection travels back to the input. Similarly, there will be a reflection if two lines with different characteristic impedance are connected together.

$$RL = 20 \log_{10} \left| \frac{Z_T - Z_0}{Z_0 - Z_T} \right| \qquad \text{Definition of the return loss}$$

$$VTC = \frac{2 \cdot Z_T}{Z_0 + Z_T} \qquad \text{Voltage transmission coefficient}$$

The phase velocity of propagation is the product of wavelength and frequency.

$$V_P = \lambda \cdot f = \frac{2 \cdot \pi \cdot f}{\beta} = \frac{\omega}{\beta}$$

Phase velocity

$$V_P \xrightarrow[\omega \to \infty]{} \frac{1}{\sqrt{L \cdot C}}$$

$$V_G = V_P - \gamma \frac{dV_P}{d\gamma} = \frac{d\omega}{d\beta} \qquad \text{Group velocity}$$

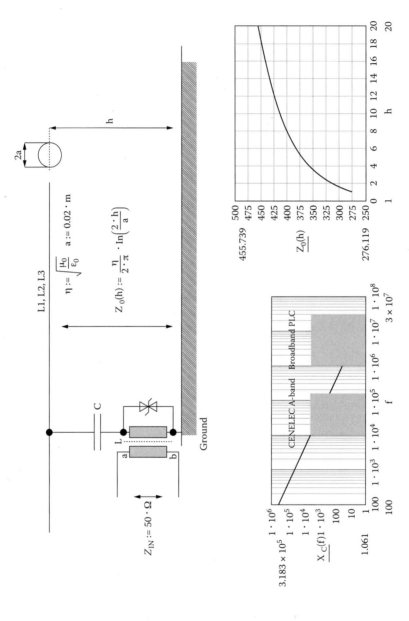

FIGURE 7.17 Typical characteristic impedance of overhead lines.

Communication Systems for Control and Automation

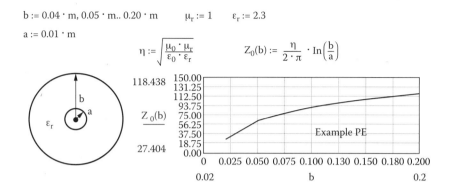

FIGURE 7.18 Typical characteristic impedance of cables.

	Paper/oil	PVC	PE	VPE	EPR
Insulation impedance	10^{15}	10^{11}–10^{14}	10^{17}	10^{16}	10^{15}
$\tan \delta * 10^{-3}$	3	10–100	0.1	0.5	2–3
ε_r	3.5	3–5	2.3	2.3–2.5	2.7–3.2

The mistermination can cause ripples on the input impedance trace of a uniform line, as well as impedance fault on the transmission line. The frequency of the ripple will be related to the distance down the line from the test end.

General Comments Regarding MV Cables. Every conductor penetrated by an alternating current is surrounded by a sinusoid electromagnetic field. Another conductor in this field sees, depending on whether we have to deal with an open circuit or a closed circuit, the induction of, respectively, a voltage or a current.

We have to distinguish between two different types of laying techniques. One is the laying of the cables in a triangular form and one is the laying of the cables in a plane (see Figure 7.19).

It is very common that the triangular type of cable is protected against mechanical stress via a metal armoring around all three cables. The armoring can be of two different types, a metal armoring not insulated against the surrounding soil and a metal armoring insulated via a plastic sheath against the surrounding soil.

Cables and wires often have to satisfy very different requirements throughout their route. Before deciding the type of cross section, therefore, one must examine their particular electrical function and also climatic and operational factors influencing the system reliability and the expected communication parameters. Figure 7.20 shows the construction of up-to-date high-voltage cable sufficient for cable screen communication.

In general, it can be said that every cable with an insulated cable screen is sufficient for this type of communication medium. If there are armored cables in

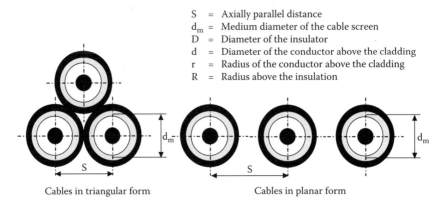

FIGURE 7.19 Cable deployment in the trench.

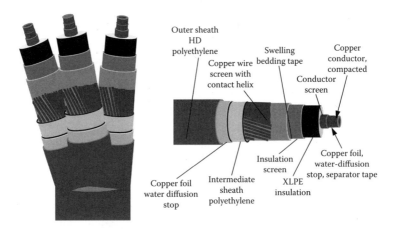

FIGURE 7.20 Example of a 3-core medium-voltage cable.

use, there should be no short circuits between the cable screen and the cable armoring behind the first 100 m after the signal injection point.

Methods of Signal Coupling. Coupling technologies for medium-voltage lines are a very crucial factor. They have to be cheap and small to fit into the substations. In general, it would be possible to use the same technologies as used for PLC systems; however, this would increase the cost per communication point in a tremendous way, and MV PLC would never be competitive to other technologies. So, when speaking about coupling technologies for MV system, we also always speak about a compromise.

There are three basic ways of coupling signals to MV cables:

- Capacitive via the core as shown in Figure 7.21
- Inductive via the screen (intrusive and nonintrusive)
- Inductive via the core

Communication Systems for Control and Automation 315

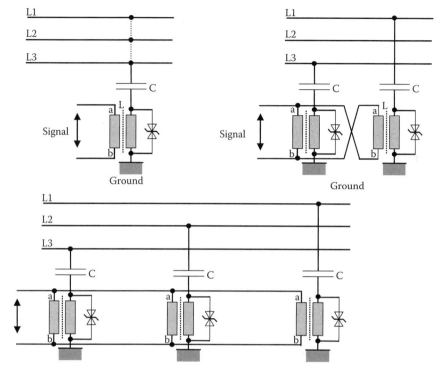

FIGURE 7.21 Capacitive coupling methods to MV lines.

The first method injects a voltage on the core, whereas the inductive coupling always injects a current to the line. Other approaches like directional coupling or transformer coupling to the core are either inefficient or too expensive.

Inductive coupling can be a very powerful method when one knows exactly what the power grid is like and if there also is a profound understanding of the real signal propagation paths. There are three major types of inductive coupling used for signal coupling.

Capacitive Coupling to the Core. Capacitive coupling is the most common approach. It can be used for overhead lines and cables. Cables are rather critical in the lower frequency band below 50 kHz due to the low characteristic impedance of about 25 ohm.

Figure 7.22 shows inductive coupling with intrusive connection to the screen where the signal flows through the insulated cable screen with return via the general mass of earth.

Figure 7.23 shows inductive coupling with nonintrusive connection to the screen where the signal flows through the insulated cable screen with return via the general mass of earth. When compared to intrusive coupling, nonintrusive provides a poorer communication path.

Figure 7.24 shows inductive coupling with nonintrusive connection to the core where the signal flows through the insulated cable core with return via the

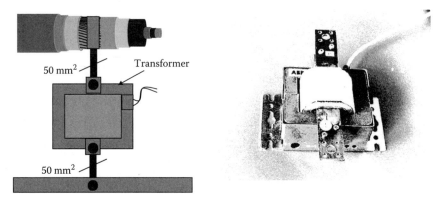

FIGURE 7.22 Intrusive coupling to the screen.

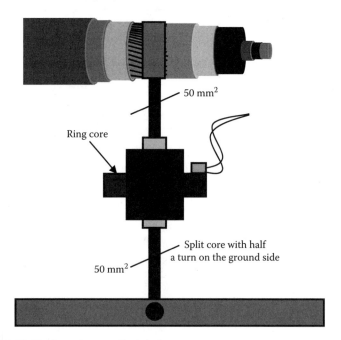

FIGURE 7.23 Nonintrusive coupling to the screen.

general mass of earth. This method works on all cables but has the big disadvantage that either the coupling is very weak or the coupler shows a strong dependency to the current in the conductor due to magnetic saturation of the core.

All inductive coupling methods have one big disadvantage in that they are load dependent, the worst being the nonintrusive coupler to the core. Beside the cable type, the method of grounding of the substation has a major influence to the screen. In Figure 7.24, an example of a short-driven vertical rod shows the HF dependency of such a grounding system.

Communication Systems for Control and Automation 317

FIGURE 7.24 Nonintrusive coupling to the core.

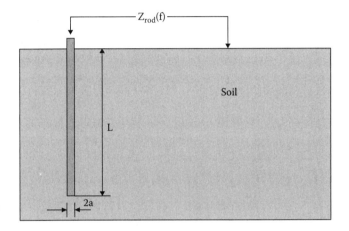

FIGURE 7.25 Illustration of a vertical ground rod of circular cross section.

The surge impedance of a single short rod of radius driven into the ground at depth of L, see Figure 7.25, is given in the equation below:

$$Z_{Rod}(f) = \frac{1}{2 \cdot \pi \cdot \sigma_g \cdot 1}\left(\ln\left(\frac{\sqrt{2} \cdot \delta_g(f)}{\gamma_0 \cdot a}\right) - j \cdot \frac{\pi}{4}\right)$$

$$\mu_0 = 4 \cdot \pi \cdot 10^{-7} \cdot \frac{H}{m}$$

$a = 0.03 \cdot m$ Groud rod radius ($1 \gg a$)

$\gamma_0 = 1.781$ Constant

$$\delta_g(f) := \frac{1}{\sqrt{\pi \cdot f \cdot \mu_0 \cdot \sigma_g}} \quad \text{Skin depth in the soil}$$

Depth in meter $(1 < \delta_g(f))$

$$\sigma_g := \frac{1}{100 \cdot \Omega \cdot m} \quad \text{Soil conductivity } \sigma_g > \omega(f) \cdot \varepsilon$$

The long term in the expression for impedance is usually of the order of 10, so that the surge impedance is predominantly resistive and relatively independent of frequency. This impedance formula is based on the transmission line model of the buried cable.

The equivalent circuit shown in Figure 7.26, despite some fluctuations in the value $Z_{\text{Ground Return}}$, is the value of $Z_{\text{Core Return}}$ that has the most influence on the transmitted signal. Simulations and measurements showed that between a high-resistive terminated cable (switch open) and a low-resistive terminated cable, the signal strength can vary in a range of up to 60 dB. But it is also obvious that in the case of a low $Z_{\text{Ground Return}}$ the variation of the value of $Z_{\text{Core Return}}$ does not matter.

It is difficult and requires much expertise to say in advance which will dominate; hence, a site survey with measurements is required. However, the simplicity and the cost of the inductive solution will justify these additional measurements.

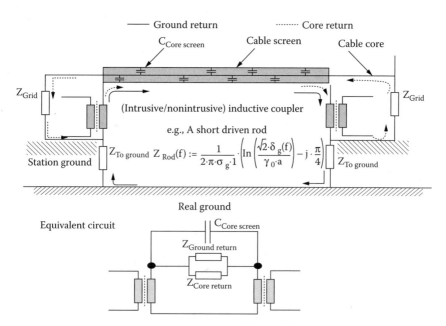

FIGURE 7.26 Signal propagation model for inductive couplers.

Communication Systems for Control and Automation

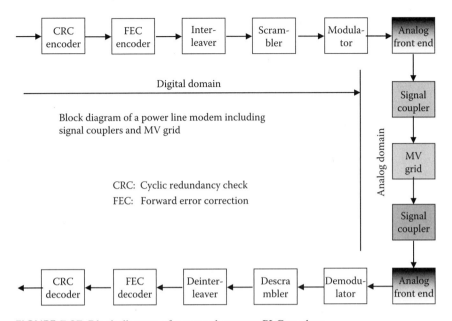

FIGURE 7.27 Block diagram of a general-purpose PLC modem.

Modulation and Coding for Power Line Carrier Systems. There is neither a common modulation nor a coding method applied to power line communications. All types of spread spectrum modulation as well as orthogonal frequency division multiplexing (OFDM) are very common. But also, more simple approaches are very successful. The goal is to set up a reliable communication system under all possible noise and attenuation conditions. Due to the fact that this is not a book about communication, the dedicated functions integrated in a state-of-the-art PLC modem will be explained based on Figure 7.27.

The goal of modulation is to transfer a bit stream into a waveform, which can be transmitted over a channel. PLC uses only bandpass types of modulation due to the fact that the lower frequencies are occupied by the power frequency and its harmonics. Baseband modulation means that all frequencies are used from 0 Hz to the maximum required frequency.

The carrier frequency can be modulated (changed) in three different ways:

1. The amplitude
2. The frequency
3. The phase

Modulation technologies are, in general, characterized by its bandwidth efficiency and its power efficiency. Also, whether we deal with a constant envelope or a nonconstant envelope type of modulation is important. Examples of constant envelope types of modulation are frequency shift keying (FSK), phase shift keying (PSK) and minimum shift keying (MSK). Examples of nonconstant envelope types are amplitude shift keying (ASK), quadrature amplitude modulation (QAM) and multicarrier modulation (MCM), e.g., orthogonal frequency division multiplexing or discrete multitone modulation (DMT).

Spread spectrum types of modulation are modulation types using a bandwidth much larger than the data rate. There are three main types used for PLC systems:

Frequency hopping	FHSS	We distinguish between fast hopping and slow hopping systems. A fast hopping system changes the frequency more than once per bit, a slow hopping one sends several bits per frequency.
Direct sequence	DSSS	DSSS systems mix the modulated signal a second time in a pseudo noise pattern. The result is the signal on the line looks like noise.
Chirp	Chirp	The base modulation is mixed with defined changing frequency. This technology is widely used for radar applications.

The advantage of all these systems is a high immunity against narrowband jammers and nonlinearities in the channel can be resolved. The big disadvantage is the complex receiver required and the long time for synchronization.

There are two major types of coding. One is the so-called source coding; the other is the channel coding. Source coding is used for data compressing and voice coding. Channel coding is used to improve the performance of noisy communication channels. Channel codes are again divided in forward error correction (FEC) codes and error detecting codes. Typical examples of FECs are block codes, convolutional codes, concatenated codes and Trellis codes. The Viterbi algorithm often decodes convolutional codes.

Most of the FEC algorithms require at least a doubling of the transferred data. This means, for a small amount of faulty data packages, an advanced repetition mechanism can be as efficient as the coding. Another problem is telegram prolongation due to coding on power line channels. The noise characteristic is very often of a form that longer telegrams have a severely lower chance to succeed than short telegrams.

Site Survey and System Evaluation. The objective of the site survey in the medium-voltage and low-voltage network is to evaluate the intrinsic communication parameters of the grid. More than a decade of the accumulated experience in making such surveys worldwide has provided sufficient evidence to support the claim that not every line or network is fit for either narrowband or broadband PLC applications.

It is true that it is possible to calculate the communication parameters of the line or the grid. However, the lack of data about the current state and condition of the line or network renders calculation-based decision about potential of the network to be used for the narrowband or broadband communication unreliable and risky. On the contrary, once undertaken, a survey of the current network gives sufficient and reliable evidence to support the findings of whether the line or the grid is fit to use for communication purposes.

Communication Systems for Control and Automation

A survey will also provide an indication as to the feasibility of conditioning the line or grid to improve its communication performance. The survey results, analysis and recommendation are unbiased and as such are independent from the technology provider. However, where the technology provider is selected in advance, a survey is conducted to identify the optimal use of the line, grid and associated devices and systems, taking into account the specifics of the technology intended for use.

The survey with its findings, analysis and recommendation provides to the company management an invaluable insight into the feasibility of the line or grid usage as the communication medium with the following additional benefits:

- Eliminates the need for expensive, open-ended experiments resulting from the lack of the evidence to make a qualified decision to enter into the full-scale system installation. There are several identified cases where a survey has not been done prior to the system installation and where the suitability of the network or some of its components has been identified during the contract execution.
- Eliminates the need to make several small-scale installations to gain the experience and evidence to select the communication technology provider.
- Provides hard evidence for the utility to make the right choice in selecting the communication technology appropriate to satisfy its needs to communicate for effective and cost-efficient network management.
- Establishes the potential of the line or the network to be offered to a third party for the use in applications requiring broadband communication.
- Generates an opportunity for the network operator to set up the business case and accurate investment planning based on hard evidence.
- Enables an unbiased selection of the most appropriate communication technology provider.

Required Data for the Preparation of a Survey. To execute the line or the network survey, one should differentiate the specifics of the line and the network. For low-voltage mains, only basic preparation work is required:

1. Accurate network documentation
2. Guaranteed undisturbed access to the selected measurement points

Surveys in the medium-voltage grid are more challenging. It is required that the network operator prepares the network for the survey and provides full support to the survey team such as

1. Providing an up-to-date single-line diagram for the network under consideration, including:
   a. Identified overhead line segments and cable segments

b. Length of each line segments
   c. Identified cable joints if known
   d. Cable type used with the cross section details
   e. Location of normally open points in the distribution network
   f. Voltage level
   g. Type of the network neutral point earthing:
      i. Solid
      ii. Over earthing resistor and the value of the resistor
      iii. Compensated
      iv. Combined from the above with the details of the control scheme and its time settings
   h. Substation and switchgear layout and the details of the switchgear earthing and cable sheet earthing
2. Installation of the coupling devices at the measurement points in the network that will be selected and supplied by the surveillance company
3. Unrestricted and undisturbed access to the measurement points during the measurement period
4. Person trained and authorized to perform device switching operations and network reconfiguration
5. Transport to the site and to the agreed place of accommodation
6. Fault isolation switching strategy for the network under consideration

Required Measurements. The measurements required for a site survey are different in accordance with the coupling method used. Also, the network topology and the application of the communication network are very crucial for the effort required at this very early stage. The following types of measurements can be performed on power lines:

- Attenuation
- Signal-to-noise ratio (SNR)
- Noise (joint time frequency domain)
- Impulse response
- Group delay
- Impedance

Basic attenuation measurements are necessary to learn about the general PLC communication performance. It is highly recommended to check important automation points under all normally used network conditions. When using inductive couplers the so-called three-point check is essential, as shown in Figure 7.28 through Figure 7.30.

The idea behind the three-point measurement is the evaluation of the screen/ground-based communication quality. The result of the measurements can be used as a good reference for further DLC system planning when using intrusive inductive couplers.

Communication Systems for Control and Automation

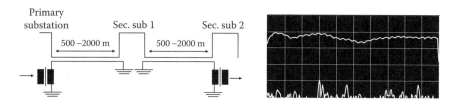

FIGURE 7.28 Three-point measurement 1 + measured transfer function.

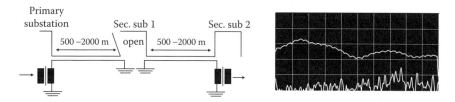

FIGURE 7.29 Three-point measurement 2 + measured transfer function.

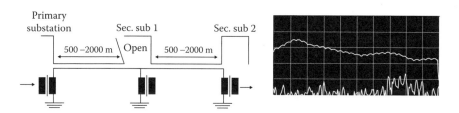

FIGURE 7.30 Three-point measurement 3 + measured transfer function.

The expected results are

Measurement	Result
1	Sufficient communication quality SNR > 30 dB
2	Attenuation of the signal at least by 20 dB compared to measurement 1
3	Increase of the signal compared to measurement 2

It can be assumed that the communication parameters within a dedicated area and the same type of cable show similar behavior. The measurement should be repeated for dedicated cable types if the outer insulation is different, or for dedicated areas if soil changes, e.g., from a water flood area towards a granite mountain.

Tools for Site Measurements. For PLC the measurement of the suitability of the circuit for communication is important because most of the time the channels are time variant and undefined. The proper measurement tools normally provide the following functions:

- Signal attenuation/transfer function
- SNR
- Impulse response
- Relative group delay
- Noise measurement (either in the frequency domain or in the joint time frequency domain)
- Impedance measurement

Such tools are available for measurements between 10 and 150 kHz as well as for the broadband range between 1 and 30 MHz.

Basics of the PLC Measurement Tool. For our application, it is desired to process and analyze the signal in the frequency domain. In the analog world, this can be easily accomplished by using a spectrum analyzer. Mathematically, this process can be duplicated by taking a Fourier transform of a continuous-time analogue signal. The Fourier transform yields the spectral content of the analog signal.

The output of an analog to digital converter (ADC) provides discrete quantified samples of the continuous input x(t). The discrete Fourier transform (DFT) transforms the discrete input time domain samples in the discrete frequency domain samples. If x(n) is a sequence of N input data samples, then the DFT produces a sequence of N samples x(k) spaced equally in frequency, for example:

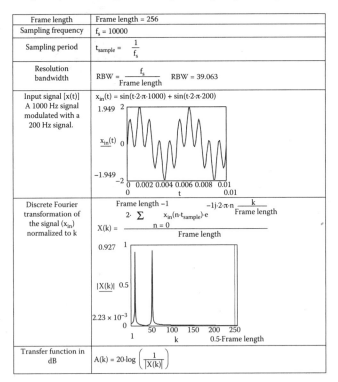

Communication Systems for Control and Automation

The DFT can be viewed as a correlation or comparison of the input signal to many sinusoids evaluating the frequency content from the input signal. For example, a 1024-point DFT requires 1024 input samples from the sinusoid signal and 1024 points from a sinusoid. Sinusoids of 1024 different frequencies equally spaced from $-f_s/2$ to $+f_s/2$ are used.

Each pass of the DFT checks the sinusoid against the input signal to see how much of the frequency is present in the input signal. This is repeated for each of the 1024 frequencies. In other words, the DFT is a cross-correlation between the input signal x(t) and 1024 internally stored sinusoid signals.

The measurement tool is based on a data acquisition card sampling the input signal with maximum 500 kHz. The sampled values are collected to a bundle of values (frames). The number of values used is the frame length in samples. The sampling frequency divided by the frame length gives the resolution bandwidth. The standard set-up is 500,000 samples and a frame length of 1024 samples results in a resolution bandwidth (RBW) of 488 Hz.

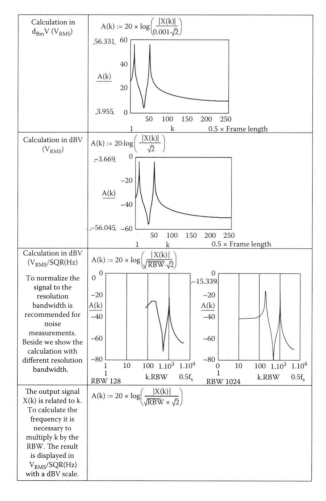

Continued.

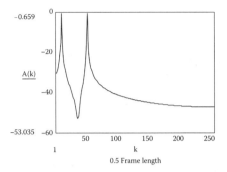

The plot of transmitted signal versus noise can be measured according to Figure 7.31.

The DLC measurement tool injects a signal between 10 kHz and 110 kHz. Between every consecutive sweep period is an idle time of 3 seconds. A typical measurement is shown in Figure 7.32. The top curve is the maximum received signal level; the lower curve is the current noise level. The injected signal level is in the average +20 dBV (peak voltage).

The resolution bandwidth of the above measurement is 488.28 Hz. It is defined by the sampling frequency divided by the frame length (500K samples/ 1024 = 488.28 Hz).

When measuring the impulse response of a line we use a pseudo noise signal on the signal injection point and a data acquisition tool at the ejection point.

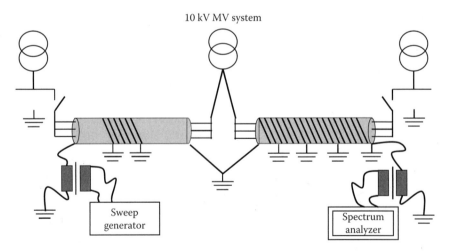

FIGURE 7.31 Measurement of the signal transfer function.

Communication Systems for Control and Automation

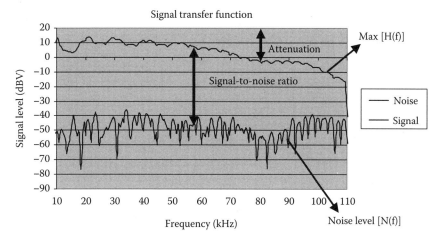

FIGURE 7.32 Typical signal-versus-noise measurement (SNR).

- Use digital pseudo noise PN sequence as stimulus
  - Advantage: crest factor = 1 means low amplitude for high RMS
  - Compared to a real dirac pulse with crest factor = infinite

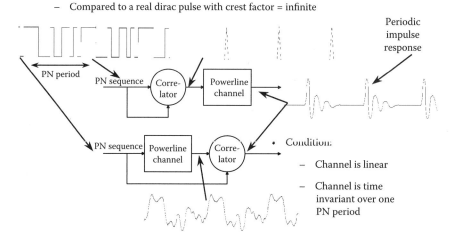

FIGURE 7.33 Measurement of line input response.

The impulse response measurement is an optional measurement only required if the SNR is very good, but the bit error rate (BER) is still very high.

The time-scale in the measurement in Figure 7.34 is 2 µs. It means that the total impulse delay between the signal injection and the signal ejection point is

$$50 \cdot 2 \cdot \mu s = 1 \cdot 10^{-4} \cdot s$$

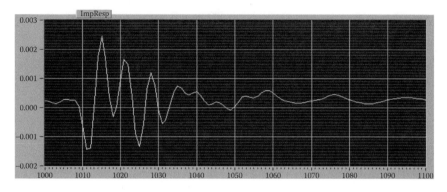

FIGURE 7.34 Impulse measurement result.

The group delay is a relative figure defining the relative propagation velocities of the different frequencies over the communication channel. The group delay can be calculated via the impulse response and looks like Figure 7.35.

The intention of the impedance measurement is to measure the source impedance at the signal injection point (Figure 7.36).

The noise measurement is performed in the joint time frequency domain (JTFD). The JTFA is, according to our experience, one of the most powerful tools for noise analysis.

Short Introduction to Power Line Noise. We need to be concerned about three different types of noise on the power line:

- Continuous-wave interference
- Impulsive noise
- White noise

Impulsive noise results from switching of inductive loads and can produce impulses saturating any receiver for periods of tens or even hundreds of microseconds. These impulses have very rapid rise times and are virtually impossible to filter out completely. The impulses are generally periodic with twice the power line ground frequency, 100 or 120 Hz, and many occur during each half cycle, due to the switching of various loads.

Worse yet, these impulses are capable of ringing the power line itself. Because the network and its attached loads possess both inductance and capacitance, they may resonate at a frequency that depends on the instantaneous load, producing decaying wave forms lasting several cycles at frequencies in the communications band. To the modem, this looks like continuous wave jamming at frequencies that cannot be predicted in advance because they vary with load conditions.

Finally, white noise can be an issue, particularly when operating at high levels of receiver gain that can be required (Figure 7.37 and Figure 7.38).

Low-Voltage Power Line Carrier. Coupling to LV networks is much simpler than for MV, and LV DLC is used predominantly for remote meter reading or

Communication Systems for Control and Automation 329

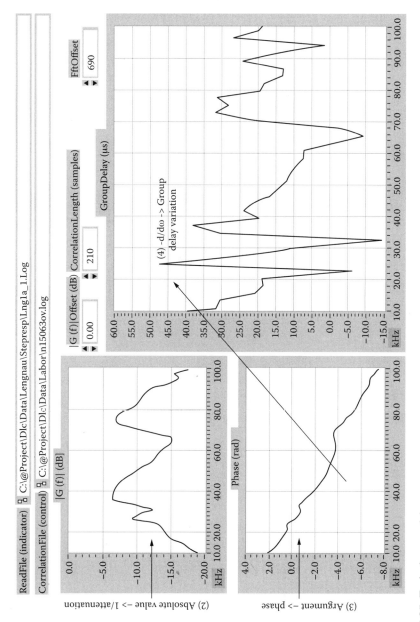

FIGURE 7.35 Typical group delay.

330 Control and Automation of Electric Power Distribution Systems

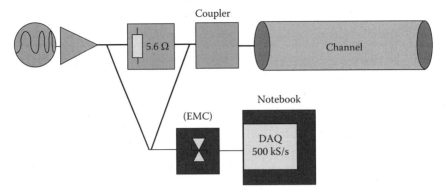

FIGURE 7.36 Setup for the impedance measurement (10 to 110 kHz).

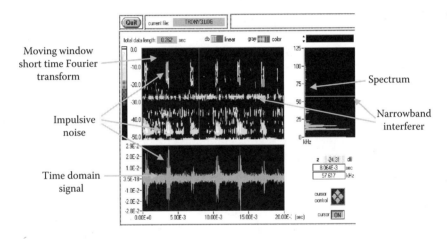

FIGURE 7.37 Interpretation of the JTFA measurement window.

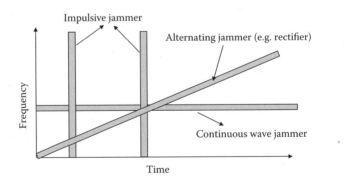

FIGURE 7.38 Interpretation of a JTFA-plot (white noise covers the overall square).

Communication Systems for Control and Automation 331

service disconnection. Meter reading requirements can be achieved with one-way communication with relatively low data transmission rates over an extended time period. Signals are transmitted from the meter to a collection point, usually the MV/LV transformer, where continuation of the communication may change media to a higher capacity wire or wireless system. In some cases, MV DLC is used if installed on the source side of the distribution transformer.

7.5.4 Summary of Communications Options

Type	Advantages	Disadvantages
Wire-Based Systems		
Telephone		
PSTN (analog)	Availability of network	No real-time applications
	Voice and data communications	Reliability depends on PSTN condition
		High operating costs
PSTN (digital)	2^8 subscribers per connection	No real-time applications
		High operating costs
Low-cost fiber optics	Well protected against electromagnetic interference	Unit price installation costs
	High data rates	
DLC	Linked to distribution network, which is owned by the customer	One to one with electric network
	Very small operating costs	Voice communications not generally supported
		Installation (capacitive) requires a de-energized state.
Cable TV	High data rate	Coverage and availability of system
	High load capacity	
Wireless		
Radio		
Conventional radio systems	Only very small infrastructure required	Only suitable for small SCADA systems
	Cheap systems available	Only for rural areas, line of sight
	Adaptable to existing protocols	Usually no integrated interfaces
	Simple frequency management	
	Suitable for real-time applications (short response times)	Robustness against jamming
	Owned by electric utilities	Only polling mode possible
	Flexible deployment	Availability of a frequency in which to operate

Type	Advantages	Disadvantages
Trunked radio	Already installed network available	Limited capacity for real-time data communications
	Cheaper than mobile phone usage	Infrastructure costs if not established for other applications
	Usually owned by electric company	
Packaged data radio network	Availability of the connection, good coverage	Limited real-time capabilities
	High capacity	Only data communications
	Optimized for high data throughput	Very high infrastructure costs
		High operating costs (system dependent)
Low-cost packet data radio	Low costs, no infrastructure required	Very limited data throughput
	Owned by the electricity subscriber	Only data communication
		No standards available
Cellular		
Cellular mobile phone networks	Good coverage in urban areas and along populated transportation routes	No real-time applications
		High operating costs
		Possible data overload by other users
Pagers		
Two-way pager	Cheap	Time delay
	Easy to install	Availability
		Optimized for short data messages
Point-to-multipoint terrestrial microwave systems	System owned by electric company	High installation costs
	High data rate	Only line-of-sight coverage
	Real-time communications (short response time)	
	DCEs compatible to many subsystems	
Satellite		
High-orbit geostationary satellite microwave systems	Coverage	High installation costs
	Customer has own HUB	Relatively large antennas
	High data rates	
Low-orbit satellite telephone system	Coverage	
	Easy installation	No real-time applications

Communication Systems for Control and Automation

Type	Advantages	Disadvantages
	Small antennas	Not fully operational globally
		Users very dependent on the service provider
		High operating costs
Low-orbit satellite telephone system	Easy installation	
	Small antenna	Small data rates
	Cheapest satellite system	Only data communications

7.6 DISTRIBUTION AUTOMATION COMMUNICATIONS PROTOCOLS

Communications protocols define the rules and regulations for the data transmission between communication devices. In simple words, a communication protocol would be the "language" employed between the transmitter and the receiver. Figure 7.39 shows the communication protocols to be described in this section. The origins of each communication protocol are included, as well as their basic structure.

7.6.1 MODBUS

Modicon, a subsidiary of Schneider Automation™, created MODBUS® in 1978 due to its necessity of transferring control signals between programmable logic controllers. The MODBUS protocol defines a message structure that the controllers will recognize and use, regardless of the networks in which they communicate, i.e., the communication media. It describes the process a controller uses to request access to another device, how it will respond to requests from the other devices, and how errors will be detected. MODBUS establishes a common format for the layout and contents of the message field.

Basic Structure, Layers. The MODBUS protocol is used to establish master-slave communication between intelligent devices, and it has two types of serial transmission modes, ASCII and RTU. For ASCII serial transmission mode, each 8-bit byte in a message is sent as two ASCII parameters, where as for RTU each 8-bit byte in a message is sent as two 4-bit hexadecimal characters. Different advantages and drawbacks are obtained by each serial transmission mode. One advantage of the RTU mode is that it has a bigger character density, and therefore,

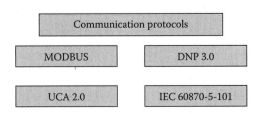

FIGURE 7.39 Distribution automation communication protocols.

Start	Address	Function	Data	LRC check	End
1 Char (:)	2 Chars	2 Chars	n Chars	2 Chars	2 Chars (CRLF)

FIGURE 7.40 ASCII message frame.

Start	Address	Function	Data	CRC check	End
T1-T2-T3-T4	8 bits	8 bits	n × 8 bits	16 bits	T1-T2-T3-T4

FIGURE 7.41 RTU message frame.

it can transmit more information for the same baud rate. ASCII's advantage is that it allows time intervals of up to 1 second to occur without causing an error.

Basic Structure, Framing. Figure 7.40 shows the message frame structure of the MODBUS protocol for ASCII serial transmission mode.

ASCII frames start with a colon (:) character (ASCII 3A hex) and ends with a carry return-line feed (CRLF) pair (ASCII 0D and 0A hex). The permitted characters for all other fields are hexadecimal 0...9, A...F. The devices connected in the network that communicates using MODBUS in ASCII transmission mode monitor the network bus, continuously looking for the colon character. When a colon is received, each device decodes the next field, which is the address field, in order to determine if this field contains its own address and then take appropriate actions. Within a message, time intervals up to 1 second can elapse between characters. If greater intervals occur, then the receiving device assumes that an error has taken place.

Figure 7.41 shows the message frame structure of the MODBUS protocol for RTU serial transmission mode.

RTU frames start with a silent interval of at least 3.5 character times. The permitted characters for all other fields are hexadecimal 0...9, A...F. The devices connected in the network that communicates using MODBUS in RTU transmission mode monitor the network bus continuously, including during silent intervals. When the first field arrives, which is the address field, each device decodes it in order to determine if this field contains its own address and then take appropriate actions. Following the last transmitted character, a similar interval of at least 3.5 character times marks the end of the message.

The address field contains two characters (ASCII) or 8 bits (RTU). The valid addresses for the slave devices are in the range of 0...247 decimal. Because 0 is used for broadcast addresses that all slave devices recognize, the slave devices are addressed in the range 1...247. A master device addresses a slave device by placing its address on the message's address field. The slave devices respond to the master device by placing its own address in the address field in order to let the master know which slave is responding.

The function field contains two characters (ASCII) or 8 bits (RTU). Valid codes are in the range of 1...255 decimal. When a message is sent from a master to a slave device, the function code field tells the slave what kind of action to perform. These actions may take the form of reading ON/OFF states, reading the

Communication Systems for Control and Automation 335

data contents of a certain register, writing to designated coils or registers, and so on. When a slave responds to the master, it uses the function code to indicate either a normal (error-free) response or that some kind of error occurred (called an exception response). For normal responses, the slave simply echoes the original function code, whereas for exception responses the slave returns a function code that is equivalent to the original with its most significant bit set to a logic 1. For exception responses, the slave data field contains a unique code that tells the master what kind of error occurred.

The data field is constructed using sets of two hexadecimal digits in the range of 00 to FF hexadecimal. This can be made from a pair of ASCII characters or from one RTU character. In the messages sent from a master to the slave devices, the data field contains additional information that the slaves must use to take the action defined by the function code. This can include items like discrete and register addresses, the quantity of items to be handled, and the count of actual data bytes in the field. For normal responses, the data field of the message sent from the slave to the master contains the data requested. As mentioned in the previous paragraph, for exception responses, the slave data field contains a unique code that the master can use to determine the next action. In some cases, the data field can be nonexistent or have zero length due to the simplicity of the requested action in which the function field alone provides all the information.

The error checking field method depends on the MODBUS serial transmission mode in question. For ASCII, the error checking field contains two ASCII characters. The error checking characters are the result of a longitudinal redundancy check (LRC) calculation that is performed on the message contents, exclusive the beginning colon to the terminating CRLF characters. The LRC characters are appended to the message as the last field preceding the CRLF characters. For RTU, the error checking field contains a 16-bit value implemented as two 8-bit bytes. The error-checking field is the result of a cyclical redundancy check (CRC) evaluation that is performed on the message contents. The CRC field is appended to the message in the last field.

Basic Structure, Checksum. Serial networks having the MODBUS protocol use two kind of error checking, parity checking and frame checking, in order to detect transmission errors. Parity checking (even or odd) can be optionally applied to each character of the message. Frame checking (LRC or CRC) is applied to the entire message. Both the character check and the message frame check are generated in the master device and applied to the master contents before transmission takes place.

The user implements parity checking with options such as even parity checking, odd parity checking or no parity checking at all. If either even or odd parity checking are selected, the 1 bits are counted in the data portion of each character. A parity bit will be transmitted in order to permit the slave device to check for any transmission error. The parity bit will be set to a 0 or 1 to result in an even or odd total of 1 bits. Before the message is transmitted, the parity bit is calculated and applied to the frame of each character. The receiving device counts the quantity of 1 bits and compares its result with the number attached to the frame.

An error is set if the numbers are not the same; however, the parity checking can only detect and set an error if an odd number of bits are picked up or dropped in a character frame.

As it was seen in Figure 7.40, the messages in ASCII mode include an error-checking field based on an LRC method. The LRC field is 1 byte and contains an 8-bit binary value, and it is applied regardless of any parity checks. The transmission device calculates the 8-bit binary number and attaches it to the message. The receiver device receives the message, calculates the LRC and compares its calculation to the 8-bit value that it received into the LRC field. If the values are not equal, a transmission error results. The LRC is calculated by adding together successive 8-bit bytes of the message, discarding any carries and two's complementing the result. LRC calculation excludes the colon character and the CRLF pair at the end of the message.

As it was seen in Figure 7.41, the messages in RTU mode include an error-checking field based on a CRC method. The CRC field is 2-byte, contains a 16-bit binary value and is applied regardless of any parity checks. The transmission device calculates the 16-bit binary number and attaches it to the message. The receiver device receives the message, calculates the CRC and compares its calculation to the 16-bit value that it received into the CRC field. If the values are not equal, a transmission error results.

The CRC first step is the preloading of a 16-bit register to all 1s. The process then starts applying successive 8-bit bytes of the message to the register contents. The generation of the CRC includes all 8-bit bytes of each character; the start bit, the stop bit and the parity bit are excluded in the CRC calculation. During the CRC generation, each 8-bit character is added using the exclusive OR function (EXOR) to the register contents. Then, the result is shifted in the direction of the least significant bit (LSB) with a zero filled into the most significant bit (MSB) position. The LSB is extracted and examined. If the LSB was a 1, then the register is exclusive ORed with a preset, fixed value. If the LSB was a 0, no exclusive OR takes place. This process is repeated until eight shifts have been completed. After the eighth shift, the next 8-bit byte is exclusive ORed with the register current value, and the process repeats for eight more shifts as previously described. The CRC value is the final contents of the register, after all the bytes of the message have been applied. When the CRC value is appended to the message, the low-order byte is appended first, followed by the high-order byte.

Basic Structure, Function Codes. Figure 7.42 shows the function codes that are supported by MODBUS. The codes are listed in decimal format.

7.6.2 DNP 3.0

DNP 3.0 provides the rules for substation computers and master station computers to communicate data and control commands. Westronic, Inc., now GE Harris, created distributed network protocol, DNP, in 1990. DNP protocol was then changed from a proprietary protocol to a public domain protocol, and in 1993 the DNP 3.0 Basic 4 protocol specification document was released to the general

Communication Systems for Control and Automation 337

Function code	Name
01	Read coil status
02	Read input status
03	Read holding register
04	Read input register
05	Force single coil
06	Preset single register
07	Read exception status
08	Diagnostics
09	Program 484
10	Poll 484
11	Fetch comm event counter
12	Fetch comm event log
13	Program controller
14	Poll controller
15	Force multiple coils
16	Preset multiple registers
17	Report slave ID
18	Program 884/M84
19	Reset communication link
20	Read general reference
21	Write general reference
22	Mask write 4x register
23	Read/write 4x registers
24	Read FIFO queue

FIGURE 7.42 MODBUS function codes.

public. In November 1993, the DNP Users Group, a group composed of utilities and vendors, acquired the ownership of the protocol. In 1995, the DNP Technical Committee was formed, having the responsibility to recommend specification changes and further developments to the DNP Users Group.

Basic Structure, Layers. DNP is a layered protocol and consists of three layers and one pseudo layer. The International Electrotechnical Commission (IEC) denominates the layering structure as Enhanced Performance Architecture (EPA). Figure 7.43 shows the context of EPA applied to the DNP 3.0.

The Application Layer responds to complete messages received from the Pseudo Transport Layer and builds messages based on the need of the user data. The built messages are then passed to the Pseudo Transport Layer in which they are fragmented. The Data Link Layer receives the fragmented messages from the Pseudo Transport Layer and send them down to the Physical Layer, where finally the messages are sent. When the amount of data to be transmitted is too big for a single Application Layer message, multiple Application Layer messages may be built and transmitted sequentially. These messages are independent of each other, and there is an indication in all messages, except of the last one, that more messages

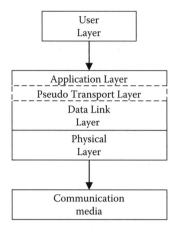

FIGURE 7.43 DNP layer architecture.

are on the way. For multiple Application Layer messages, each particular message is called a fragment; therefore, a message may be either a single-fragment message or multifragment message. It would be important to mention that the Application Layer fragments from master devices are typically requests, whereas in the case of slave devices they are typically responses to those requests. A slave device may also send a message without a request, which is called an unsolicited response.

The Pseudo Transport Layer, as mentioned before, segments the Application Layer messages into multiple and smaller frames for the Link Layer to transmit or, when receiving the Pseudo Transport Layer reassemble the frames into longer messages to be received by the Application Layer. For each frame, it inserts a single-byte function code that indicates if the Data Link Frame is the first frame of the message, the last frame of the message or both (for single-frame messages).

The Data Link Layer has the responsibility of making the physical link reliable. The Data link layer incorporates error detection and duplicate frame detection in order to increase the data transmission reliability. The Data Link Layer sends and receives packets, which are called frames. The maximum size of a data link frame is 256 bytes.

The Physical Layer is primarily concerned with the physical media over which the DNP protocol is being communicated. This layer handles states of the media such as clear or busy, and synchronization across the media, such as starting and stopping, among others. The DNP protocol is normally implemented over simple serial physical layers such as RS-232 or RS-485.

Basic Structure, Framing. A frame is a portion of a complete message communicated over the Physical Layer, and its structure may be divided in header and data segments, as shown in Figure 7.44.

Header	Data

FIGURE 7.44 DNP frame.

Communication Systems for Control and Automation

| Sync | Length | Link control | Destination address | Source address | CRC |

FIGURE 7.45 DNP header segment.

| Block 1 | | | Block n | |
| User data | CRC | ... Other blocks ... | User data | CRC |

FIGURE 7.46 DNP data segment.

Figure 7.45 shows the header segment, which contains important information such as the frame size, the master and remote device addresses, and data link control information. From Figure 7.45, the header subsegments will be briefly described.

Every header starts with two sync bytes or starting octets that help the remote receivers determine where the frame begins. The length specifies the number of octets remaining in the frame, which does not include the CRC error checking octets. The minimum value for the length is 5 octets, and the maximum value is 255. The link control octet, or frame control octet, is used between sending and receiving link layers in order to coordinate their activities.

The destination address and source address, as their names indicate, refer to the device that should process the data and the device that sent the data, respectively. DNP allows up to 65,520 individual addresses, and every DNP device should have a unique address attached to it. Three DNP addresses are reserved for particular applications such as "all-call-message" in which the frame should be processed by all devices. Destination and address fields are sized for 2 octets, in which the first octet is the least significant bit and the second one is the most significant bit. The CRC is a 2-octet field that helps in the cyclic redundancy check tasks.

The Data segment, shown in Figure 7.46, is commonly referred to as the payload and it contains the data coming from the DNP previous layers. User data fields contain 16 octets of user-defined data except the last block of a frame, which contains 1 to 16 octets as required. A pair of CRC octets are included, each 16 octets of data in order to provide a high degree of assurance that transmission errors can be detected. The maximum number of octets in the data payload is 250, not including the CRC octets.

Basic Structure, Objects. DNP employs objects in order to let the slave devices know what kind of information is required. Before describing the objects defined in DNP, some definitions will be presented.

In DNP 3.0, the term *static* is used with data and it is referred to as the current value. Then, static binary input data refers to the present ON or OFF state of a bi-state device. Static analog input data contains the value of an analog signal in the exact moment it is transmitted. The term *event* in DNP 3.0 is related to something significant happening. An event occurs when a binary input changes from ON to OFF, or an analog value changes by more than its configured deadband limit.

DNP may present the data in different formats. Static, current value, analog data can be represented by variation numbers as follows:

1. A 32-bit integer value with flag
2. A 16-bit integer value with flag
3. A 32-bit integer value
4. A 16-bit integer value
5. A 32-bit floating-point value with flag
6. A 64-bit floating-point value with flag

The flag is an octet that contains information such as whether the source is on-line, the value contains a restart value, communications are lost with the source, the data is forced and the value is over range.

Event analog data can be represented by these variations:

1. A 32-bit integer value with flag
2. A 16-bit integer value with flag
3. A 32-bit integer value with flag and event time
4. A 16-bit integer value with flag and event time
5. A 32-bit floating-point value with flag
6. A 64-bit floating-point value with flag
7. A 32-bit floating-point value with flag and event time
8. A 64-bit floating-point value with flag and event time

From the diverse formats in which the static and analog data are presented, it can be seen that if both the static analog and analog event data use the same data format, the user cannot distinguish which is which. DNP implements the object numbers in order to avoid confusion, then static analog values are assigned to object number 30 and event analog values are referred to as object number 32. Having this implementation, static analog data can be presented in one out of six formats, and event analog values may be presented in one out of eight formats. All valid data formats in DNP 3.0 are identified by the object number and variation number. DNP 3.0 object groups for SCADA/DA implementations are shown in Figure 7.47.

For each object group, one or more data points exist. A data point is a single data value of the type specified by its object group.

Basic Structure, Function Codes. Application Layer fragments begin with an Application Layer header followed by one or more object header/object data combinations. This Application Layer header is subdivided into application control code and application function code. The application control code contains information regarding:

- Whether the fragment is single-fragment or multifragment
- Whether the application fragment confirmation is requested
- Whether the fragment was unsolicited

and contains a rolling Application Layer rolling number that helps the detection of missing or out-of-sequence fragments.

Communication Systems for Control and Automation

Object group	Object representation	Object number range
Binary input	Binary (status or boolean) input information	1–9
Binary output	Binary output or relay control information	10–19
Counters	Counters	20–29
Analog input	Analog input information	30–39
Analog output	Analog output information	40–49
Time	Time in absolute or relative form in any resolution	50–59
Class	Data classes or data priority	60–69
Files	Files of file system	70–79
Devices	Devices (rather than point) information	80–89
Applications	Software applications or operating system processes	90–99
Alternate numeric	Alternate or custom numeric representations	100–109
Future expansion	Future or custom expansion	110–254
Reserved	Permanently reserved	0 and 255

FIGURE 7.47 DNP object groups for SCADA/DA applications.

The Application Layer function code indicates the actual purpose of the message, i.e., what the slave device should do. DNP 3.0 only allows a single requested operation per message, and the function code applies to all objects included.

The function codes available in DNP 3.0 are shown in Figure 7.48.

For Figure 7.48, function codes 3, 4, and 5, select, operate, and direct operate, respectively, the applicable relay operation code fields are shown in Figure 7.49.

The Application Layer object header contains the required information to specify an object group, a variation within the object group, and a range of data points within the object variation. A request message fragment only contains as object headers the object groups, variations, and point ranges that are requested to read. A read response fragment message would contain, in addition to the object header, the requested object data.

Basic Structure, Report by Exception. For each object group, there are data points that contain change data. Change data is referred to as only those data points that have changed within a particular object group. For example, if object group number 1 represents binary inputs, and object group 2 represents binary input change data, when a data point in group 1 changes, a change event for the same data point is created for object group 2. The reports in which only the changed data are included are called report by exception, or RBE, in DNP 3.0.

DNP classifies object groups and data points within them into classes 0, 1, 2, and 3. Class 0 represents all static, not changed, event data. Classes 1, 2, and 3 represent different priorities of change event data. For each change data point, a time can be associated with the change, and each detection of a data value that changes is considered a change event. DNP 3.0 defines scanning for class data to the request actions that involve different change event data with different classes.

Function code	Function name
0	Confirm
1	Read
2	Write
3	Select
4	Operate
5	Direct operate
6	Direct operate — no acknowledgment
7	Immediate freeze
8	Immediate freeze — no acknowledgment
9	Freeze and clear
10	Freeze and clear — no acknowledgment
11	Freeze with time
12	Freeze with time — no acknowledgment
13	Cold start
14	Warm start
15	Initialize data
16	Initialize application
17	Start application
18	Stop application
19	Save configuration
20	Enable spontaneous messages
21	Disable spontaneous messages
22	Assign classes
23	Delay measurement

FIGURE 7.48 DNP function codes.

Code	Indication
1x	Pulse on
3x	Latch on
4x	Latch off
81x	Trip (pulse)
41x	Close (pulse)

FIGURE 7.49 Relay operation code fields.

7.6.3 IEC 60870-5-101

The IEC 60870-5 is a general protocol definition developed by the International Electrotechnical Commission Technical Committee 57. The IEC 60870-5 is a series of standard documents that consists of base standard sections and companion standards. The companion standards are integrated by selections of sections taken from the base standards in order to acquire a specific configuration. This section will briefly describe the IEC 60870-5-101 profile, which is a messaging structure for RTU-IED communication.

Communication Systems for Control and Automation 343

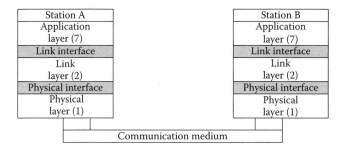

FIGURE 7.50 IEC layer architecture.

Basic Structure, Layers. The basic reference model contains seven layers. However, the simplified reference model, which is used in the IEC 60870-5-101 profile, has fewer layers, and it is called the enhanced performance architecture (EPA) model. Central stations as well as outstations perform their local application tasks called application processes, and the communication between the central and outstations is based on the communication protocol.

Figure 7.50 shows the EPA model applied to the communication process between two stations. Based on this figure, the communication process between station A and station B starts with the acceptance of the application data at the top of the station A layer stack. The application data pass down through all layers, collecting the required data needed to control the working of the protocol, until they emerge at the bottom. The message is sent at the bottom part of station A through the communication media and it is received by the bottom part of station B. The message now passes up all layers, and by doing that, all control data are dropped until the original application data are received by the top layer and passed to the application processes in station B.

Basic Structure, Framing. Figure 7.51 shows the IEC messaging structure derived from the layering arrangement. All data fields shown in the figure consist of octet strings of 1 or more octets. The application service data unit (ASDU) is a block of data being sent from the application processes in one station to the application processes in another station. For the IEC 60870-5-101, the ASDU is equal to the application protocol data unit (APDU), because no application protocol control information (APCI) is added.

The link layer adds its own link protocol control information (LPCI) to the APDU to form the link protocol data unit (LPDU). The LPDU is transmitted as

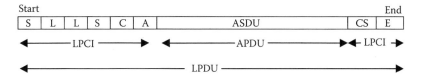

FIGURE 7.51 IEC messaging structure.

a contiguous frame with no idle line or gaps between asynchronous characters. The IEC 60870-5-101 frame may be divided into a header and a body. The header consists of the S+L+L+S characters, and the body contains the remaining characters. The LPCI is formed by

$$LPCI = S + L + L + S + C + A + CS + E$$

where S = Start character with a fixed defined bit pattern, L = Length character, specifying the length in octets of the ASDU + C + A, C = The link control character, A = The link address field, CS = The check sum character, and E = End character with a fixed defined bit pattern.

Basic Structure, Application Layer Provisions. The IEC 60870-5-101 defines two sets of provisions for the application protocol, application functions and application service data units.

Among the application functions are the following:

- Station initialization
- Data acquisition by polling
- Cyclic data transmission
- Acquisition of events
- General interrogation
- Clock synchronization
- Command transmission
- Transmission of integrated totals
- Parameter loading
- Test procedure
- File transfer
- Acquisition for transmission time delay

The general arrangement for the different types of ASDU suitable for the application is shown in Figure 7.52. The legend for each data field is as follows:

T = Type identification (1 data octet).
Q = Variable structure qualifier (1 data octet). Indicates the number of information objects or information elements.
C = Cause of transmission (1 or 2 data octets). Causes include periodic/cyclic, spontaneous, request/requested, activation, etc.
CA = Common address (1 or 2 data octets). Distinguishes the station address/station sector address.

FIGURE 7.52 IEC 60870-5-101, application service data units structure.

Communication Systems for Control and Automation 345

OA = Information object address (1, 2 or 3 data octets).
IE = Set of information elements.
TT = Time tag information object.

7.6.4 UCA 2.0, IEC 61850

The Electric Power Research Institute (EPRI) started the Utility Communication Architecture project in November 1998, as a part of the Integrated Utility Communication (IUC) program. Version 1.0 of UCA was based on discussions with 14 electric utility companies. UCA 2.0 evolved from UCA 1.0, and it is generally divided in UCA 2.0 for real-time database exchange and UCA 2.0 for field devices, which is the one to be briefly described in this section. EPRI sponsored research activities having as an objective the development of object models of common field devices. The Manufacturing Message Specification (MMS) forum working group and the Substation Integrated Protection, Control and Data Acquisition project were developed based on this research activity. The Generic Object Models for Substation and Feeder Equipment (GOMSFE) contains the results of the previously described projects in a conglomerated fashion.

Basic Structure, Layers. The UCA 2.0 protocol is organized based on the Open System Interconnection reference model, and seven layers integrate the communication protocol. In UCA 2.0, the real-time data acquisition and control applications employ the Application Layer standard ISO/IEC 9506, Manufacturing Message Specification, which services include reading, writing, reporting of variables, and event management.

Basic Structure, Objects and GOMSFE. UCA 2.0 for real-time device access has developed detailed object models that identify the set of variables, algorithms, and so on, required to support the basic functionality of each device class. These object models have named variables instead of point lists. When the objects are accessed by MMS, common data formats and variables are associated to the object model.

There are two main levels of field device object models, the basic, which is an elementary modeling of the field device, i.e., a switch control, and the specialized, which allows several degrees of definition and application, i.e., breaker control or breaker reclose control.

Each field device object model includes a description of the field device function or application, a functional block diagram, and the object model. The object model is a model of a device function or application that receives control commands (binary and analog), setting changes (binary and analog) and indication data (binary and analog) from other objects. The object model maintains relevant data (parameters, settings) and indication data (binary and analog). The object model outputs control commands and indication data.

The object model components are

- Configuration parameters: Values that determine the setup of the device and are not expected to change often. Parameters include any datatype (visible string, bitstring, etc.), binary values, analog values.

- Settings: Values that determine the operation of the device and can change often. Settings include any datatype (visible string, bitstring, etc.), binary values, analog values.
- Operation: Values that represent the actual output decisions or commands of the model to perform its functions. Operations include binary control and set points.
- Status: Represents the indication or values directly concerned with the functions of the device. Status include binary status values (boolean or bitstring), analog values.
- Associated parameters: Values associated with the function of the model. Associated parameters include any datatype, status values and analog values.

7.7 DISTRIBUTION AUTOMATION COMMUNICATIONS ARCHITECTURE

7.7.1 CENTRAL DMS COMMUNICATION

This distribution grid can be monitored, supervised, controlled and automated by systems consisting of SCADA, remote terminal units, and substation control.* The communication architecture must overlay the distribution hierarchy and connect all the monitoring and secondary control equipment by different links (optical, microwave, local networks, telephone line, radio of DLC, etc.). The design of the communication system for controlling and automation of a distribution system, particularly when applying DLC, must consider the hierarchy of this primary system and the heterogeneity of the monitoring equipment.

A heterogeneous architecture comprises a number of different SCADA, RTU and IED types, using different communication protocols and application protocols. Traditionally, the SCADA level of the telecontrol network has been connected via direct wires or point-to-point links from its front end to the remote terminal units. RTUs may also be bussed on the same communication link. RTUs can themselves serve as control units for sub-RTUs.

For the integration of heterogeneous systems and subsystems into an architecture, the following two aspects must be distinguished:

1. The application view, consisting of objects on which application functions are performed
2. The communication protocols and their conversion

How DA communications are deployed depends greatly on the utility's objectives for DA itself. With the division of utilities into separate entities, more distribution companies are using distribution management systems in which the distribution automation functions are integrated. The integration of substation and

---

* Substation control system (SCS), also called substation automation (SA).

Communication Systems for Control and Automation

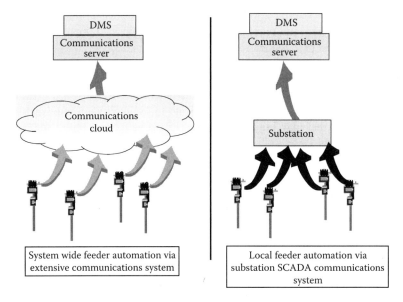

FIGURE 7.53 Distribution automation and the communication media.

feeder automation equipment into the electric utility DMS system depends on several factors, which include

- The communication protocol — the field unit must be capable of communicating with the legacy system or adhere to a standard protocol for immediate or future integration
- Communication media available with the associated infrastructure, copper circuits, radio, microwave, or optical fibers
- Automation strategy, if this is the chosen one for a particular feeder or is a general deployment for the system

Communication to large- and medium-sized substations requires high volume and frequent transmission of data. This data link is treated as part of the utility's WAN* communication infrastructure employing microwave, fiber optics or dedicated lines. In contrast, communications to small substations and feeder devices as required at the low end of the control hierarchy for feeder automation, although requiring less data traffic, has many more destinations distributed across the entire service area.

Two common approaches to communication for the integration of the distribution feeders into the SCADA/DMS are shown in Figure 7.53:

- Central DMS system — The left picture shows a system-wide implementation suitable when the communication media, usually wireless, is available throughout the service territory. The communication system

---
* Wide area network.

is used to interface to all feeder equipment such as reclosers and sectionalizers and retrieve the information from the remote field units for central processing.
- Local intelligent controller — The picture on the right shows a local hierarchical implementation suitable for either incremental implementations or for selected regions of the electric utility network. For this case, local radios would communicate to an intelligent node such as an RTU that may initiate local automation as well as report status to the central DMS system.

At the SCADA level, one can assume that spontaneous traffic originates from the SCADA system and that each SCADA system controls a number of slave stations (RTUs), which respond to requests of the SCADA.

7.7.2 Polling and Report by Exception

In the typical SCADA system, the end device is polled periodically for information such as status, amps, etc. Some systems operate on a report-by-exception principle, in which stations spontaneously report events (state changes) when such occur. Although this scheme is today little used, this may change in the future. In general:

1. Polling of slaves is justified if the slave has little or no application-dependent processing.*
2. As the slaves become more powerful, they are able to perform local functions (protection, reclosing) and generate events. In this case, a report by exception could use the available bandwidth better.

The two modes can be mixed in one system. For example, report by exception could occur during a quiet interval of polling. Some advantages and disadvantages of each are listed below.

Polling Network

Advantages	Disadvantages
Management of this system is no big problem, and the costs are relatively low. The same infrastructure may also be used for voice communication on manually assigned channels. It also allows the use of traditional polling protocols.	Continuous polling without precalculation at the remote side produces a large number of calls and a very heavy data load. Therefore, renting or leasing public systems is not very economical. Controlling thousands of RTUs in high-density locations will require many frequencies.

Continued.

---

* WANs are not well suited for a polling operation, because access to devices requires a connection opening operation unless one is able to maintain a large number of open connections. This can slow down response time to unreasonable delays.

Communication Systems for Control and Automation

Advantages	Disadvantages
	Alarms are delayed until the unit is polled again. Therefore, this system is not recommended for a large amount of remote units in a high-density area.

Event-Driven Network

Advantages	Disadvantages
Allows a single channel to be shared by a relatively large number of remote units and is therefore very suitable for very tight (RTUs/km^2) areas.	The infrastructure costs for the communication network itself is higher than for a polling system.
Typically shorter time delays for short message data transfer like alarms.	Continuous polling and centralized cyclically measurement updating produce a very high overhead and data load.
In comparison to the polling system, the telemetry infrastructure is becoming cheaper (no intelligent node controllers required) but the functions and costs are transferred to the communication infrastructure.	Delay time, capacity and overload forecasts are sometimes quite difficult to establish and may change heavily during the live time.

7.7.3 Intelligent Node Controllers/Gateways

Between the various communication hierarchies, there is a need for intelligent node controllers (INCs) to handle the traffic between the uplink and the downlink of the communications infrastructures. These can be between different carrier technologies, with approximately the same speed, but also between carrier technologies with very different speeds such as a high-speed LAN interface to a radio system. Furthermore, the same INC equipment is available to support the routing and path organization in systems that may reroute on loss of the primary path. The need to support protocol conversions and packaging of the information optimally according to the requirements of the communication infrastructure implies a flexible, programmable but also cost-effective platform for the low-end INC activities, up to a high-end INC/communication server for handling multiple computer links and fast data transfers. The implementation of these controllers serves also as a platform to support application functionality, which may be distributed across the grid. Such applications range from automation support for the automatic reconfiguration of the grid after a disturbance to the control of loads according to a local schedule downloaded from a higher level center.

7.7.4 Interconnection of Heterogeneous Protocols

Because different protocols may be used at the feeder level from that of the SCADA system, it may be necessary to interconnect different protocols. This is typically accomplished in two methods: protocol conversion and encapsulation.

Protocol Conversion. Classically, subnetworks with different protocols are interconnected by a gateway, i.e., in OSI terms by a Level 7 connection. The drawback of this approach is that the conversion from one protocol to the other requires knowledge of the semantics of the transported application data.

Encapsulation. The encapsulation method assumes that the front end and the RTUs communicate over a telecontrol protocol and that their requests and responses have to transmit over different subnetworks with different protocols. In this case, the requests are forwarded over the WAN with protocol W as transparent data: the intermediate network ignores in principle the frame contents. However, the transparency does not extend to the addressing scheme, because all frames must be fitted with an address depending on the application address.

7.8 DA COMMUNICATIONS USER INTERFACE

All distribution automation applications must include an interface between the remote feeder equipment and the user, which indicates the changes that the distribution system structure will have, due to maintenance and outages, among others. Normally, FA is a component of the total data acquisition system, which reports information back to the central control master station where integrated HMI for the entire network is maintained. In some cases, FA is polled by primary substation RTUs or SA that, in turn, pass on.

7.9 SOME CONSIDERATIONS FOR DA COMMUNICATIONS SELECTION

The distribution automation communication options for electric utilities were briefly described previously. As it was indicated, the technologies are generally divided into wire and wireless. A number of technical and economical issues must be evaluated before selecting a communication technology.

For the communication technology selection, it is very important to understand the utility goal. Among the issues to take into consideration are

- The number of remote units to be integrated in the scheme
- The amount of information to be retrieved
- The frequency in which the data are retrieved during a certain period of time
- Monthly costs or initial capital costs
- Maintenance on items such as communications network, protocol conversions, field batteries, etc.

Some communication technologies are more appropriate than others, depending on the specific application. The optimal communication technology for a wide-area deployment is not the same as the one that is optimal for a particular distribution feeder application. As an example below, it is possible to use the

Scenario/number of units	10 units	50 units	100 units	200 units	400 units	1000 units
A (190 bytes)	Unlic. radio	Satellite	Satellite	Satellite	Satellite	Satellite
B (956 bytes)	Unlic. radio	Unlic. radio	Satellite	PPSR	PPSR	PPSR
C (2,048 bytes)	Unlic. radio	Unlic. radio	PPSR	PPSR	PPSR	PPSR
D (30,626 bytes)	Unlic. radio	Unlic. radio	PPSR	PPSR	PPSR	PPSR
E (68,346 bytes)	Unlic. radio	Unlic. radio	Unlic. radio	Unlic. radio	CDPD	CDPD

FIGURE 7.54 Optimal communication technology options.

communications cost, data requirements, and number of units to generate a table of options. Figure 7.54 is only an example that shows the most cost-effective technology option after five years for different technologies. The different scenarios are for increasing amounts of information that the utilities retrieve from the remote field devices. The approximate byte amounts per month for each scenario are shown in parentheses. The amount of field units provides an idea of the distribution automation application size.

As can be seen in Figure 7.54, the unlicensed radio communication technology is the optimal solution for "small" distribution automation applications, up to 10 remote devices. For scenario A, which retrieves the least amount of information and for applications involving 50 or more field units, satellite communication technology is the optimal solution. The utility or vendor can generate a similar example based on costs and needs to determine the optimal technology.

7.10 REQUIREMENTS FOR DIMENSIONING THE COMMUNICATION CHANNEL

7.10.1 CONFIRMED AND NONCONFIRMED COMMUNICATION

The aim of this section is to give guidelines and to show ways of how to calculate the required communication speed or throughput for distributed automation and control systems. In a world where everybody thinks in megabits per second, there is a tendency to think that the higher the speed, the better the performance.

For control systems, the parameter of interest is not the communication speed but the reaction time. The reaction time itself is defined by the time it takes to make something happen (nonconfirmed command), Figure 7.55, or the time it takes until it is confirmed that something happened (confirmed command), Figure 7.56.

7.10.2 CHARACTERIZATION OF COMMUNICATION SYSTEMS

There are two major types of communication systems. One type has a continuous always-on link between the central unit and the remote terminal unit; the other one has to set up a link between the central unit and the remote terminal unit prior to the data exchange. A typical example for a system with the requirement to set up a link between the central station and the remote terminal unit is the

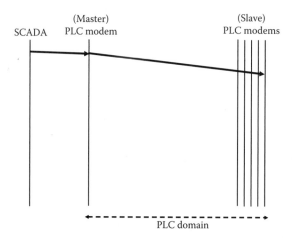

FIGURE 7.55 Nonconfirmed command.

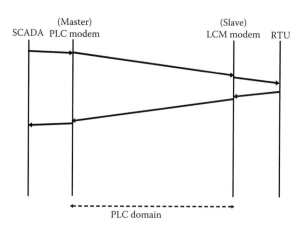

FIGURE 7.56 Confirmed command.

public switched telephone system. An example for an always-on system is a power line-based system. The following example will show how such systems differ when not using the communication speed but the reaction time.

We assume that both systems have a net data rate of 1200 bps, and the complete data exchange between master (central unit) and slave (remote terminal unit) has 100 bits. So, for the always-on system, it takes about 90 ms to do the job. For the telephone system, it is first required to dial the line, and to get a line assigned and so on. This procedure takes at least 2 seconds. That means it takes about 2.1 seconds to do the job, resulting in an overall data rate of about 50 bps.

Both systems are labeled with 1200 bps. If we transmit a long data sequence, it is also the case, but if we use it for automation purposes where we do not have to transmit long data file, it is a misleading parameter.

Communication Systems for Control and Automation

The next important parameter to characterize a communication system for automation purposes is the way a dedicated device gets access to the communication channel. We distinguish here between two major access technologies (all other technologies are a mix between these two technologies):

- One technology uses a central master responsible for controlling the access to the communication channel. Again using as example the plain old telephone system, it would mean that the central station dials one remote terminal unit after the other and is asking for updates.
- The other technology uses equally prioritized units distributed all over within the system. That means everybody is able to inform somebody about a change of state or an emergency as soon as it happens. When again using the PSTN example, it just means that everybody is allowed to call everybody. Another well-known example of such a type of system is the Ethernet.

So, what is the difference of such systems in respect of performance? It is obvious that equally prioritized systems generate a far lower average communication load than a system with a central arbitration unit. Yet, it can cause unpredictable reaction times in case there is something like an emergency. In such a situation, everybody tries to use the communication system, and finally the system is overloaded and nobody gets access. That is a well-known scenario during natural disasters, but is not the case when using a central arbitrated system. Even its average performance is slow compared with an equally prioritized one; when there is a disastrous situation, the performance can be far better — it is dependent on the application what type of arbitration method may be the most efficient.

7.10.3 COMMUNICATION MODEL

Figure 7.57 shows the general setup of every supervision, control, data acquisition, and automation system, completely independent of the used communication medium.

7.10.4 CALCULATION OF THE REACTION OR THE RESPONSE TIME

When doing an overall performance calculation, the major effort is the identification all occurring delays in the system under consideration. There is no universal model but the diagrams in Figure 7.57, Figure 7.58, and Figure 7.59 will give some general ideas about the principles involved.

Very often, the necessary information regarding the system internal timing is not very well documented or available to the customer. From the customer point of view, the best is to measure the time it takes to transmit some information or to enable a command by doing some measurements. When measuring the absolute delay and the delay jitter it takes to send a command, a good approximation of the total system performance is possible.

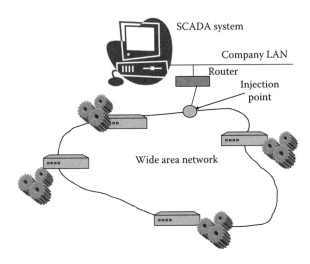

FIGURE 7.57 General communication model.

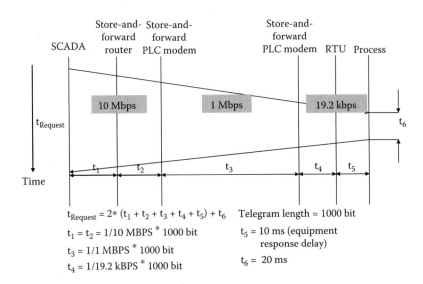

$t_{Request} = 2* (t_1 + t_2 + t_3 + t_4 + t_5) + t_6$ Telegram length = 1000 bit

$t_1 = t_2 = 1/10$ MBPS * 1000 bit

$t_3 = 1/1$ MBPS * 1000 bit

$t_4 = 1/19.2$ kBPS * 1000 bit

$t_5 = 10$ ms (equipment response delay)

$t_6 = 20$ ms

FIGURE 7.58 Timing illustration.

Communication Systems for Control and Automation 355

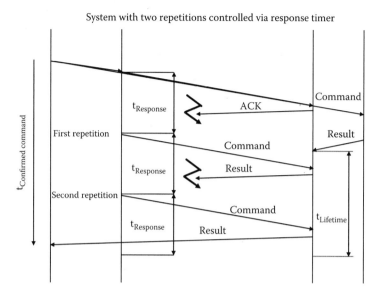

FIGURE 7.59 Timing illustration with additional delay.

8 Creating the Business Case

8.1 INTRODUCTION

This chapter will develop a procedure to create the business case for justifying distribution automation. Although the main emphasis will be on extended control, general principals for justifying substation automation (SA) will be explored. We will be relying on concepts developed in earlier chapters, particularly those covering distribution systems, reliability assessment, fault location, and automation logic. After a detailed method has been explained, it will be illustrated with two case studies in the final chapter.

Creating a business case considers the combination of hard and soft benefits — those that can be economically quantified and those that are intangible but influence the perception of a utility's performance. Hard benefits can be divided into investment savings (CAPEX) and reduction in operating costs (OPEX). Some authors categorize the justification into tangible and strategic benefits. The availability of hard data to estimate benefits can also be the major differentiator between hard and soft benefits. Benefits are also direct and indirect, direct being derived directly from the application being justified, whereas indirect are achieved through another application that depends on data from the direct implementation. The categories that benefits can be divided in any analysis are shown pictorially in Figure 8.1.

The influence of indirect benefits emphasizes the importance of a DMS within the total enterprise IT structure of the utility and the need for realistic yet seamless integration with the other enterprise IT applications (GIS, NAM, CIS/CRM, WMS, CMS, ERP,* etc.).

The soft benefits should not be overlooked because they can impact other enterprise activities indirectly. Also, important benefits may produce only minimal hard quantification. Further DA functions implemented produce different types of benefit across specific parts of the network. In all cases, functional benefits have to be quantified into an economic value. A general method for evaluating the worth of DA will be based on the concept of a benefit opportunity matrix

---

* Geographical information system, network asset management, customer information system/customer relationship management, work management system, computerized maintenance management, enterprise resource management.

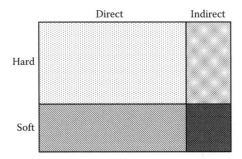

FIGURE 8.1 Diagrammatic representation of hard, soft, direct, and indirect benefits.

and definitions of generic benefits. This foundation will be extended to specific evaluation of substation and feeder automation (FA).

8.2 POTENTIAL BENEFITS PERCEIVED BY THE INDUSTRY FOR SUBSTATION AUTOMATION

Utility industry experience suggests that benefits should be expected from substation and feeder automation. This perspective was obtained from a number of industry surveys completed over the past 10 years. The surveys point to where utilities expect benefits. It will act as a guide during the exploration of how benefits are generated.

8.2.1 INTEGRATION AND FUNCTIONAL BENEFITS OF SUBSTATION CONTROL AND AUTOMATION

The fundamental concept surrounding DA is that of integration of the components into a system. In Reference 5, Tobias presented a subjective evaluation of integration benefits for substation automation defining three distinct levels of integration*:

1. Intelligent device level
2. Switching device level
3. Substation level

The results of this evaluation [5] are summarized in Figure 8.2, giving benefits for both the manufacturer and the utility (user). The benefits from the individual levels can be extrapolated to infer an increase in benefits with deeper integration.

The perceptions of the industry were shown in a global survey conducted by Newton-Evans Research in the mid-1990s. The survey indicated (see Figure 8.3) that apart from lack of funds, the most significant obstacle to substation automation implementation was economic justification. Over 100 respondents prioritized

---

* The same integration philosophy will also apply to feeder devices considering automation ready as the ultimate level of integration.

Creating the Business Case

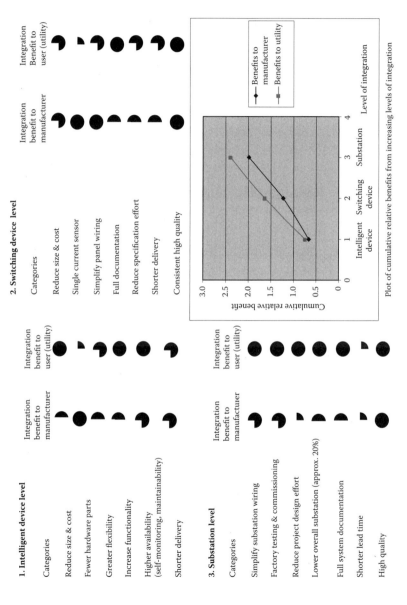

FIGURE 8.2 Subjective evaluations of relative benefits for different levels of SA device integration.

the financial benefits derived from SA. The two benefits ranked highest were operational: to reduce response time for problem fixing and to lower O& M costs.

The remainder of benefits accrued from reduction in capital investment of substations falling into two categories, either from reduction in project timescales or from smaller and more flexible control equipment. It is relatively easy to determine the benefits from new designs because these can be costed and compared with former designs. The operational benefits are sometimes, however, subjective and more difficult to quantify.

The most highly ranked benefits are derived from operational efficiencies through speed of data access and system flexibility. The results are similar whether considering retrofitting automation to an existing system or building a new automated substation

The survey assumes that SCADA control of substations is in place and the benefits are the difference between wiring of existing noncommunicating relay and auxiliary contacts to RTUs in comparison to a LAN-based SA installation.

8.2.2 SCADA vs. SA

The traditional method for remote control of substations is via RTUs as part of the SCADA system. Direct benefits of this control are derived from being able to operate switches remotely and from monitoring more precisely the power system state. This will be considered the base case for evaluating the benefit contribution of remote control and monitoring. True substation automation as described earlier* will only provide small incremental benefits to power system operation. The major benefits accrue from improvements in O&M costs and capital expenditure for new substation builds or expansion. The cost of retrofitting SA to legacy substation control may not produce benefits due to the cost of replacing traditional protective relays. This is illustrated in Figure 8.4.

This figure shows the annual operation costs for base case control (steeper gradient — dashed line) and for a fully automated substation. The initial savings for implementing SA directly for a new substation over the base case solution (traditional RTU, noncommunicating protective devices) is shown by savings "A." Cost "D" shows the addition of retrofitting SA to an existing substation. The benefits of each option are shown by the shaded areas, illustrating that for new substation builds there are immediate benefits, whereas for retrofitting there must be a period of time before the retrofitting costs are repaid.

8.2.3 Economic Benefits Claimed by the Industry

Typical economic benefits published in a variety of texts claim substantial economic benefits. In an analysis of a typical four-panel substation where each panel costs approximately $25,000, ABB [6] suggest that the same functionality can be provided by one panel for about the same cost, therefore providing a first-time cost saving of 75%.

---

* Chapter 2, Section 2.9.2.

Creating the Business Case

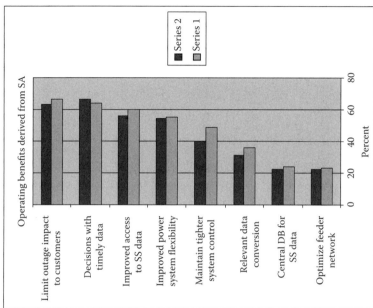
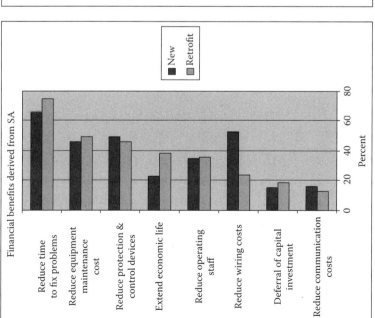

FIGURE 8.3 1997 Survey of Financial and Operational Benefits Derived from Substation Automation (Courtesy of Newton-Evans Research Company Inc.).

362 Control and Automation of Electric Power Distribution Systems

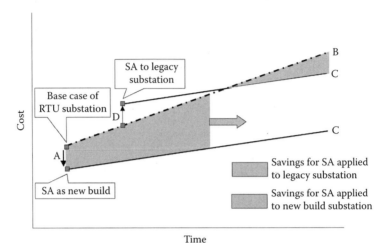

A = Saving in construction cost at new build (CAPEX)
B = Operating cost of non automated substation (OPEX)
C = Operating cost of automated substation (OPEX)
D = Cost of SA retrofit to existing substation (CAPEX)

FIGURE 8.4 Illustration of economic benefits for substation control and automation.

General Electric in its marketing literature* quantifies the savings from substation automation as shown in Figure 8.5. The explanation of these benefits in the literature suggests that some feeder automation functions may have been considered to contribute to the CAPEX savings using remote feeder switching.

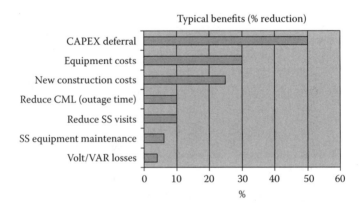

FIGURE 8.5 Typical substation automation benefits in percentage cost reduction (General Electric).

---

* www.Geindustrial.com/pm GE Substation Automation.

Creating the Business Case 363

KEMA Consulting [9] give a sample case where the benefits from continuous on-line diagnosis of equipment, remote control of feeder breakers, feeder capacitor bank monitoring and voltage control produce a benefit/cost ration of 2.14.

Black & Veatch [11] present a business case for SA in a medium-sized utility having a total of 116 substations (16 grid stations, 40 medium- and 60 small-sized primary distribution stations). The 4.3 benefit/cost ratio was derived from the following annual benefits:

- Reduced time to find/fix problems
- Reduced O&M (SCADA, metering, protective relaying and recorders)
- Remote operation
- Predictive transformer maintenance
- Transformer load balancing
- Reduced crew travel time
- Reduced training
- Asset information/drawing management

The important conclusion is that the majority of benefits result from reduced operation and maintenance costs derived from reduced personnel and re-engineered business processes. In order to capture these benefits, a utility must be prepared to make the appropriate changes and reassignments in their organization.

8.3 POTENTIAL BENEFITS PERCEIVED BY THE INDUSTRY FOR FEEDER AUTOMATION

Surveys of utilities intending to implement automation on their feeders have determined perceived priority functional needs to be performed by automation. The priority of these functions is shown in Figure 8.6. It will be used to guide identification of opportunities for benefits in more detail and to build the procedure for creating the business case.

The publications reviewed covering solely feeder automation lacked substantial quantification of monetary benefits; however, all make a strong business case on the value to the utility. The benefit that predominates is quality of service improvement based on reducing SAIDI/customer minutes lost (CML). Feeder automation has proved flexible and reliably delivered in excess of the planned benefits of reducing CML by 33% on MV circuits [17,18]. Achieving satisfactory levels of perceived SAIDI is crucial to maintaining customer loyalty [20], and increased economic value is possible by phasing device automation with asset replacement programs. An added benefit [21] is the possibility of on-line loading data that enables network reinforcement to be completed just in time. This paper, in addressing simple remote control compared with simple isolation and restoration automation, concludes that the partial automation function delivers a 25% improvement increased customer satisfaction. The automation intensity level (AIL) selection is crucial to a positive benefit-to-cost ratio, as is the selection of

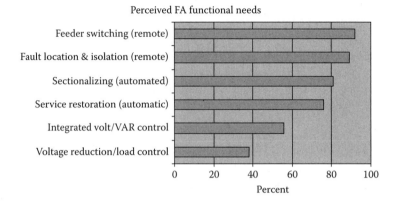

FIGURE 8.6 Perceived needs of feeder automation reported by the utilities in a recent survey (Source: Newton Evans).

candidate feeders [19]. None of the above references explicitly delineate the stated benefits in monetary terms; however, Reference 11 outlines examples of the application of a Total Power System Integration model, which has been used to assess potential benefits. The results of a study in Reference 13 show an overall positive benefit-to-cost ratio of 2.29.

The remainder of this chapter provides a review of benefit assessment methods and procedures for their quantification.

8.4 GENERIC BENEFITS

Exhaustive work completed under EPRI project EL-3728, "Guidelines in Evaluating Distribution Automation," [8] will be used as a starting point and framework for developing the benefits analysis methodology. This work identified seven types of generic benefit that apply to the DA functions. The seventh covered automatic meter reading which although not covered in this book,* will be included for completeness.

Type 1: Capital Deferred Delayed in Time (Years). The benefits derived in this type are as a result of applying DA functions, which allow the delay of a capital purchase for another time in the future. The benefits are quantified in terms of present value of revenue requirement (PVRR) of investment delayed.

> Benefit = (PVRR of equipment needed without DA over planning period) − (PVRR of equipment needed with DA over the period starting from the time when the specific DA equipment was implemented).

This is usually seen as the same primary equipment needed but with a delayed installation time. This type of capital deferral is achieved by the ability to switch

---

* Demand side management/load control and automatic meter reading, although often included as DA, are not covered in this text.

loads between adjacent substations, thus delaying the need to install additional transformers until the load grows sufficiently that adjacent substation capacity is insufficient (see Chapter 3).

Type 2: Capital Displacement — Same Year. This generic benefit reflects the displacement of traditional hardware with intelligent hardware used to implement DA. The use of digital protection modules instead of conventional electromechanical relays is a common example. The utility must evaluate the costs carefully for this type of benefit because, depending on how the first cost of the new equipment or the residual value of the conventional asset is assessed, the benefits can be negative. Another example is where the need for a line switch is determined and implemented by an automated switch costing more, rather than a manual device. However, the other DA benefits cannot be achieved without this expenditure for the DA-controlled equipment.

> Benefit = PVRR [base system hardware (purchase price + installation cost) – distribution automation system hardware (purchase price + installation cost)] over the period starting from the time when the specific DA equipment was implemented.

Type 3: Operation and Maintenance — Hardware Dependent. This benefit is based on the assumption that the digital (IED) equipment implemented for DA is more reliable and less expensive to maintain. A typical example in substation automation is the flexibility and remote interrogation capabilities provided by communicating IEDs, which potentially allow remote setting changes and troubleshooting, thus reducing worker-hours and substation inspection visits.

> Benefit = PVRR [base system hardware O&M requirements – DA system hardware O&M requirements] over the period starting from the time when the specific DA equipment was implemented.

Type 4: Operation and Maintenance — Automation Function Dependent. This benefit category is non-hardware dependent, being the result of an improvement in a process that results from the installation of DA. Typically, the effort needed to collect data from substations and feeder locations is greatly reduced with the implementation of the DA data logging function, which is unrelated to the legacy system installed. Benefits from remote device operation are also obtained.

> Benefit = PVRR [base system O&M requirements affected by the DA function – DA system O&M requirements] over the period starting from the time when the specific DA equipment was implemented.

The above generic benefits will be used as the foundation for assigning one or more of these benefits to each DA function, where a more specific relationship will be developed for calculating the benefit.

Type 5: Capital Deferred — Displaced in Time Due to Demand Reduction. Any DA function that reduces demand releases the need for additional upstream capacity in the power system, impacting the transmission and generation. Loss minimization and VAR optimization functions will improve the demand losses in the network, thus reducing peak capacity requirements.

Benefit = PVRR [(peak generation costs/kW + transmission costs/kW)(total kW reduction)] over the period.

Type 6: Operational Savings — Reduction in kWh Due to Demand Reduction. This benefit is the energy equivalent of Type 5 and allows the annual savings in operating expense to be related to a DA function. The reduction in demand losses can be transferred using a loss factor to the savings in annual energy losses.

Benefit = PVRR [(total kW reduction)(hours in operation)(appropriate generation and transmission cost/kWh)] over the period.

Type 7: Operational Savings — Reduction in kWh Due to Displacement in Time (Energy).* Although similar to Type 6, this benefit is derived solely from the implementation of load management and automatic (remote) meter reading:

Benefit = PVRR [(total kWh reduction)(hours in operation/yr)(% of kWh shifted from on to off peak)(differential fuel cost from on to off peak/kWh)] over the period.

This saving can be formulated to represent the improvement in accuracy, elimination of theft and manpower efficiencies as a consequence of improved meter reading, resulting from implementation of an entire new metering program.

In the absence of typical values for generation and transmission costs used by utility management, it will be necessary to conduct detailed system studies to develop the true generation and transmission capacity and energy costs for the last three benefit types. Also, to develop precisely the needs of distribution network capacity over the period in question on which capacity released through automation and other operating benefits can be based, a detailed distribution planning study will be necessary. All these studies should be carried out using application software dedicated to the purpose.

Summary. The generic benefits above can be broadly grouped into:

- Capital and O&M benefits related to hardware (Types 1, 2, and 3)
- Non-hardware-related resulting from DA functions (Type 4)
- Capacity- and energy-related benefits derived from reduction in essentially upstream capacity requirements (Types 5, 6, and 7)

The categorization of prospective benefits serves to describe where benefits can be derived. The development of the business case should concentrate on those

---

* Automatic meter reading (AMR) and AMR systems are not treated in this book.

functions that will reap the highest benefit and impact the business most. For example, the benefits related to generation and transmission will not affect the business of a distribution utility unless a capacity credit is a variable in the rate structure or penalties exist for exceeding supply demand limits. A screening method may produce sufficient benefits, thus avoiding the need for detailed studies that would be necessary to squeeze out benefits in all categories.

8.5 BENEFIT OPPORTUNITY MATRIX

The opportunity matrix is used as a simplified overview of the DA functions treated, the location of the implementation, and the anticipated benefit category. The benefit categories used are expanded versions of the generic benefits that target the specific areas affected (Table 8.1).

The direct anticipated benefits are now grouped according to the DA function implemented and the area in which the benefits will be obtained to form the opportunity matrix of anticipated benefits (Table 8.2).

The final column is provided to indicate where operational improvements will result that may not be quantifiable in terms of savings and could be classified as a soft benefit.

The opportunity matrix acts as a guide for identifying the anticipated benefits for candidate DA functions.

8.6 BENEFIT FLOWCHART

In order to develop the economic side of the business case, functional benefits, once identified, have to be quantified in economic terms. The benefit flow diagram introduced by KEMA [7] provides an overall view of the process for any candidate automation function and shows the steps to convert functional benefits into monetary benefits. A general diagram for DA is shown in Figure 8.7.

Diagrams can be developed for each DA function and compared to ensure that benefits are not double-counted, because typically the implementation of increased automation results in incremental benefits in similar categories.

8.7 DEPENDENCIES, AND SHARED AND UNSHARED BENEFITS

8.7.1 Dependencies

Implementation of any one DA function requires a minimum installation of hardware and the associated infrastructure. Once installed, other DA functions can be added at small incremental cost and deliver significant additional benefits. The development of sufficient returns is a matter of adding more functions that use the initial infrastructure investment and that deliver benefit. It is usual that DA, to be justified, has to be a combination of functions, each contributing to the overall economic payback.

TABLE 8.1
Summary of Direct and Indirect Benefits Areas Anticipated from DA Implementation

Direct Benefits

Investment Related

Deferred
- Supply system capacity
- Distribution substation additions
- Distribution substation transformer addition/replacement
- Feeder bay/gateway/primary feeders

Displaced or reduced
- Conventional SCADA and RTUs
- Conventional meters and recorders at substations
- Conventional controls (capacitors, regulators)
- Conventional protection
- Substation control and supervisory wiring

Operations and Maintenance Related

Interruption Based
- Increased revenue due to faster service restoration (kWh saved)
- Reduced crew time to locate fault and restore service
- Customer based
- Reduced customer complaints
- Value to customers of improved reliability

Operational Savings and Improvements
- Reduced trips to substations and feeder switches (switching and data gathering)
- Improved voltage regulation
- Reduced substation and feeder losses
- Repair and maintenance savings
- Reduced manpower for meter reading at substations
- Faster generation of switching plans
- Improved detection of equipment failures
- Faster restoration of supply due to remote feeder reconfiguration

Indirect Benefits

- Capability to change digital protection settings remotely
- Improved data and information for distribution engineering and planning

- Equipment loading data provides improved asset management
- Network status data improved perceived quality of customer satisfaction — "shine"

Exploring the development of a DA solution as described in earlier chapters, there are two streams of automation focused on different levels of control:

- Substation control and automation
- Feeder control and automation

The hardware infrastructure for each of these streams consists of a master control center, adaptation or provision of the primary equipment for automation/remote control and the communication system linking the master with the remote units. The largest dependent element between common DA functions in this infrastructure is the communication system. Although the master central control is common to both SA and DA, today's computer hardware can be easily sized

TABLE 8.2
DA Function/Anticipated Benefit Opportunity Matrix

Automation Function	DA Area SA*	DA Area FA†	Investment Related[a]	Interruption Related	Customer Related	Operational Savings	Improved Operation
Data	✔	✔	✔			✔	✔
Data monitoring	✔	✔	✔			✔	✔
Data logging	✔	✔				✔	✔
Integrated volt/VAR control	✔	✔	✔		✔	✔	✔
Bus voltage regulation	✔		✔				✔
Transformer circulating current control	✔		✔				
Line drop compensation	✔		✔				✔
Substation reactive control	✔		✔			✔	
Feeder remote point voltage control (vs. regulator control)		✔	✔		✔		✔
Feeder reactive power control (capacitor switching)		✔	✔			✔	
Automatic reclosing	✔	✔	✔	✔			✔
Substation							
Remote switch control	✔		✔	✔	✔	✔	✔
Digital protection with communicating IEDs	✔		✔				
Load shedding	✔					✔	
Load control	✔		✔	✔		✔	
Cold load pickup	✔	✔	✔	✔		✔	✔
Transformer load balancing (adjacent substation capacity)	✔	✔	✔			✔	✔
Feeders							
Remote switching[b]		✔	✔	✔	✔	✔	✔
Fault location		✔		✔		✔	✔
Fault isolation		✔	✔	✔		✔	✔
Service restoration		✔	✔	✔	✔	✔	✔
Reconfiguration		✔	✔			✔	✔

[a] Deferred and displaced.
[b] Remote feeder switching is also termed *feeder deployment switching* and refers to all switches outside the substation along primary (MV) circuits.
* SA — Substation Automation; † FA — Feeder Automation.

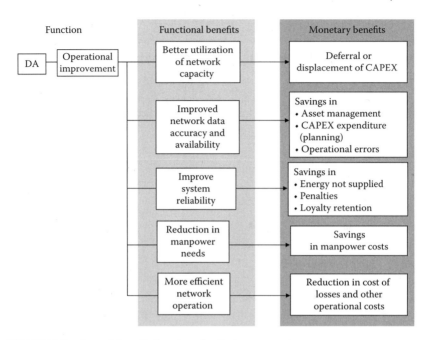

FIGURE 8.7 General benefit flowchart for feeder automation.

or extended without significant cost to allow increasing software functionality and the number of control points. The selection of the communication system may vary for SA and DA. The former is often justified based only on the needs of the SCADA/SA benefits. Communications to the potentially large number of remote feeder switches, particularly for high AIL, may require a different medium and even a combination of different types due to the varying environment and terrain across the service area. The DA communication system can take many forms as described in Chapter 7, of which three typical combinations are listed below:

- Implemented as an extension of the SA communication network
- Implemented as an independent, systemwide network requiring large initial investment
- Incremental deployment through a more restrictive coverage targeted at specific locations

The third option represents low-power radio or per-location lease/payment of the communication link typical of mobile phone technology. The benefit of incremental deployment is that the large initial investment is avoided. Any small investment can be allocated or expensed on a per-location basis.

Once the hardware platform and infrastructure for automation is in place, automation functions can be added. Figure 8.8 shows the dependencies in DA implementation with the two control streams.

Creating the Business Case

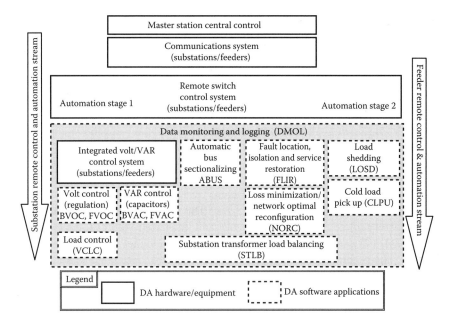

FIGURE 8.8 Hardware platform/DA infrastructure and DA software applications dependencies. BVOC — bus voltage control, FVOC — feeder voltage control, BVAC — bus VAR control, FVAC — feeder VAR control, VCLC — voltage conservation load control, ABUS — automatic bus sectionalizing, FLIR — fault location isolation and service restoration, NORC — network optimal reconfiguration, STLB — substation transformer load balancing, LOSD — load shedding, CLPU — cold load pick up, DMLO — data monitoring and logging.

Clearly, implementation of remote switching is the foundation for all DA where remote control of substation circuit breakers is traditional SCADA and the addition of remote control of feeders outside substations represents extended control or FA. This supports one of the major business goals of a DMS, which is to improve control response in the network. Once switches are remotely controlled, other DA functions can be added as software applications that depend on the ability to quickly open and close switches. The degree of monitoring and data logging possible and also the level at which other applications can be implemented will depend on the automation stage, because stage 1 as defined in Chapter 1 will not provide communication of analogue values.

8.7.2 Shared Benefits

Many DA functions contribute benefits that are shared, and thus double-counting should be avoided. Also, there are functions and subfunctions defined, first by the necessary control infrastructure and second by the role within the power delivery system. For example, remote feeder switching is a foundation function from which all others are dependent even, to some extent, integrated volt/VAR control. In a substation, automatic bus sectionalizing (ABUS) requires remote control of the switches to give the operator status information and control over disabling the process. The fault isolation location and service restoration (FLIR) function is dependent on remote switching to be effective, as is the network optimal reconfiguration (NORC) application. Further, transformer substation load balancing (STLB) can be considered a subfunction of NORC because it requires transferring of load between sources by feeder switches to balance substation transformer loadings. Voltage conservation load control (VCLC), cold load pick up (CLPU) and load shedding (LOSD) also produce dependent benefits. The extensive work done in the EPRI project discusses these dependencies and the interrelation of the benefits over many pages of text and tables. This has been summarized in Figure 8.9. This diagram shows the two main functions of remote feeder switch control and integrated Volt/VAR control plus all the DA software functions and subfunctions divided into SA and FA applications. The generic benefits derived from each function are shown in the middle of the figure with arrows depicting the source of the benefit. Dotted lines indicate major dependencies and point to the outer ring of the particular generic benefit to convey the notion of a shared benefit. Benefits from the data monitoring and logging online (DMLO) function are not shown in order to avoid further complexity.

The main DA functions deliver benefits through direct improvements in the areas of capital deferral and O&M. The main sharing of benefits occurs for upstream plant catered for by generic benefit Types 5 and 6. Loss reduction in the network components reduces the need for generation and transmission capacity and also the cost of energy whether purchased or self-generated. The contribution of each DA function will be governed by the operating constraints and Thevenin's law; thus, each function as implemented will only deliver marginal improvements. The functions sharing contributions to generic benefit Types 5 and 6 are shown in Figure 8.10 using a benefit flowchart.

Minor shared benefits occur between the feeder voltage and VAR control functions related to improved maintenance (generic benefit Type 3) of new control equipment that potentially is used by both functions within integrated volt/VAR control. Also, there is a dependency between voltage conservation load control and cold load pick up when used in combination to pick up load faster, thus reduce the energy not supplied. This is a potential reduction in operating costs within generic benefit Type 4.

8.7.3 UNSHARED BENEFITS FROM MAJOR DA FUNCTIONS

Now that the discussion on shared benefits has been completed, the development of the business case returns to where the main payback is generated. DA benefits are mainly derived from two major functions:

Creating the Business Case

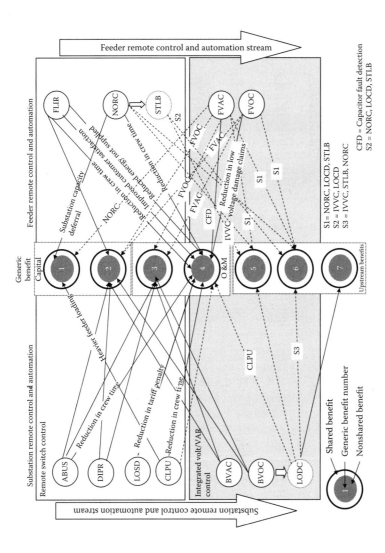

FIGURE 8.9 DA functions and subfunctions show the derivation of benefits and dependencies. Generic benefits are capital deferral; capital displacement; operation and maintenance — hardware dependent; operation and maintenance — software dependent; capital deferral due to demand (capacity) reduction; operational savings due to kWh reduction; and operational savings due to reduction in kWh due to time displacement.

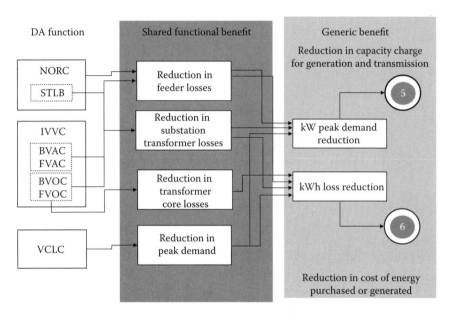

FIGURE 8.10 Benefit flow diagram for shared benefits for NORC, IVVC, and VCLC functions.

- Remote switch control
- Integrated volt/VAR control

Remote switch control enables many other subfunctions to be introduced that add incremental benefits by improving the operator's decision process, thus improving the accuracy of actions and the speed of their implementation.

The level of benefits can vary depending on the stage of automation adopted* because simple remote control offered without analogue measurements but in combination with local automation (reclosers and autosectionalizers) will give different results than a stage 2 totally integrated centrally controlled implementation using analogue measurements. Volt/VAR control is similar, where using local control must be compared with the potential for improvements given by a totally integrated implementation.

Remote Switch Control Benefit Flowchart. The different areas where benefits can be calculated are shown in the benefit flowchart in Figure 8.11, which will form the basis for detailed calculations later in this chapter. The flowchart is not specific to either SA or FA, and thus, when making the detailed calculation, care should be taken to include the benefits appropriate to the particular SA and FA function implementation as part of remote control.

Integrated Volt/VAR Control Benefit Flowchart. The integrated volt/VAR control is a narrower application than remote switch control, thus benefits are

---
* See Chapter 1.

Creating the Business Case

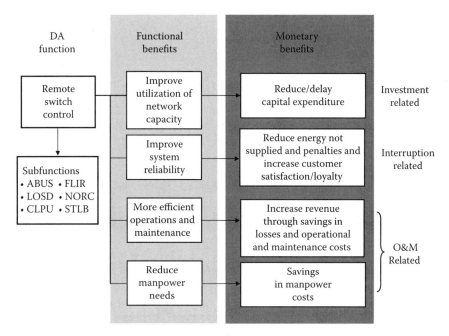

FIGURE 8.11 Overview benefit flowchart for remote switch control showing where main direct unshared benefits are obtained.

derived from a more restricted area. Also, there is the potential that benefits are shared because for example, power factor correction improves voltage regulation, thus relieving the voltage control function.

The operational benefit of automation over traditional control methods (capacitor switch control and voltage regulation via tap changers) is marginal and results from finer real-time control and integration of the functions, only possible with a system approach using the DMS network model to optimize control actions. Maintenance savings from new tap changers and modern substation capacitor bank control are expected. Benefit flowchart, Figure 8.12 depict these benefits.

Data Monitoring and Logging (DMLO) Benefits. Data monitoring and logging is a byproduct of implementing remote control. It is almost inconceivable that implementing a system for this function alone could be justified. The system data that potentially become available increase the visibility of the operators in areas where they have had to use experience and knowledge built up over many years. Load information usually taken annually by extensive field surveys and limited real-time current readings at some points in the network, usually at grid substations, can be remotely read with DA. These additional data provide vastly increased visibility for the operator and management. It provides information for improved network capacity planning, engineering and asset management activities outside the direct operation of the network. Such accrued benefits should be considered indirect as shown in Figure 8.13.

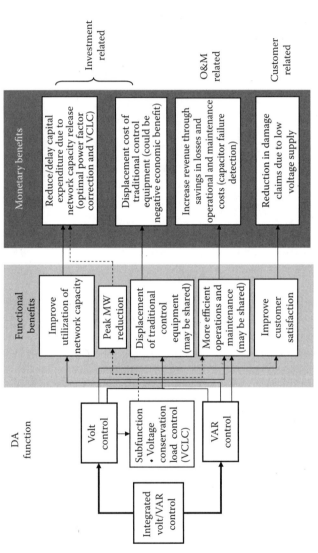

FIGURE 8.12 Overview benefit flowchart for integrated volt/VAR control showing where main direct unshared benefits are obtained.

Creating the Business Case 377

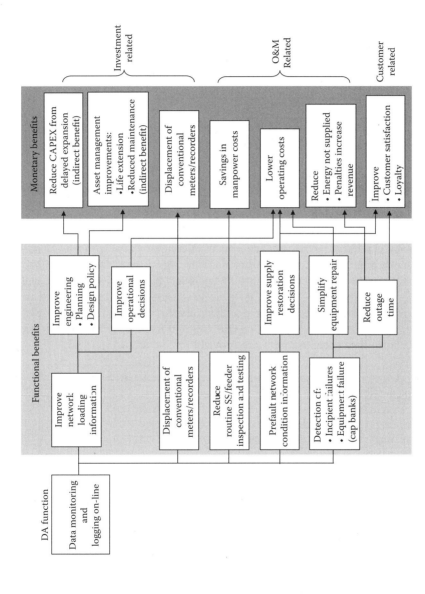

FIGURE 8.13 Benefit flowchart for data monitoring and logging on-line.

Incremental benefits from this DA function as an independent function are difficult to quantify because many advanced functions for which benefits are derived depend themselves on the data available from the monitoring function, which is usually an integral part of implementing DA. For example improvement in customer satisfaction through the DMLO function should be considered as integral with the FLIR function when it is implemented. If FLIR is not implemented, then the DMLO function provides the operator invaluable information for manual decision making necessary for fault location, isolation and supply restoration. Monitoring of the health of equipment increases the benefit derived from knowing when a DA function is able to contribute benefits. For example, VAR control will not reap benefits if capacitors are not functioning; thus, DMLO could be considered to contribute to energy-related benefits derived from loss reduction by reducing the time for capacitor repair.

8.7.4 Benefit Summary

The level of automation implemented, as discussed in earlier chapters, allows successively more information to be made available to the operator. Also, the deployment of different types of automated device affects the resulting system performance in terms of interruptions seen by certain customers, and thus, the interruption frequency is reduced. Increasing automation, in contrast to remote control under operator decision, will reduce the restoration time significantly. Once feeder automation is implemented, benefits from substation load balancing can be made with Feeder Automation Stage 2.* When analogue data measurements become available, improved engineering, operating and asset management decisions are possible, each reaping benefits to the business. The previous sections in this chapter have indicated that the main hard benefits for DA are derived from the following areas:

- Capital deferral or displacement of network assets associated with each function
- Savings in personnel
- Energy related savings
- Other operation and maintenance savings
- Customer-related benefits

The following sections of this chapter develop a number of expressions for estimating the benefits in the above categories. The expressions derive the benefits on an annual basis, the assumption being that these annual values can be applied on a longer-term basis by using the appropriate carrying charges and period to suit each individual case. Even though the majority of formulas presented are for determining the individual benefits of respective functions on a savings minus DA implementation cost basis, it is usual to calculate a total implementation cost for the proposed system and then compute the aggregate savings expected. This

---
* Defined in Chapter 1.

Creating the Business Case 379

not only reduces the probability of double-counting benefits but also provides visibility in the magnitude of savings that must be obtained and the consequent depth of analysis that will be required should initial screening methods not deliver the required payback.

8.8 CAPITAL DEFERRAL, RELEASE, OR DISPLACEMENT

Various DA functions are instrumental in allowing control of the network configuration, voltage and loads, which will result in the deferral of system investment expansion or the release of existing upstream system capacity. The latter may not directly benefit the distribution utility unless reflected in the supply contract.

The installation of any DA function will displace existing conventional control equipment or, if not installed, can be used to compare the cost of the conventional approach with provision for automation. The derived benefit may be negative and reflects the cost of automation equipment to be covered by other benefits.

8.8.1 Deferral of Primary Substation Capital Investment

Remote feeder switching at the normally open point (NOP) allows load to be transferred from the normal supply point to one or more adjacent substations. This eliminates the need to install capacity at each substation to cover transformer contingencies. Overall system capacity is, therefore, deferred. Chapter 3 discusses the planning aspects of determining the capacity deferred on the basis of load growth, transfer capacity and substation firm capacity step size. The economic benefit is derived from the difference between the timing and capacity that would have been installed with and without remote switching. On a present worth basis, there are three approaches:

1. Benefit in present worth of annual revenue requirements over the planning period = (substation expansion step cost + installation cost) × carrying charge × PWF + O&M annual costs × PWF) for period d to N of the capacity requirement without remote switching × (substation expansion step cost + installation cost) × carrying charge × PWF + O&M annual costs × PWF) for period c to N of the capacity requirement with remote switching.
2. A more simplified analysis using only the present worth and timing of the capital investment over the FA planning period.
3. Evaluating the present worth of benefits over the shorter amortization period used for an automation project.

These different evaluations are compared in Table 8.3 using the example in Section 3.2.9 of Chapter 3.

Screening Method. A simplified method that provides an estimate on a system basis is proposed for screening purposes. It can be used at various levels

TABLE 8.3
Substation Deferred Capacity Due to Feeder Automation at the NOP

Capital					Method 1 — PW of Annualized Costs					Method 2 — PW Cap	
					No NOP Automation		Automation				
Without DA	With DA	Period	Transf PW	PW Fac	PW $USD	Transf PW $USD	Automation $USD	Total $USD	PW $USD	Without	With
200,000	50,000	1	18,879	0.934579	17,644		12,195	12,195	11,397	186,916	46,729
		2	18,879	0.873439	16,489		12,195	12,195	10,651		
		3	18,879	0.816298	15,411		12,195	12,195	9,954		
		4	18,897	0.762895	14,402		12,195	12,195	9,303		
	100,000	5	18,879	0.712986	13,460		12,195	21,634	15,425		71,299
		6	18,879	0.666342	12,580	9,439		9,439	6,290		
		7	18,879	0.62270	11,757	9,439		9,439	5,878		
		8	18,879	0.582009	10,988	9,439		9,439	5,494		
		9	18,879	0.543934	10,269	9,349		9,439	5,134		
	100,000	10	18,879	0.508349	9,597	9,439		18,879	9,597		50,835
200,000		11	37,357	0.475093	17,938	18,879		18,879	8,969	95,019	
Totals					150,533	18,879			98,092	281,934	168,863
DIFF									52,441		113,072

Creating the Business Case

of resolution from the complete system down to operating districts or areas defined by similar spare capacity or load growth. The selection of the margins for capacity must be carefully considered on the basis of the size, number of substations and feeder structure to represent the planning criteria used by the utility for substation expansion. The method develops the amount of spare substation capacity available then, for a given growth rate, determines how many years' delay is possible before additional substation capacity is required. The value of economic benefit is the difference between the cost of substation investment after the delay period less the investment for the portion of the DA scheme in the base year that delivers the benefit, all on a present worth basis.

The average substation contingency capacity ($ASCC_0$) in the base year is given by the expression:

$$ASCC_0 = [(TANC \times ACRF/100) - (TANC \times ACCP/100)] \times [1 - CPM/100]$$

where TANC = total area normal capacity (= Σ substation normal capacities in the area). ACRF = average contingency rating factor in percent (the average increase in normal capacity of transformers in the area that can be sustained for the period of the outage). This will depend on the planning policies for substations; however, under FA remote switching time is short, to assess the available capacity. Conservatively the rating should remain at normal. ACCP = average contingency capacity provision in percent of total installed substation transformer capacity. This designates the loss of capacity as a result of contingencies. It should approximate the loss of one average substation transformer in the area. For relatively small areas, it is possible to assess the value in MVA depending on the number of substations and transformer sizes. However, if an entire system were to be screened a percentage of the total installed capacity would seem more appropriate. CPM = Capacity Planning Margin in percent. This reflects the capacity margin that will allow sufficient lead time for increased substation capacity to be installed. It is dependent on whether allowance is needed for an additional transformer in a substation or an entirely new site. The value is dependent on substation sizing and design policy, as well as load growth in the area.

The average substation released capacity ($ASRC_0$) in the base year is the difference between the average substation contingency capacity ($ASCC_0$) and the area maximum peak demand ($APMD_0$).

$$ASRC_0 = ASCC_0 \times APMD_0$$

$ASRC_0$ must be $>> 0$ for there to be any benefit derived from substation capacity deferral.

The number of years N of substation capacity delay is the number of years the load will take to grow to meet the available average contingency capacity or the load plus the released capacity:

$$N = \frac{\log[(APMD_0 + ASRC_0)/APMD_0]}{\log(1 + x/100)},$$

where $APMD_0$ = area peak maximum demand in base year, $ASRC_0$ = average substation released capacity, and x = exponential growth rate in %.

The economic benefit resulting from remote feeder switching to transfer load from one substation to another following loss of a substation transformer is given as

$$\text{Benefit} = (ASCS \times Acost \times CC) \times PVF_0 \times [(DAC \times PFV_0) + (ASCS \times Acost \times CC) \times PFV_N,$$

where ASCS = average substation capacity expansion step per substation in percent of normal capacity. This percentage must be developed for a set of sample adjacent substations and then applied to the total area normal capacity. Acost = average linear cost per MVA for substation expansion, DAC = distribution automation cost associated with the load transfer benefit, CC = annual carrying charge associated with the particular equipment, and PVF = present value factor; PVF_N for year substation capacity expansion will have to be made, PFV_0 for the year that DA is implemented or the substation capacity would have to have been added — the base year.

The application of this screening method is shown in the following example, using the expressions developed above and the data in Table 8.4, the following results are obtained.

The average substation release capacity amounts to zero and thus no substation deferral possible for example 1, whereas in example 2, a delay of 4 years is possible because the released capacity is approximately 235 MVA given the assumptions on the various operating margins.

It is worth examining the performance of the screening method for the two examples to draw some conclusions. The sensitivity of the method to variations in load growth, capacity planning margin and average contingency capacity provision is shown for each example in Figure 8.14.

Clearly, in example 1, the screening method in too sensitive to changes in these parameters and gives excessive and unrealistic deferral periods for low load to capacity ratios (LCRs) to be useful. It should not be used for examination of small areas, which should be done using traditional planning techniques as suggested at the beginning of this section. The large-system example gives realistic results; however, the deferral of substation capacity by DA only takes place at relatively high load-to-capacity ratios because at low levels sufficient capacity exists within each substation to make transfer between substations unnecessary. Reviewing the results, a rule of thumb could be suggested that substation deferral possible will be between 2 and 5 years for normal growth and capacity margin policies as shown by the application area designated in the figure.

Creating the Business Case

TABLE 8.4
Table of Typical Data Used in the Examples for the Substation Deferral Calculation

		Example 1 — Small Area Adjacent Group	Example 2 — Large System
Area peak maximum demand, MVA	APMD	118	1385
Exponential load growth	X	2%	4%
Number of substations		3	30
Substation capacities		30	—	—
	60 (2 × 30)	2	20
	60 (1 × 60)	—	—
	90 (3 × 30)	1	5
	120 (2 × 60)	—	5
Total area normal capacity, MVA	TANC	210	2250
Load to capacity ratio		.56	.62
Average contingency rating %	ACRF	100	100
Average contingency capacity provision in % of TANC	ACCP	30% of TANC Equivalent to loss of one 30 MVA substation transformer	10% of TANC Equivalent to loss of four 60 MVA transformers
Capacity planning margin in %	CPM	20	20
Average linear substation capacity cost per MVA (Acost)	ALCC	$40,000	$40,000

8.8.2 Release of Distribution Network Capacity

Loss Reduction from VAR Control (VAC). The deferral of distribution network capacity resulting from automated VAR control is assumed minimal because traditional capacitor controls will apply the VAR correction at time of system peak. The benefit from DA is the accuracy and continuous nature of the control. This provides energy savings, not demand savings. However, literature on DA stresses VAR control as an important function. This is correct if there has been no power factor correction on the system, and such compensation is considered part of the DA implementation. It is traditional to justify the cost of the total compensation equipment against the capacity and loss savings in the planning phase. In the case that the compensation is considered part of the DA implementation, the benefits are directly related as follows:

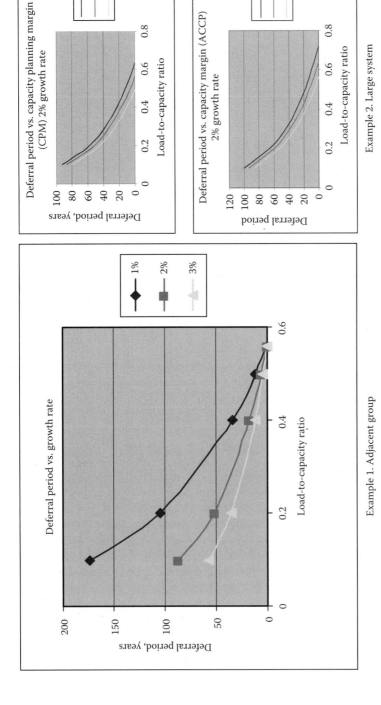

FIGURE 8.14 Sensitivity analysis for the two example cases of the screening method showing that the method is not realistic for small adjacency groups due to the sensitivity to input parameters, producing unrealistic deferral periods. *Continued.*

Creating the Business Case

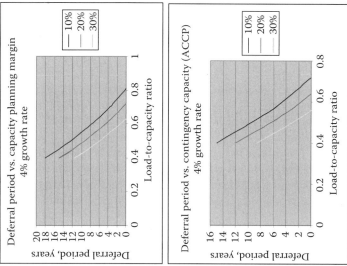
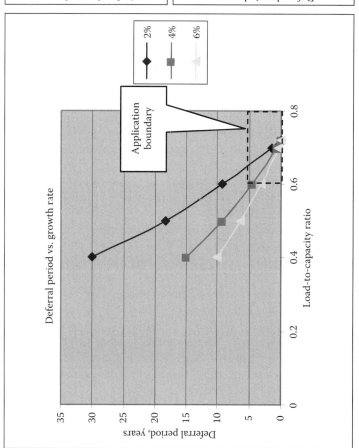

FIGURE 8.14 *Continued.*

Reduction in Demand.

$$\text{Current}_{(after)} = \text{current}_{(before)} \times (\text{power factor}_{(before)}/\text{power factor}_{(after)})$$

Reduction in Losses.

$$\text{Losses}_{(after)} = \text{losses}_{(before)} \times [(\text{power factor}_{(before)}/\text{power factor}_{(after)})^2 \times 1]$$

Avoidance of Increased Feeder Capacity due to Cold Load Pick Up Function. Cold load pick up refers to the increase in load that arises upon restoration of supply. The precise definition is important because it usually refers to the stable (nontransient, both electrical and mechanical) load increase due to noncoincidence when picking up after a cessation of power. Normally it does not include transient loads such as motor starting that lasts for a few seconds following re-energization.

The increase in load after an outage is dependent on end use load types and their characteristics as defined by their change in coincidence over time, the weather sensitivity of the loads and the 24-hour load profile. To determine the potential increase in load, the time of day, the duration of the outage and the weather sensitivity parameters must be considered. Although detailed studies using an end use model are possible, they are complex, and in the absence of actual measurements on representative feeders, in all probability, operators have gained experience with the actual system reaction to outages at various times of the year and day. Consequently, they have developed an estimate of the percentage increase in load for typical outage durations that should be compensated for in the re-energization plan (typical values 2–30%).

The avoidance of tripping due to CLPU upon re-energization can be accomplished by a number of strategies:

- Sizing or limiting the normal loading of the feeder to accommodate the increased load due to CLPU
- Logical sequential switching of the network to bring back load in steps under the LODS function
- Application of voltage conservation load control to reduce the effective load
- Temporary adjustment of protective device settings to prevent tripping for the duration of the CLPU peak

The latter three strategies can be implemented as an integrated DA function that would avoid the need for capital expenditure to increase feeder capacity. Thus, the benefit in delayed feeder capacity would offset the expense of implementing any or all of the DA CLPU functions.

This benefit is calculated in a similar manner as before for substations but must incorporate the required released capacity resulting from load reduction or potential for remote switching.

Creating the Business Case 387

8.8.3 RELEASE OF UPSTREAM NETWORK AND SYSTEM CAPACITY

Upstream capacity can be released by reducing the downstream loading. Load reduction can be achieved either directly from voltage conservation load reduction or by a DA function that reduces losses. It is assumed that such control action would be most beneficial at system peak, thus reducing peak capacity charges. A general expression for this benefit is given below based on the kW load reduction at annual network peak.

Annual benefit from release of generation and transmission (G&T) capacity resulting from a DA control function:

$$\text{Benefit} = (\text{kW load reduction}) \times (\text{generation peak capacity}^* \\ \text{charge/kW} + \text{transmission peak capacity charge/kW})$$

In the case of direct load control via load shedding, the expression for kW load reduction is

$$\text{KW reduction} = (\text{kW peak load} \times \text{kW LODS target load}).$$

In the case where demand side management (end use load shedding) is implemented, then the following expression applies that describes the degree of end use device control:

$$\text{kW reduction} = (\text{average controllable load per customer in kW customer} \\ \times (\text{number of customers under influence of load control})$$

In the case of peak load reduction via voltage conservation load control, the expression for kW load reductions

$$\text{kW reduction} = (\text{voltage change above normal regulation}) \times (\% \text{ reduction of load in kW per voltage reduction}) \times (\text{distribution load under voltage control}).$$

A typical value for both (voltage change above normal regulation) and (% reduction of load in kW per voltage reduction) ranges between 1 and 2.

In the case of VAR control, the expression for kW load reduction is:

$$\text{kW load reduction} = (\text{number of distribution circuits}) \times \\ (\text{average incremental reduction in losses in kW from automation per circuit}).$$

In the case of network optimal reconfiguration or substation transformer load balancing, the expression for kW load reduction is

---

* G&T capacity charges as reflected at the distribution level.

NORC Function.

kW load reduction = (total distribution load in kW with potential for load balancing) × (average % loss of associated circuits) × (% reduction in losses as a result of the DA function)

STLB Function.

kW load reduction = (number of substation transformers with potential for load balancing) × (average kW loss/transformer) × (% reduction in losses as a result of the DA function)

The number of substation transformers with potential for load balancing refers to those substations under the direct load balancing function or where switching of the feeder network is possible. The potential for loss reduction due to optimal feeder reconfiguration of a nonoptimized network has been shown in many planning studies to be approximately 10%, and typical distribution network losses range between 5 and 8%.

8.8.4 DISPLACEMENT OF CONVENTIONAL EQUIPMENT WITH AUTOMATION

The installation of automation equipment (digital protective relays, electronic volt/VAR controllers, etc.) potentially displaces conventional protection/control equipment for substation and feeder switches/circuit breakers, voltage regulators, shunt capacitors and data loggers. The expression for the benefits takes the same form.

Annual benefit from displacement of conventional protection and control with automated control by type is:

Benefit = [(number of pieces of equipment to be displaced or retrofitted by automation by type) × (capital cost + first-time cost for installation for conventional equipment) × (annual carrying charge)] − [(number of automated controls/automation ready devices) × (capital cost + first-time cost for installation for automation) × (carrying charge)]

8.9 SAVINGS IN PERSONNEL

These savings fall into three categories, each having different weight depending whether delivering benefits for SA or FA:

- Reduction in substation operators and control center operating levels
- Reduction in inspection visits to substations/feeder devices
- Crew time savings (CTS)

Creating the Business Case

The first two generally are the most significant benefits derived from traditional substation SCADA and SA, whereas crew time savings is very dependent on the automation intensity level and the network complexity factor (NCF). Crew time savings could loosely be classed as reduction in field staff.

8.9.1 Reduction in Substation/Control Center Operating Levels

The calculation of manpower savings is simply the difference between staffing before and after implementing the DA system.

Annual benefit = (manpower costs before DA − manpower costs after DA)

The manpower costs should be those associated with the function implemented or that corresponding to the organizational change made as a result of implementing real-time control/automation. Specifically, the manpower skill classifications could change. As an example, the cost reduction for consolidating manual control rooms with one modern DMS implementation is as follows:

$$\text{Annual benefit} = [(\Sigma\ NS \times OR_b \times MHr_b \times NSH_b) \times (\Sigma\ NS \times OR_a \times MHR_a \times NSH_a)]\ PVF_N$$

where NS = number of staff per rate classification, OR_b = manpower rates/classification used before automation, MHR_b = hours per shift before automation, NSH_b = number of shifts before automation, OR_a = manpower rates/classification used after automation, MHR_a = hours per shift after automation, NSH_a = number of shifts after automation, PVF_N = present value factor, and N = year.

8.9.2 Reduction in Inspection Visits

The ability to monitor sites remotely following the implementation of DA reduces the need to make as many site visits as in the past, where visits were necessary to gather loading data for engineering and planning and also to inspect assets for maintenance purposes.

$$\text{Annual benefit} = \\ [(\text{number of sites visited/year})(\text{MHR/site})(\text{number of visits/year})(\text{IR}) \\ \text{before DA} \times (\text{number of sites visited/year})(\text{MHR/site}) \times \\ (\text{number of visits/year})(\text{IR})\ \text{after DA}]\ PVF_N$$

where MHR/site = the duration of the visit at each site in worker hours, IR = inspectors rate per hour (currency per hour), PVF = present value factor, and N = year.

Inspection visits for asset life assessment will still be required but should be substantially reduced due to condition monitoring. In calculating the reduction in site visits, the utility should consider the types of visit undertaken before and after DA implementation.

8.9.3 Reduction in Crew Time

Development of a Crew Time Savings Expression for Interruption-Related Benefits. The ability to estimate crew time savings is fundamental to many of the cost-benefit calculations because logically, benefits are derived by moving from a manually operated environment to one of remote control/automation. Savings in crew travel time occur in both interruption- and investment-related benefits.

A basic relationship for calculating annual crew time savings for interruption-related benefits has been proposed in the EPRI work as follows:

$$\text{Benefits} = \text{fn} \,[(\text{number of faults on the feeder, number of switches to be operated, switching time per switch, crew hourly rates}) \times (\text{time for automated system to switch, control operator hourly rate})] \, \text{PVF}$$

$$\text{Benefit} = \lambda(L)[(\text{MNST/Fault})(\text{CR}) \times (\text{FAST/Fault})(\text{OR})]\text{PVF}_N$$

where λ = feeder annual outage rate/unit circuit length, L = circuit length, MNST = manual switching time, which includes travel time to switch location, CR = Crew hourly rate (including vehicle cost), FAST = feeder automation switching time, OR = operator hourly rate (including control room overhead costs), PVF = present value factor, and N = Year.

Individual relationships were developed by the EPRI work for the three distinctive phases, fault location, fault isolation, and supply restoration of operation following a fault. These are summarized in Figure 8.15.

It should be noted that the only difference between the three relationships is the need to determine the number of switches to be operated for each condition. However, the time for field crews to perform switching should reflect the travel time, which could be different in each of the conditions and time-consuming to determine for every location. Explicit enumeration of the switching operation for all feeders and possible faults in a network is unrealistic and should be limited to when detailed analysis is required on selected candidate feeders. A more general method for screening FA benefits is required.

Crew travel time for CTS associated with interruption-based benefits is a function of the feeder length, the feeder configuration and structure, the fault location, the number and location of switching devices and, when remote control is implemented, the location and number of automated switching devices. The network complexity factor and automation intensity level described in Chapter 3 will be used in developing a relationship for crew travel time and the resulting crew time savings that can be used as the basis of a general screening method. This method will account for source feeder circuit breakers, which would be

Creating the Business Case

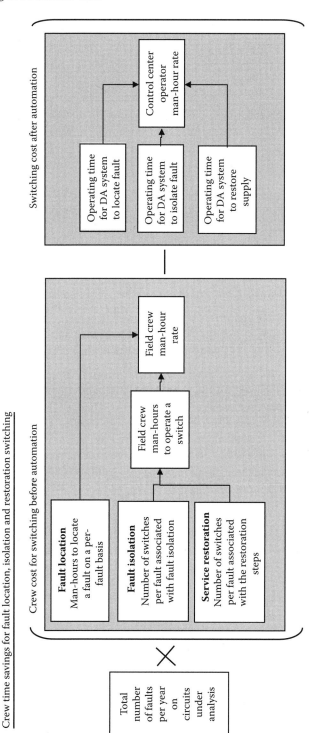

FIGURE 8.15 Annual cost-benefit calculations for crew time savings associated with an interruption.

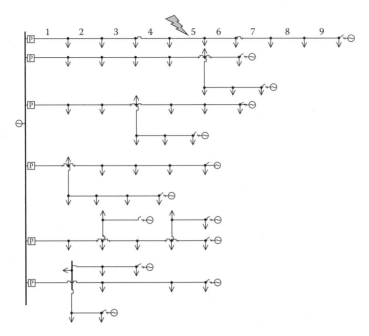

FIGURE 8.16 Generic feeder configurations used to develop an empirical network complexity factor (NCF).

automated (remote controlled) under traditional SCADA as well as successive increases in FA (increased value of AIL).

Crew Distance Traveled for Circuits without Extended Control. For every fault on the most basic distribution circuit (without any form of automation), the utility will need to dispatch a crew to locate the fault, disconnect the faulted section, carry out switching to restore healthy sections and prepare for fault repairs. Associated with these activities will be the cost of the time taken by the crew. Part of this is traveling time, including reaching the site, inspecting the circuit to find the fault and traveling between switching points; the remainder is time to operate switchgear. It is relatively straightforward to calculate this crew traveling distance and show that, if the switchgear can be made remotely operable, then much of the traveling time can be eliminated. The savings thus made can provide a significant economic benefit to the utility.

In Chapter 3, we introduced some model circuits, all of identical circuit length, reproduced in Figure 8.16 again for convenience. We shall now consider the crew travel time needed, for illustration purposes, for a fault on the fifth segment of the straight circuit, based on the given assumptions.

We shall assume the following:

- The circuit is divided into a number of equal-length segments, in the example, nine.

Creating the Business Case

- The crew always starts the fault location process at the source substation.
- Where there is a tee connection, the crew decides to inspect the line, downstream of that tee, which represents the greatest circuit length, and hence the highest probability of containing the fault.
- When the fault is found and it is a permanent fault, the crew disconnects the fault at the nearest upstream disconnection point and then travels back to close the source circuit breaker to restore supply up to that disconnection point.
- If alternative supplies are available and the fault is permanent, the crew then disconnects the fault at the nearest downstream disconnection point and supply up to that point is restored by closing the normally open point.
- Where a disconnection point is a switch that can be operated live, it can be either manually controlled, in which case the crew has to travel to the switch to be able to operate it, or it can be SCADA controlled, in which case the crew does not have to travel to the switch.
- A normally open point can be either manually controlled, in which case the crew has to travel to the switch to be able to operate it, or it can be SCADA controlled, in which case the crew does not have to travel to the switch.

The calculation would involve working out the traveling distance, and hence time, for each possible fault position and then taking an average to represent faults at any average point on the circuit. For the example, the crew would start at the source substation:

1. Travel forward along the route of the circuit, finding the fault after 4.5 segments.
2. Return 1.5 segments to open the upstream disconnecting switch.
3. Return 3 segments to the source substation to close the circuit breaker.
4. Travel forward for 6 segments to open the downstream disconnecting switch.
5. Travel forward for 3 segments to close the normally open switch.
6. Return 4.5 segments to the point of fault to initiate repairs.

This gives a total traveling distance of 22.5 segments for a fault in the fifth segment. Examining each possible fault location on this circuit gives a traveling distance ranging from 18.5 to 26.5 segments, with a total of 202.5 and an average of 22.5 segments. The reason why the average value is the same as the value for the fifth segment is that the fifth segment is in the center of a uniform model.

Repeating the calculations for each of the other circuits gives the following results in Table 8.5, noting that this table includes the network complexity factor for each circuit as described and calculated in Chapter 3.

These results have been used to develop an empirical formula to give the straight line in Figure 8.17.

TABLE 8.5
Parameters for Calculation of Crew Travel Times

Circuit Number	Average Traveling Segments	Distance Traveled as Multiple of Circuit Length	Network Complexity Factor
1	22.5	2.5	1.0
2	17.5	1.9	1.5
3	15.5	1.7	2.0
4	12.7	1.4	2.5
5	12.7	1.4	3.0
6	10.0	1.1	3.5

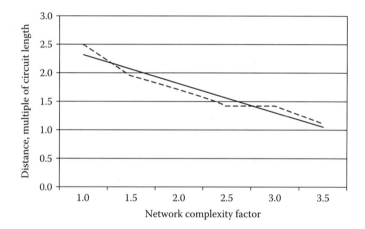

FIGURE 8.17 Approximate relationship between NCF and crew distance traveled as multiple of circuit length.

The empirical equation for the best-fit straight line is

$$D(m) = 2.77 - 0.5 \times NCF,$$

where $D(m)$ is the crew distance traveled, as a multiple of the circuit length, and NCF is the network complexity factor.

Crew Distance Traveled for Circuits with Extended Control. The calculations for the set of nonautomated circuits can be extended to cover the situation where extended control (automation) has been added to the switchgear. The results for different automation strategies are given in Table 8.6, where the AIL or degree of automation is highest in the left-hand column case 3. The automation intensity levels used in this example are described as follows:

Creating the Business Case

TABLE 8.6
Crew Travel Distance (Dm) as a Function of Automation Intensity Level

Circuit Number	Network Complexity Factor	Crew Distance Traveled as Multiple of Circuit Length for Stated Extended Control on Network			
		Manual Control AIL 0%	Source CB	Source CB and NOP	All Switches AIL 100%
1	1.0	2.5	1.7	0.9	0.5
2	1.5	1.9	1.1	0.9	0.5
3	2.0	1.7	1.2	0.7	0.4
4	2.5	1.4	1.2	0.7	0.3
5	3.0	1.4	0.9	0.6	0.4
6	3.5	1.1	0.9	0.6	0.3

- The column "Source CB" is where extended control has been added to the feeder circuit breaker in the substation at the start of the circuit, providing the facility to close the circuit breaker from a control room, and thereby removing the need for the crew to travel to the source substation to operate the circuit breaker. Therefore, crew travel distance has been reduced.
- The column "Source CB and NOP" is when, in addition to the source circuit breaker, extended control has been added to each of the normally open points. This removes the need for the crew to travel to the normally open point.
- The column "All Switches" is when, in addition to the source circuit breaker and the normally open point, extended control has been added to each of the line switches on the circuit. This removes the need for the crew to travel to each of the line switches. In addition, each of the line switches has been fitted with a fault passage indicator.

The AIL increases from left to right by column, the last column "All Switches" representing 100%, which in practice can seldom be justified.

The empirical equations for the best-fit straight line (Figure 8.18) for the model circuits now become

For no extended control — AIL 0%,

$$D(m) = 2.77 - 0.50 \times NCF.$$

For extended control of source breaker only,

$$D(m) = 1.77 - 0.27 \times NCF.$$

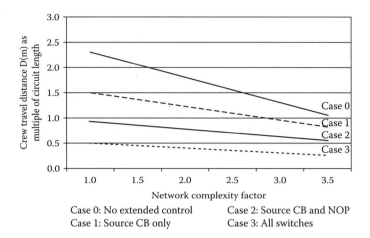

FIGURE 8.18 Plot of empirical relationship for crew travel distance as a function of NCF and AIL.

For extended control of the source breaker and NOPs,

$$D(m) = 1.10 - 0.16 \times NCF.$$

For extended control of all switches — AIL 100%,

$$D(m) = 0.59 - 0.095 \times NCF.$$

From this, it can be seen that crew distance traveled decreases as NCF increases. Because the network of identical length is packed into a smaller area rather than a long single line and because there are more NOPs, the travel distance to the nearest suitable NOP is reduced. Crew distance traveled also decreases as the automation intensity level increases because extended control removes the need to travel to the switch before operating it. This situation will be dramatically changed when circuit length is introduced; for example, a rural feeder may have a low NCF, which is countered by additional length.

Calculation of Annual Savings from Reductions in Crew Distance Traveled. It is extremely important to note that the savings in traveling distance during fault isolation switching will be matched by the same savings in switching after the fault has been repaired and the circuit is being restored to its normal condition.

So far, only the crew distance traveled, expressed as a multiple ($D(m)$) of the actual circuit length (L), per fault has been considered. The actual crew distance (ACD) traveled will, therefore, be

$$ACD = D(m) \times L.$$

Creating the Business Case

Now, the monetary cost of the crew distance will depend on the average speed of travel (speed) and the hourly cost of the crew and their transport (hourly cost).

$$\text{Cost per fault} = \text{ACD} \times \text{hourly cost/speed}$$

Now for any circuit, the number of faults per year will be the product of the length of that circuit and the annual fault rate for that circuit, or

$$\text{Faults} = \text{length (L)} \times \text{fault rate } (\lambda) \text{ per year.}$$

So, the annual cost will be

$$\text{Annual cost} = D(m) \times L \times \text{hourly cost} \times L \times \lambda/\text{speed}$$

or

$$\text{Annual cost} = D(m) \times L^2 \times \text{hourly cost} \times \lambda/\text{speed}$$

And taking into account the additional cost of restoration switching,

$$\text{Annual cost} = 2 \times D(m) \times L^2 \times \text{hourly cost} \times \lambda/\text{speed}.$$

From this formula, it is clear that the annual costs depend on the factor $D(m)$, which in turn is a function of both the NCF and the automation intensity level AIL. The total annual cost of fault switching, including postrepair restoration switching can be examined by applying the formula. Assuming the following basic circuit parameters and applying them to the expressions for $D(m)$ developed previously,

$$\text{Length of the circuit (L)} = 20 \text{ km}$$

$$\text{Hourly cost} = \$100$$

$$\text{Fault rate } (\lambda) = 18 \text{ faults per 100 km per year}$$

$$\text{Speed} = 10 \text{ km per hour.}$$

The annual crew costs of fault location isolation, and postrepair supply restoration switching are given in Table 8.7, where the savings as a function of AIL can be clearly seen. These savings can, of course, be capitalized over a number of years at a specified cost of finance to compare against the capital cost of installing the automation.

Special Case 1. Adding Source Autoreclosing to Overhead Networks. So far, the model has not taken into account the difference between overhead line networks and underground networks. The major difference, as discussed in Chapter

TABLE 8.7
Annual Costs in kUSD of Fault Switching, Including Postrepair Restoration Switching

Circuit Number	Network Complexity Factor	Annual Costs in kUSD			
		No Control AIL 0%	Source CB	Source CB and NOP	All Switches AIL 100%
1	1.0	3.3	2.2	1.4	0.7
2	1.5	2.9	2.0	1.2	0.6
3	2.0	2.5	1.8	1.1	0.6
4	2.5	2.2	1.6	1.0	0.5
5	3.0	1.8	1.4	0.9	0.4
6	3.5	1.5	1.2	0.8	0.4

3, is that overhead line networks experience transient faults and permanent faults. The model, so far, assumes that all faults cause the source circuit breaker to trip and lock out, irrespective if the fault is transient or permanent. By adding autoreclosing to the source circuit breaker, transient faults would not cause the source circuit breaker to lock out, thereby removing the need for a crew to be dispatched to patrol the line. Because approximately 80% of faults on overhead lines are transient, we can simulate in the model the effect of implementing the source autoreclose function by reducing the fault rate, currently set at 18 faults per 100 km per year, to 20% of this value, or 3.6 faults per 100 km per year, giving the results in Table 8.8.

The difference between Tables 8.7 and 8.8 shows the savings that can be made by adding source autoreclosing. For example, where the NCF is 1, then 3.3×0.7 (= 2.6) kUSD can be saved each year by adding source autoreclosing.

TABLE 8.8
Annual Costs in kUSD of Fault Switching, including Postrepair Restoration Switching, for Source Autoreclosing

Circuit Number	Network Complexity Factor	Annual Costs in kUSD			
		No Control AIL 0%	Source CB	Source CB and NOP	All Switches AIL 100%
1	1.0	0.7	0.4	0.3	0.1
2	1.5	0.6	0.4	0.2	0.1
3	2.0	0.5	0.4	0.2	0.1
4	2.5	0.4	0.3	0.2	0.1
5	3.0	0.4	0.3	0.2	0.1
6	3.5	0.3	0.2	0.2	0.1

Creating the Business Case

The capitalized value of these saving needs to be considered against the cost of providing the reclosing function, but it is likely to be economically attractive.

Special Case 2. Underground Circuits in City Areas. In general, underground circuits are shorter in length and of a higher NCF than overhead circuits; hence, the crew traveling distances are less significant and the savings made by adding extended control are reduced. In addition, because of lower fault rates, the number of faults per year is lower. It follows that the economic benefit of extended control with respect to crew time savings on underground systems will, in general, be small. The major exception to this would be in the most densely populated cities where average traveling speeds can be very low, perhaps exacerbated by traffic controls not functioning because of the electrical fault. At the same time, distribution substations in cities can be difficult to access. For example, it may be necessary to get a janitor to provide access to the substation outside normal working hours. If these factors are prevalent, detailed calculations will show whether the addition of extended control may be justified or not.

Special Case 3. Critical Length for Overhead Circuits. Examination of the $D(m)$ expression shows that the annual costs, and hence the annual savings, depend on the square of the circuit length. It is, therefore, interesting to see whether a quick rule of thumb can be derived to identify a length of circuit where crew time savings would always justify the addition of extended control.

Suppose that the cost of the full automation with FPI for the straight circuit (NCF = 1) was 40 kUSD, that the project had a life of 20 years and the cost of finance was 5%, then to be economic, the investment would need to save 3.21 kUSD per year.

For no extended control on the NCF = 1 circuit, the value of $D(m)$ is 2.5, and for the full extended control, the value is 0.5. If the sustained fault rate is 10 faults per 100 km per year, the crew cost 100 USD per hour and the average speed of travel is 10 km per hour, the equation, therefore, becomes

$$3210 = 2 \times (2.5 - 0.5) \times L^2 \times 100 \times 10/100 \times 1/100$$

where the solution for L is 28 km. Therefore, any line of length greater than 28 km would justify the addition of full automation on the basis of the savings in crew time alone.

Special Case 4. Radial Circuits. By definition, the radial circuit does not have a switched alternative supply and, hence, no normally open point(s) where extended control could be considered. Figure 8.17 includes crew traveling time for the operation of normally open points, so the curve for radial circuits would be lower in magnitude, although some savings would still be made for the extended control of midpoint switches.

Quick Crew Time Savings Estimation Tool. We have only considered a single feeder in deriving the relationships for crew time savings. The actual assessment of financial value of these savings depends on many variables related to the utility and the characteristics of the network. The financial elements, although many (e.g., hourly crew rate, vehicle costs, carrying charges), are

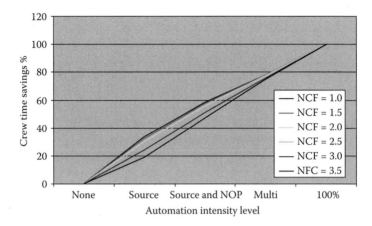

FIGURE 8.19 Percentage crew time savings as a function of AIL and NCF.

common to the network as a whole. The network characteristics though, vary from feeder to feeder, and hence a method that eliminates the need to explicitly describe every feeder would simplify the evaluation of benefits.

A comparative method is proposed where the savings are expressed as a percentage so that, if a utility knows how much this element of crew time costs each year before automation or at a particular AIL (e.g., feeder source circuit breaker under SCADA control), then an immediate percentage reduction can be estimated for feeders represented by NCF as AIL is increased. The envelope of savings is given in Figure 8.19.

For example, from the graph we can see that for a feeder of standard total unit length,

- By adding control of the source breaker only, a saving of 34% can be made for simple networks (NCF = 1.0) and 19% for complex networks (NCF = 3.5).
- By adding control of the source breaker and the normally open point, a saving of between 47% and 59% can be made, depending on the type of network.
- By adding control to each and every switching device, even if an unlikely situation in reality, there would be no need for post fault restoration and fault location switching (i.e., the saving would be 100%).

This method can now be used to screen an entire network or zone comprising many circuits or perhaps even the whole utility using weighted data. Suppose a utility comprised of 1000 circuits could be broken down as in Table 8.9.

Then we can calculate the distance multiple, $D(m)$, from the empirical formulas and hence, knowing the number of circuits of each length and NCF, the actual distance traveled. If we know the savings that can be made, in percentage terms,

Creating the Business Case

TABLE 8.9
List of Feeder Categories in Terms of NCF and the Percentage Savings Resulting from Increasing AIL

Circuit Specific Data				No Automation		% Savings per Circuit with Selected Extended Control			km Savings Made in Actual Crew Distance		
Counts	Type	NCF	km	D (m)	ACD km	Source	Source and NOP	Multi	Source	Source and NOP	Multi
50	OH	1	5	2.27	568	34	59	78	193	335	443
30	OH	1	10	2.27	681	34	59	78	232	402	531
50	OH	1	15	2.27	1703	34	59	78	579	1004	1328
75	OH	1	20	2.27	3405	34	59	78	1158	2009	2656
50	OH	2	5	1.77	443	32	57	78	142	252	345
50	OH	2	10	1.77	885	32	57	78	283	504	690
25	OH	2	15	1.77	664	32	57	78	212	378	518
25	OH	2	20	1.77	885	32	57	78	283	504	690
45	OH	3	5	1.27	286	24	51	76	69	146	217
30	OH	3	10	1.27	381	24	51	76	91	194	290
40	OH	3	15	1.27	762	24	51	76	183	389	579
30	OH	3	20	1.27	762	24	51	76	183	389	579
50	UG	1	5	2.27	568	34	59	78	193	335	443
30	UG	1	8	2.27	545	34	59	78	185	321	425
50	UG	1	12	2.27	1362	34	59	78	463	804	1062
75	UG	1	15	2.27	2554	34	59	78	868	1507	1992
50	UG	2	5	1.77	443	32	57	78	142	252	345
50	UG	2	8	1.77	708	32	57	78	227	404	552
25	UG	2	12	1.77	531	32	57	78	170	303	414
25	UG	2	15	1.77	664	32	57	78	212	378	518
45	UG	3	5	1.27	286	24	51	76	69	146	217
30	UG	3	8	1.27	305	24	51	76	73	155	232
40	UG	3	12	1.27	610	24	51	76	146	311	463
30	UG	3	15	1.27	572	24	51	76	137	291	434
Total	1000				20,568				6492	11,714	15,964

Counts = number of feeders per category, ACD = actual crew distance traveled.

* Taken from Figure 8.12.

then the actual savings made in crew travel time for each circuit can be calculated and the total for all circuits can be compared to the total with no extended control to derive a percentage. When this percentage is applied to the utility estimate for actual crew costs, then the potential savings can be very quickly estimated.

In this example, we can see that average savings of 32%, 57% and 78% can be made for increasing automation intensity levels and, if the utility estimate for present crew costs is taken as, say, 2 MUSD, the potential savings are indeed significant.

8.9.4 CALCULATION OF CREW TIMES SAVINGS ASSOCIATED WITH INVESTMENT- AND OPERATION-RELATED SAVINGS

Crew time savings is also involved when feeder reconfiguration is used to achieve benefits from the following:

- Loss reduction due to dynamic optimal reconfiguration (operation-related savings)
- Capital deferral of substation capacity (investment-related savings)

These crew savings are purely dependent on the number of switching operations that have to be carried out and their frequency, which for the latter case is substation transformer failure rate dependent. Loss minimization is dependent on the load characteristics of the network. In both cases, it is questionable whether there is a savings in crew time because neither function would have been practical without remote control of switches, particularly the NOPs, and only a cost covering implementation of automation seems relevant. However, the benefit derived from these two areas is expressed below:

$$\text{Annual benefit} = (NRS)(NSS) \left[(MNST)(CR) \times (STCO)(OR) \right] PVF_N,$$

where NRS = number of reconfiguration sets per year, NSS = number of switches per reconfiguration step, MNST = manual switching time/switch (includes travel time to the switch location), CR = crew hourly rate (including vehicle cost), STCO = switching time taken by control room operator/switch reconfigured, OR = operator hourly rate (including control room overhead costs), PVF = present value factor, and N = year.

This assumes that the capital cost of implementation is carried by another function, and thus only an incremental benefit of that investment should be considered.

Whereas the estimate of CTS for interruption-based benefits requires consideration of many switching options dictated by the fault location, CTS for operation- and investment-related benefits requires fairly static descriptions of the switches to be operated. Although simpler in concept, the reconfiguration sets (switching plans) must be determined by thorough engineering analysis and planning.

8.9.5 REDUCED CREW TIME AND EFFORT FOR CHANGING RELAY SETTINGS FOR CLPU

In the event that the utility solves CLPU trips by manually changing the protection settings at primary substations, then DA will allow settings to be changed remotely for CLPU incidents, thus removing the need to visit substations to perform the task.

Creating the Business Case

Annual benefit = (total number of feeder interruptions per year)
× (time to change protection setting including travel time) × (manpower rate)

= (feeder fault rate/unit length/year) × (total network circuit length)
× (time to change setting in hours) × (manpower rate/hour)

8.10 SAVINGS RELATED TO ENERGY

Energy revenue is reduced as a result of an interruption or from losses in the network (technical and nontechnical). Reduction in outage time reduces energy not supplied (ENS), and optimal equipment loading of the network also reduces technical losses. Nontechnical or commercial losses are not considered in this book and are more associated with energy balancing from metering on the network and at the customer.

8.10.1 REDUCTION IN ENERGY NOT SUPPLIED SAVINGS DUE TO FASTER RESTORATION

Unless there is a continuous alternative supply available, any fault on a power system that causes protection to operate will cause loss of supply to one or more customers. We have already described in Chapter 6 how this loss can be characterized in terms of reliability indices such as SAIFI and SAIDI. The effect on network reliability as a function of the automation strategy (protection and AIL) has been discussed in preceding chapters, and thus the values of indices used in the benefit calculation must reflect network performance before and after DA implementation. Chapter 3 also describes the dependence on the magnitude of energy loss on the feeder load characteristic resulting from the combination of customers supplied from the feeder. The annual energy not supplied (AENS) is given by

$$\text{AENS} = \text{SAIFI} \times \text{CAIDI} \times \text{annual load factor} \times \text{annual peak demand of the particular load.}$$

This is converted into the annual revenue lost (ARL) by applying the appropriate monetary value for that energy. In terms of a benefit to justify distribution automation, as discussed later in Section 8.12, this could be as little as the utility's profit on a kWh sold, the customer's perceived cost of lost power, or even the energy-based penalty, should it be imposed.

$$\text{ARL} = \text{AENS} \times \text{cost of energy}$$

If we now consider, as an example, a typical urban underground network, we can use the data for SAIDI (SAIFI × CAIDI) to calculate the kWh lost as a result of faults according to Table 8.10.

TABLE 8.10
Parameters from a Typical Urban Underground Network Comprising 12 MV/LV Substations of 300 kVA Capacity Operating at a Power Factor of 0.9 and an Annual Load Factor of 0.5*

Substation	SAIDI	kWh Lost
7	0.712366	96
9	1.174950	159
12	1.174950	159
15	2.582590	349
16	1.174950	159
19	2.582590	349
23	2.582590	349
24	2.582590	349
25	2.582590	349
28	2.582590	349
29	2.582590	349
33	0.712366	96
Total		3109

* On average, each substation serves 80 customers.

If, for example, the difference between selling price and purchase price (profit margin) for each unit was $0.02, then the utility would lose just $62 per year as a result (3109 kWh × $0.02). This is, of course, not a large sum, but for networks with higher values of SAIDI, which would typically be long overhead systems, the loss would be correspondingly higher. The significance of energy not supplied in justifying any automation scheme is very limited unless the value of unsupplied energy is increased by either penalties or loss of customer retention (customer loyalty).

$$\text{Annual benefit} = ALR_{NA} - ARL_{DA}$$

where ALR_{NA} = annual lost energy without automation and ALR_{DA} = annual lost energy with distribution automation. The speed of restoration provided by the DA functions remote feeder switching, FLIR and CLPC (VCLC) is reflected in an improvement to the reliability index SAIDI. Calculation of the respective benefit contributions of each of the DA functions must be considered to avoid double-counting.

8.10.2 Reduced Energy Revenue Due to Controlled Load Reduction

Load Reduction Due to Voltage Control for Load Conservation. Load reduction can be achieved by controlled lowering of the system voltage to achieve a target load level, provided the target is achievable within the legal operating voltage limits (114–126 V in the United States, and in the U.K., 230 V, +6–10%). The degree of load reduction per customer is directly related to the voltage dependency of the loads and the percentage of that load type at the load points on the feeder. This information is difficult to determine and is amassed by special field studies into the dynamic nature of loads. A typical value is 20% of AMDD* for the voltage-dependent portion but will vary considerably for loads with air-conditioning.

The annual reduction in energy due to voltage-controlled load reduction is

Value of ENS = (voltage change above normal regulation) ×
(% reduction of load in kW per voltage reduction) ×
(distribution load under voltage control) × (8760) × (energy cost in $/kWh).

Typical values for voltage change would range from 1 to 2 V and the load variation is probably in the same order in percent.

Load Reduction Due to LODS. Direct load reduction or load shedding is implemented either by a set of predefined switching actions that reduce the peak loads in steps or by a smart switching algorithm that determines candidate switches from associated load values. Direct load control of end use devices is also possible (demand-side management). The resulting reduction in revenue is estimated as follows:

Value of ENS = (average controllable load per customer in kW/customer) ×
(number of customers under influence of load control) ×
(number of hour/year load control is active) × (energy cost in $/kWh)

8.10.3 Energy Savings Due to Technical Loss Reduction

Certain FA functions are directed at reducing network losses by operating the existing network in a more optimal manner. These losses can be reduced in the following way:

- Improved volt/VAR control of the feeder regulators and shunt capacitors
- Load Balancing between transformers and interconnecting feeders of adjacent substations

8.10.3.1 Loss Reduction from Feeder Volt/VAR Control

Loss Reduction on Distribution Feeders (Transformer Core Loss) Due to Improved Volt Control. Automated remote control of feeder voltage regulators usually maintains the voltage at the set points with a tighter tolerance than traditional methods. This can be even more effective when operating within the total DA integrated volt/VAR control function that controls substation regulation

---

* AMDD — Average Maximum Diversified Demand.

and compensation equipment. Maintaining tight voltage tolerances potentially reduces distribution transformer no-load losses (core losses) that would increase as a powered function of voltage increase.

Annual benefits from potential maintenance of voltage at a set point on a per feeder basis are

Benefit = (total number of distribution transformers per feeder) ×
(average kW core loss per transformer) ×
(% change in core loss for a per unit voltage change) ×
(per unit voltage excursion change between traditional and automated set point voltage control) ×
8760 × (utility's hourly cost of energy).

Typical values are as follows: average kW core loss per transformer ranges from 0.5 to 0.2% of name plate rating for small (20 kVA) to large (1000 kVA) distribution transformers, % change in core loss for a per unit voltage change (3%), per unit voltage excursion change between traditional and automated set point voltage control (1.0).

The benefit from load reduction accomplished by lowering the voltage assumes that no violation of the voltage regulatory limits will occur. It must be countered with the corresponding ENS due to the voltage reduction (Section 8.10.2).

Loss Reduction on Distribution Feeders Due to Improved VAR Control (BVAR, FVAR). Automated remote control of feeder-switched shunt capacitors improves voltage regulation and reduces VAR flow and improves power factor. The benefit from loss reduction due to optimal VAR control within the integrated volt/VAR function is estimated from the following expression.

Annual benefits from power factor correction:

Benefit = (number of circuits) × (average reduction in losses due to automatic VAR control) × (number of hours per year automatic control is active) × (utility's hourly energy cost)

The average reduction in feeder losses is calculated by considering the present percentage loss level and the assumed power factor correction the installed feeder capacitors will provide. The average reduction may only approximate 100 kW and the number of active hours of control 1300 hours per year).

Benefit estimates derived on a feeder-by-feeder basis will be time-consuming to obtain a systemwide estimate, and system averages should be taken for a general estimate. The utilities cost of energy will also vary between peak and off-peak periods, and a weighted average over the loss reduction period should be considered.

In conclusion, this benefit is not for power factor correction, but for the benefit from improvement of capacitor control. If a utility has not installed capacitors

Creating the Business Case

for power factor correction, the benefits from so doing should really be set against the cost of capacitor installation. However, if capacitors are being installed as part of the automation scheme, then the full capacitor cost (primary plus remote control) should be used against the full benefits.

Loss Reduction Due to NORC and Load Balancing. Load balancing to reduce losses is achieved by reconfiguring the network to minimize losses. During network planning, the selection of NOPs is often determined to minimize losses at peak load. In terms of FA, network reconfiguration is made through operation of remotely controlled feeder switches that can allow dynamic changes to the feeder network. It is usually driven by a central loss minimization application within the DMS servers. Reconfiguration by FA will only deliver benefits when there is a need for the location of NOPs to be altered. This will only occur should the load characteristics of adjacent feeders have a significantly different daily or seasonal profiles to justify regular reconfiguration or there be permanent loss of a substation transformer or major feeder.

Annual benefits from feeder reconfiguration for load balancing:

$$\text{Benefit} = (\text{load of circuits being reconfigured in kW}) \times (\text{average network losses in \%}) \times (\% \text{ reduction in losses achieved by reconfiguration}) \times (\text{number of hours per year automatic control is active}) \times (\text{utility's hourly energy cost})$$

If only STLB is implemented, then replacing the network data to show only those transformers under the DA function modifies the benefit equation as follows:

$$\text{Benefit} = (\text{number of transformers under the DA function}) \times (\text{average load loss/transformer in kW}) \times (\% \text{ reduction in losses achieved by DA function}) \times (\text{number of hours per year automatic control is active}) \times (\text{utility's hourly energy cost})$$

Although the loss minimization application now regularly appears in DMS specifications, the authors conclude that the effectiveness of such an application for optimally planned systems is marginal given the potential inaccuracy of the data, the limited degrees of freedom within the network and the practicalities of performing repeated reconfiguration switching with the potential to interrupt customers for little value. At most, only seasonal reconfiguration maybe justified.

8.11 OTHER OPERATING BENEFITS

There are other operating benefits that develop as a result of implementing DA. Some are direct benefits and others, indirect. They can be classified under the following headings:

- Repair and maintenance savings
- Improved information on assets
- Improved customer relationship management

8.11.1 Repair and Maintenance Benefits

The introduction of modern intelligent electronic devices necessary to implement the DA functions provides increased flexibility and communication, and fewer moving parts, which potentially should reduce the need for maintenance and repair. The flexibility also simplifies the task of setting and resetting operation regimes. IEDs refer to the whole range of different protective devices providing autoreclosing, bus fault protection, instantaneous and time overcurrent protection, substation transformer protection, under frequency as well and all the control functions at remote devices such as volt and VAR control and fault passage indication.

The general benefit expression for implementing modern IEDs over traditional electromagnetic devices makes the assumption that the failure rate of the new devices is substantially lower and that the devices do not require a regular inspection and testing schedule, all diagnostics being conducted remotely.

Annual benefit = (maintenance cost of traditional electromagnetic devices − maintenance cost of modern IEDs)

= [(number of EMD devices of same type) ×
(device type annual failure rate) × (man-hours effort to correct failure) +
(yearly test schedule × number of tests for device type) ×
(man-hours effort to perform the test)] × (man-hour rate for

EMD technician) − (number of IEDs of type replacing EMD) ×
(IED type annual failure rate) × (man-hours to restore the
IED to operation) × (man-hour rate for IED technician)

8.11.2 Benefits from Better Information (DMOL)

Detection of Feeder Capacitor Bank Malfunction. Under the VAR control function, the increase in VARs that should result in switching on capacitor banks can be monitored to determine whether a bank has failed. The savings from monitoring are threefold:

- Reduction of inspection visits, the savings of which can be calculated as explained in Section 8.9.2
- Reduction in losses as a result of the time a bank is malfunctioning
- Avoidance of the capacity requirements as a result of the time a capacitor bank is malfunctioning

Each of the expressions to calculate the related benefits has been covered previously in Section 8.8, Section 8.8.2, and Section 8.10.3. It is the determination

Creating the Business Case

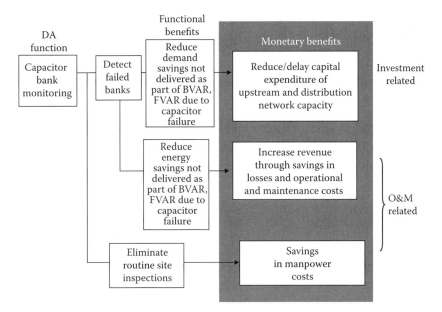

FIGURE 8.20 Benefit flowchart for capacitor monitoring as part of the DA VAR control function.

of the number of hours that a capacitor bank is malfunctioning that determines the energy savings lost and is a function of the capacitor bank failure rate and the interval between inspection visits. Switched capacitor banks forced out due to blown fuses or mechanical control failure can amount to 20% of total banks in very severe cases and lasting for months due to the inspection cycle. The potential benefits are summarized in the benefit flowchart in Figure 8.20.

Improved Operating Decisions Based on Current Information. Up-to-date information on system status and loading directly prior to a fault will enable the operator to reassemble the system to achieve maximum restoration possible in the minimum time. Further, with implementation of advanced decision tools the likelihood of an error and of overloading the remaining healthy network will be reduced. This benefit is subjective and of the soft category because it requires an estimate of the value of this information or the value of a bad operating decision that could have been avoided.

Annual benefit = (number of faults resulting from a bad operating action) ×
(estimated worth of such failures per failure) ×
(the expected % improvement from better data)

Improved Engineering and Planning Decisions Based on Improved Data. Distribution system planning has always suffered from not only lack of but also inaccuracy of data. DA provides the opportunity to improve that data model of the network, particularly of loads and the loading information. The benefits derived from such information will improve capacity expansion planning and also asset utilization and management. The worth placed on these improvements is again subjective, and this benefit should be considered not only soft but also indirect. The benefit relationship is similar to that for improved decision making, and the benefits for improved asset management could be a percentage value applied to the total network asset equity base or the annual CAPEX figure instead of the number of devices monitored. As a minimum, 5% improvement in asset utilization could be claimed as a result of better information.

$$\text{Annual benefit} = (\text{number of devices monitored}) \times (\text{estimated worth of the data per device})$$

8.11.3 Improved Customer Relationship Management

Many DA functions provide better service and information to customers, hence improving customer satisfaction. This improved "shine" is difficult to quantify because some of the improvements are perceived. For example, a utility that restores supply within an hour but is unable to provide its customers with any information on the outage is perceived to be less capable than the utility that responds to customers' trouble calls with information about the incident yet also return power in one hour. The value of a customer complaint may be assessed as a penalty value; however, the penalty has been taken as the value of ENS in a previous benefit calculation, the utility must place some value on goodwill that is in addition to the penalty. The following benefits can be considered:

Decrease in Outage Time Improves Customer Satisfaction. Remote switching combined with the FLIR function provides faster restoration times and, therefore, improves customers' goodwill.

Reduction in Low-Voltage Complaints. Claims for damage to utilization equipment due to low voltage (below statutory levels) should be reduced with improved voltage regulation control as part of the DA integrated volt/VAR function.

The savings expression for the above two benefits has a common form:

$$\text{Annual benefit} = (\% \text{ reduction in customer complaints}) \times (\text{cost per customer complaint}) \times (\text{average number of complaints per 1000 customers}) \times (\text{number of customers on circuits under the DA function control}/1000)$$

A typical value given in the EPRI report suggested the cost of a customer complaint in the United States could be assessed between $200 and $500, and the number of complaints could amount to 3–5 per 1000 customers. Reduction in outage-generated customer complaints is a function on the improvement in SAIDI as a result of FLIR and may be in the order of 5%.

Customer Loyalty/Retention. A significant number of industrial and commercial customers need a reliable supply of electricity to be able to carry on their business process. If the supply does not meet the customer needs, and that might

Creating the Business Case

be in terms of outages, voltage dips and spikes, supply voltage or harmonics, then the process must be stopped and the business will lose production or be faced with additional costs.

Consider a real example of a factory producing short half-life radioactive isotopes for medical use. The production process was such that the product was manufactured for immediate and urgent delivery directly by express carrier to the nearby airport for immediate dispatch to the customer. Any failure of the electricity supply affected the production and meant a wasted product and a cost of $70,000. The factory was located on an urban underground network that was affected by voltage dips caused by transient faults on the large overhead feeder connected to the same source substation, and it was found that its equipment was sensitive to these voltage dips.

A major factor in assessing the cost of an interruption in supply is the notice that the customer receives of that interruption. Many sensitive customers can tolerate a supply interruption if they were to receive, for example, two hours' notice. They would be able to reschedule their production, and the resulting costs of the outage, while probably not being zero, would be less than for an interruption without notice. But, of course, most outages are as a direct result of faults on the network, and these normally occur without giving notice.

The costs that such customers incur on the failure of the electricity supply to meet their needs are, of course, a cost to them directly and not a cost to the local utility. However, as the costs to customers rise, then there will come a point where they will stop taking the poor supply from the utility and generate their own independent supply. Assuming that the locally generated supply was of the correct quality, then they will move away from the local utility supply when

Cost of supply from utility + cost of supply failures > cost of local generation.

The ratio of the estimated cost of lost energy to the cost of local generation can be viewed as the customer decision ratio of retention index. It provides an indication of when customers may consider substituting the utilities supply by installing their own generation. An example below shows the results of such a study in a poorly performing area of a utility. All customers with an index greater than unity are candidates for installing their own generation. Figure 8.21 shows the number of customers who are already using their own generation instead of supply from the utility.

This would represent a loss to the utility of the supply of electricity, not just for a few hours a year as discussed above, but for 8760 hours per year. It also means that the capital investment made by the utility to connect the customer to the utility network is producing no return on the capital employed.

8.12 SUMMARY OF DA FUNCTIONS AND BENEFITS

The quantification of DA benefits has taken a circuitous route, starting with general considerations and a suggested opportunity matrix. The importance in

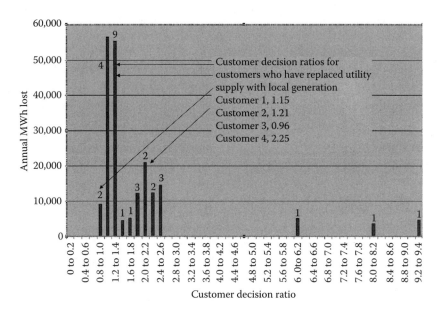

FIGURE 8.21 Customer decision ratio/loyalty showing the number of customers per band and the total annual MWh at risk to the utility for each decision ratio value.

functional dependencies and potential benefits sharing was also covered before specific benefit expressions for each of the major benefit types, resulting from a particular DA function, could be developed. The final task is to complete the work by constructing a revised benefit opportunity matrix cross-referencing all benefit expression developed in Section 8.8 to Section 8.11.

8.13 ECONOMIC VALUE — COST

Whatever the technical cost to the business of an outage in terms of duration and frequency and the associated resources needed to restore or minimize further occurrences, there is a monetary value to the event that has prevailed throughout. Thus, all functional benefits need to be converted to monetary benefits. Some of the costs are straightforward but the cost of energy to be used in the calculation of ENS is open to different interpretations, which will be explored in this section.

The cost of poor quality is quantified differently by a utility than by its customer. The larger the disparity, the more at risk the utility is of the customer taking remedial action. A utility must be aware of its customers' threshold when deciding the value of a remedial program and when converting functional benefits into monetary terms. This section concludes the discussion on calculation of benefits by looking at different values a utility may place on energy costs, particularly when evaluating energy not supplied.

TABLE 8.11
Cross-Reference Table for DA Functions and Chapter Sections Containing Benefit Expressions

	Abbreviation	Capital Deferral/ Displacement	Manpower Savings	Savings in Energy	Other Operating Savings
Remote SWitch Control	RSWC	8.8.1	8.9.1, 8.9.4	8.10.1	
Fault Location, Isolation and Restoration	FLIR		8.9.4	8.10.2	8.11.3[a]
Network Optimal ReConfiguration	NORC		8.9.5	8.10.3	
Substation Transformer Load Balance	STLB			8.10.3	
Cold Load Pick Up	CLPU	8.8.3	8.9.6	8.10.1	
LOaD Shedding	LODS				
Integrated Volt/VAR Control					
VAR Control	VARC	8.8.3		8.10.3	8.11.2
Volt Conservation Load Control	VCLC	8.8.3		8.10.3	8.11.3[a]
Data Monitoring and LOgging	DMLO		8.9.2, 8.9.4		8.11.2
IED-Based Control and Protection Equipment					8.11.1[b]

[a] CRM benefits.
[b] Repair and maintenance.

8.13.1 Utility Cost

Traditional. In assessing the benefit to cost ratio for any investment in automation, the utility calculates its margin based on standard cost accounting practices for operating costs (burdened man-hour rates, maintenance costs, energy cost, etc.) and capital costs, all on an annual basis. The cost of energy can vary from an average purchase price to the utility or as an opportunity cost of the selling price to the end user in the situation where the distribution entity still retains the supply business. The cost of losses is a direct cost, being a combination of the energy cost with a capacity investment component necessary to cover the network capacity to accommodate the losses. It is usual for most utilities to have set an energy value policy as part of their cost of service pricing for delivering the energy. In reality, the economic value of energy not delivered, by itself, using this type of evaluation is seldom sufficient to justify any network performance improvement measures in contrast to a penalty-based environment.

Penalties. In several countries where deregulation has been implemented, the regulator is designing incentives in the form of penalties, and in some cases rewards, to directly encourage utilities to improve their performance. These are output focused, and any performance improvement project can be compared with a specific economic penalty value. Under a penalty regime, the benefit-to-cost calculation is dominated by the resulting value of energy not supplied because the regulator tends to set the value closer to the customer's cost for loss of supply than that of the utilities. The major standards relate to interruptions to supply without notice and are typically either interruption based or duration based, or possibly combinations of the two. The following are typical examples:

- An interruption-based penalty for each customer who received more than a predefined number of interruptions in a year (e.g., 5).
- A duration-based penalty (1) for each customer whose cumulative duration of interruptions in a year totals more than a specified duration (e.g., 6 hours).
- A duration-based penalty (2) for each customer interruption that exceeded a defined duration (e.g., 18 hours).
- An interruption- and demand-related penalty that increases the interruption-based penalty as a function of the load actually interrupted. This would mean that the larger customers, probably not domestic customers, were associated with higher penalties.
- A duration- and energy-related penalty that increases the duration-based penalty as a function of the load actually interrupted. Such a penalty would be based on the kilowatt hours (units) of electricity that was not supplied to the customer because of the outage and would mean that again the larger customers, probably not domestic customers, were associated with higher penalties.

The key definitive parameters of most penalties are illustrated in Figure 8.22, which shows two duration thresholds:

- "A" is the duration at which a fault is statistically defined as permanent or sustained (momentary/permanent boundary). This varies from country to country and from regulatory period. A value of 5 minutes is typical, although in the U.K. the initial value of 1 minute has now been extended to 3 minutes.
- "B" is the duration at which a permanent interruption is of sufficient duration to warrant a penalty.
- The interval between "A" and "B" represents where penalties are paid on either accumulation of interruptions or as a function of actual duration and load.

Some regulatory authorities specify that the penalties are made in cash to the affected customers, whereas others apply the penalty directly to the utility, for example, through a restriction of annual operating profits that are permitted. Both provide the utility with one measure of the effectiveness with which its distribu-

Creating the Business Case 415

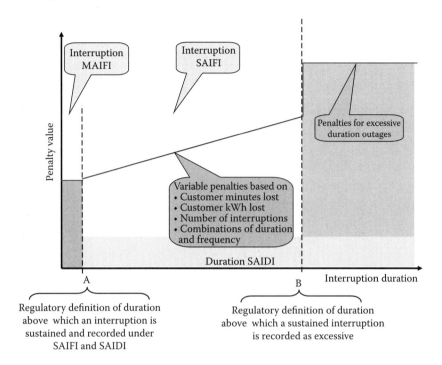

FIGURE 8.22 Summary of penalty types and defining thresholds.

tion network delivers electricity to its customers and offers some financial motivation for improvement.

Example Calculation of Penalties. The effects of these penalties can be illustrated by considering some typical values applied to a long rural network. The rural long-overhead circuit is modeled with 30 km of overhead line, controlled by one circuit breaker at the source substation. Because this circuit breaker is not fitted with autoreclosing, then all faults, whether transient or permanent, on the system cause the circuit breaker to trip.

The overhead line fault rate is modeled as 37.2 faults per 100 km per year for transient faults and 12.4 faults per 100 km for permanent faults. Repair time is 5 hours, and the three sectionalizing switches can be operated in 1 hour.

There are 12 load points, each with 80 customers and a total substation load of 300 kVA each.

With this information, the average reliability of the circuit has been calculated as follows:

- SAIDI = 36.51 hours per year
- SAIFI = 11.51 interruptions per year

Table 8.10 shows the distribution down the feeder of reliability indices and results.

TABLE 8.12
Reliability Indices for a Typical Feeder

					kWh Lost	
Substation	Customers	SAIFI	SAIDI	Load	Per Customer	Total
7	80	11.51	23.72	300	40.04	3203
9	80	11.51	23.72	300	40.04	3203
12	80	11.51	21.76	300	36.72	2937
15	80	11.51	26.66	300	44.98	3599
16	80	11.51	26.66	300	44.98	3599
19	80	11.51	48.65	300	82.09	6567
23	80	11.51	48.65	300	82.09	6567
24	80	11.51	48.65	300	82.09	6567
25	80	11.51	48.65	300	82.09	6567
28	80	11.51	48.65	300	82.09	6567
29	80	11.51	48.65	300	82.09	6567
33	80	11.51	23.72	300	40.04	3203
Average	na	11.51	36.51	na	na	na
Total	960	na	na	3600	na	59,146

Considering how the penalties discussed might apply to the sample network, we would find the following:

Interruption-Based Penalty. Because of the single protection point (the source circuit breaker), all customers have the same number of interruptions per year at 11.51. Therefore, if the penalty structure was, say, a $50 payment to each customer who had more than five interruptions in the year, the utility would need to meet the penalty of 960 × $50 or $48,000. However, the actual outages are caused by both transient faults and permanent faults. If transient faults could be dealt with by a source recloser, then the actual outages that count for this penalty would be caused by permanent faults only. The result would be a significant drop in sustained outages, with a corresponding reduction of penalties.

Duration-Based Penalty (1). If the penalty structure was, say, a $40 payment to each customer who was off supply more than 6 hours in the year, then the utility would need to meet the penalty of 960 × $40 or $38,400. If, however, the structure was changed to be a $60 payment to each customer who was off supply more than 30 hours in the year, then only the customers supplied from substations 19, 23, 24, 25, 28 and 29 would be eligible. The utility would need to meet the penalty of 480 × $60 or $28,800.

Duration-Based Penalty (2). The table giving the data for SAIDI shows the total outage for the year in question but does not break this down into different durations for different faults because most reliability software assumes that a fault in a given location of the network will always have the same restoration time, whether that restoration be by switching or by repair action. In practice,

Creating the Business Case

because switching time and repair time can depend on other factors such as severe weather conditions, which will delay switching and repair times. Consequently there will be a variation of the actual interruption durations, although this variation will not affect the average annual durations shown here.

If, however, actual measurements showed that 30 customers experienced one interruption that exceeded a certain limit, then this aspect of penalties could be calculated. Suppose the penalty was based on $75, which applied for each interruption that exceeded 18 hours, then the utility would need to meet a penalty of 30 × $75 or $2,250 per year.

Interruption- and Load-Related Penalty. This category of penalty is based on the product of the number of interruptions and customers load in kW. The relevant load could be either the maximum demand of the load or the load at the time of the fault. If we assume maximum demand, then from the table, we can see that the product is 11.51 × 12 × 300 = 41,436 kW for interruptions and, if the penalty rate was $1.50 per kW interruption, then the utility would need to meet a penalty of $62,154 per year.

The actual distribution of this penalty would depend on the load of an individual customer; for example, a customer with a load of 100 kW would produce a penalty of $1,727, whereas a customer with a load of 5 kW, probably a domestic customer, would incur a penalty of $86 per year.

If the penalty was based on the load at the time of the interruption, then either a direct measurement of the load interrupted, using automated meter reading, or additional calculations based on load profiling and the start time of the interruption would be needed. Because the load interrupted could not be more than the maximum demand, it follows that actual load-related penalties would give reduced values.

Duration- and Load-Related Penalty. This category of penalty is based on a value for a kWh not supplied. By taking the load at each substation in the model and assuming a power factor of 0.9 and a load factor of 0.5, it is possible to calculate the kWh of electricity not supplied due to faults during the year. From the table, this is shown to be 59,146 kWh per year.

At the time of writing, the Norwegian regulator has initiated the penalty rate of 38 Norwegian krone (NOK), or approximately $4.20, for a kWh not supplied to an industrial or commercial customer and 2 NOK, or approximately $0.20 for a kWh not supplied to other customers. If we use these rates, we would find that the utility would need to meet the penalty of 59,146 × $4.20 or $248k if all the customers were industrial or commercial, or if all the customers were nonindustrial or noncommercial, the utility would need to meet the penalty of 59,146 × $0.20 or $11.8k. In practice, a mix of customers would ensure that the actual penalty fell within these two extremes.

The actual penalties that a utility might incur can be very easily determined by a fault and interruption reporting scheme such as National Fault and Interruption Reporting Scheme (NAFIRS), which is operated by the U.K. electricity supply industry. When the reporting scheme that is used to calculate the penalties is controlled by the same organization on which the penalties are to be imposed, it is natural for the regulation authorities to have a random checking scheme. Such a system can be created by a dial-up reporting device, located at a few strategically placed individual customers, which reports a supply failure.

Typical Penalties. Performance-based regulation (PBR) is gaining ground across the globe because out-of-service regulation is becoming obsolete. Regulators are introducing PBR plans for distribution companies to ease burdens of the regulatory process, to improve customer service, to increase profits for utility shareholders and to lower rates to consumers.

North America. A number of examples of PBRs in North America were presented in the *Financial Times*.* The severity of the PBR varied from state to state. The emphasis is to encourage true reporting of performance statistics as part of the process as illustrated by the new reporting requirements for utilities in Wisconsin and Illinois, rather than the establishment of predefined reliability standards and penalties as implemented in California and New York. However, the natural use of this information will be for the regulator to benchmark utilities and set average performance figures. In Wisconsin, reliability statistics were to be reported annually (SAIFI, SAIDI and CAIDI) together with details in their annual report of network improvement projects, reliability programs, maintenance completed and planned, together with measures of customer satisfaction by area. In Illinois, all utilities were to report annually their reliability metrics (SAIDI, SAIFI, CAIDI and MAIFI), their network improvement budget, and measures to improve reliability. On an event basis, special reports had to be filed for any interruption over 3 hours affecting 10,000 or more consumers. The PBR approved by the California regulators for San Diego Gas & Electric in 1999 is given in Table 8.13.

Scandinavia. The regulators are introducing their own flavor of PBR. In Norway, energy-based penalties were introduced in 2001 at 38 NOK per kWh for commercial/industrial customers interrupted for all sustained interruptions (> 5 minutes) and 4.2 NOK per kWh interrupted for residential customers.

In Sweden, an entirely different approach was introduced in 2003. All distribution network investment, operational costs and reliability are compared with a synthetic grid value model (GVM) that normalized the performance of all Swedish network companies. Good reliability performance was a positive addition to the investment model. The regulator developed an average performance line for the country as a whole, and penalty incentives were set for utilities on the lower performance side of the average.

Finland uses a similar approach to Sweden for benchmarking utility performance using data envelope analysis (DEA) to develop an efficiency value for each network company instead of the GVM.

England. In England, now in its third regulatory period, a 3-year Information Incentive Program (IIP) has been introduced that will force utilities to provide significant operational and investment data to the regulator in return for a set reward-penalty incentives. The program runs for the final 3 years of the period,

---

* Davies, R., Performance-Based Reliability Regulations, Part of *Financial Times Energy*, E Source.

TABLE 8.13
San Diego Gas & Electric PBR Plan Approved by the California Regulators in 1999

	Target Value	Incentive Value $ Millions	Max Reward Equal or Fewer	Max Penalty Equal or More
		System Reliability		
SAIDI	52 minutes/year	3.75	37 minutes/year	67 minutes/year
SAIFI	0.90 outages/year	3.75	0.75 outages/year	1.05 outages/year
MAIFI	1.28 outages/year	1.00	0.95 outages/year	1.58 outages/year
		Employee Safety		
OSHA reportable rate	8.8	3	7.6	10
		Customer Satisfaction		
Very satisfied	92.5%	1.5	94.5%	90.5%
		Call Center Response		
Calls answered within 60 secs	80%	1.5	95%	65%

and the targets have been set for each distribution company to reflect the individual CAPEX and OPEX plans submitted to the regulator and the network environment.

The scheme falls into two major categories, revenue at risk (RAR) and outperformance bonus, the latter also being known as area of outperformance (A of O).

Revenue at Risk. The revenue at risk scheme balances network performance against revenue that the utility can be permitted to receive. In practice, this means that if the quality of supply does not meet the predefined targets, then the utility income is curtailed. Note should be taken that it is the actual income that is curtailed, not just the profit margin on that amount of income.

Each utility has been set a target for the average number of interruptions (SAIFI) and the average outage duration (SAIDI), which can be summarized on a national basis in Figure 8.23.

The left-hand diagram shows target line for the interruptions on a year-by-year basis, the target being tightened up for 2004/5. At the end of each year, the actual historical performance is compared to the target line:

- If the performance is better than target, then no revenue is placed at risk, but at the same time, under this part of the scheme at least, the utility does not make any gain from performance better than target.
- If the performance is worse than target, then utility revenue is curtailed, on a pro rata basis, up to a predetermined maximum, or worst case,

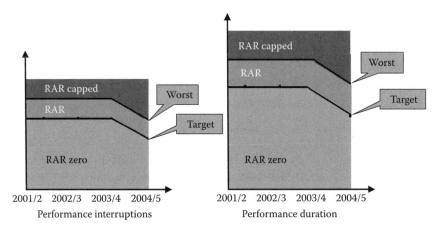

FIGURE 8.23 Graphical representation of U.K. Information and Incentive Program (IIP) on a national basis.

point. The worst case is typically 20% above the actual target level and has been applied to prevent excessive penalties being applied during periods of, for example, extremely severe weather conditions.

Over the 3-year period, the actual cumulative revenue that could be stopped if the targets are not met by all utilities amounts to nearly 55 MUSD.

The right-hand diagram shows similar situation for outage durations, where the target line and worst case levels are clearly defined. Again, the targets are tightened in the third year and the total, 3-year, RAR is nearly 140 MUSD. The performance for interruptions and duration are separately measured for each utility, and it is, therefore, quite possible for a utility to be on target for one, but not both, measurements.

Outperformance Bonus. The RAR scheme has been developed to reward utilities that provide a service that represents a significant improvement on the target values set, and this is shown in Figure 8.24.

Only if a utility meets the targets for both duration and interruptions will its out performance bonus be paid. For interruptions, this will be paid, pro rata, for a maximum improvement of up to 15% better than target. For duration, this will be paid, pro rata, for a maximum improvement of up to 20% better than target. Three-year total outperformance bonuses will be payable up to 56 MUSD for durations and nearly 24 MUSD for interruptions.

The IIP scheme, therefore, places a national maximum penalty (RAR) of nearly 200 MUSD and offers a maximum outperformance bonus of nearly 90 MUSD over a 3-year period of operation. It is intended that these economic motivators will provide an impetus for utilities to monitor network performance closely, but it remains to be seen how much they will actually encourage utilities to target investment to optimize the rewards available.

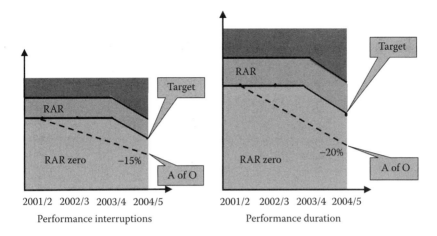

FIGURE 8.24 Graphical representation of U.K. Information and Incentive Program on a national basis showing areas of opportunity (A of O).

8.13.2 Customer Cost

The cost of an outage seen by customers bears a very different value because it reflects directly on their lives and, more importantly, on loss of production. The cost of supply failures that is seen by customers depends heavily on the actual use to which the electricity is placed. Large customers often have developed a value for their processes that are dependent on the loss of production. Certain processes have a duration threshold beyond which a total shutdown and clearing of the system is necessary at a cost substantially above that prior to the threshold being reached. Although it is difficult to generalize loss-of-supply costs, research by Allen and Kariuki has proven to be extremely useful in estimating customer costs [1, 2]. The research has been able to identify typical costs of an individual, unscheduled interruption according to the duration and the type of customer. This work presents costs normalized to the annual electricity consumption of the customer, and in Table 8.14, they are given in units of GNP per MWh annual consumption per interruption of a specified duration.

These normalized figures can be adjusted according to the local economy by using a scaling factor. Kennedy [3] gives the gross national product per capita for developing countries, OECD, and Eastern Europe/former Soviet Union and then adjusts this data to achieve the purchasing power parity (PPP, an indicator of real income) per capita (Table 8.15). We can use the above relationship to estimate the relative worth of supply interruptions in these locations. For example, the data supplied by Allen and Kariuki [1, 2] apply for the U.K., part of the OECD, so the factor by which these costs would be multiplied when considering the developing countries would be 3300/19,500 or 0.17. Further, these figures can be converted to a local currency or some accepted base such as USD using the current exchange rate (e.g., 1£ = $1.40 U.S.).

TABLE 8.14
Annual Interruption Costs in GBP for Different Outage Durations

	Interruption				Duration			
Load class	0	0.0167	0.333	1	4	8	24	Hours
	0	1	20	60	240	480	14,400	Minutes
Commercial	0.46	0.4800	1.640	4.910	18.13	37.06	47.58	
Industrial	3.02	3.1300	6.320	11.94	32.59	53.36	67.10	
Domestic	0.00	0.0000	0.060	0.210	1.440	1.44	1.44	
Large user	1.07	1.0700	1.090	1.360	1.520	1.71	2.39	
			Costs	£/MWh	annual	consumption		

Note: An interruption of zero duration represents the cost of a momentary interruption — MAIFI.

TABLE 8.15
Selected Demographic Data by Economic Region

	Population		Income, Gross National Product			
	Million	%	$ Billion	%	$ per Capita	PPP $ per Capita
Developing countries	4450	75.5	3800	13	850	3300
OECD	900	17.0	25,000	84	27,500	19,500
EE/FSU	400	7.5	800	3	2000	5500
Total	5750	100.0	29,600	100	5150	7000

8.13.3 ECONOMIC VALUE

The economic value a distribution utility puts on an interruption is vital to its business because the investment to improve the supply quality to the customer must be weighed against the benefits, some of which are subjective. An example will best illustrate the point and use the different costs discussed above. The situation in question is in a developing country, where system growth has outpaced the available investment, with a consequent detriment to the supply quality to an industrial park where one specialized industrial customer is suffering from the poor supply quality. The peak demand of the load is 2 MVA at a power factor of 0.9 and a load factor of 0.72. This gives an annual consumption of

$$\text{Annual MWh} = \text{MW demand} \times \text{load factor} \times \text{power factor} \times \text{hours in a year MWh/year}$$
$$= 2 \times 0.72 \times 0.9 \times 8760 = 11{,}353 \text{ MWh/year}$$

The reliability statistics are shown in Table 8.16.

TABLE 8.16
Reliability Level at Industrial Park Supply Point

Index	Initial Condition	Improved Condition	Improvement %
MAIFI	9.28	4.99	46.2
SAIFI	2.52	1.44	42.9
SAIDI	9.14	4.78	47.7
CAIDI	3.63	3.31	8.8

Utility Outage Value. The value of the outage at this customer for the utility is calculated as follows, given that the energy is sold at 0.05 USD equivalent in local currency per kWh at a margin of 10%.

The average annual interruption cost in terms of lost margin on sale of energy for the utility is

$$\text{Customer's annual MWh consumption} \times \text{SAIFI} \times \text{CAIDI} \times \text{margin of energy sale}$$
$$= 11{,}353 \times 2.52 \times 3.63 \times 0.05 \times 0.1 \times (1000/8760) = 59.3 \text{ USD/year}.$$

The annual revenue from sale of energy to this customer is

$$= 11{,}353 \times 0.05 \times 1000 = 567{,}650 \text{ USD/year}.$$

The annual profit at risk from loss of revenue should the customer take drastic remedial action and install their own generation is

$$\text{Customer's annual MWh consumption} \times \text{margin of energy sale}$$
$$= 11{,}353 \times 0.05 \times 0.1 \times 1000 = 56{,}765 \text{ USD/year}.$$

Customer Outage Cost. The annual outage cost to the customer is calculated as follows. By taking the average duration of an outage at this customer, CAIDI, from Table 8.16, is 3.63 hours and, from Table 8.14 for an industrial customer, the cost of an outage of this duration is between 11.94 (1 hour) and 32.59 (4 hours). The cost for the example outage is determined by linear interpolation between the two points in the table to be 18.10 (3.63 hours). Because this customer is in a developing country and it prefers to use USD as the currency base, the cost figures have to be adjusted by the factor described earlier as follows:

$$\text{Normalized interruption cost (NIC)} = \text{GBP} \times \text{cost adjustment factor} \times \text{exchange rate}$$
$$= 18.10 \times 0.17 \times 1.4$$
$$= 4.31 \text{ USD/MWh annual consumption/interruption of 3.63 hours}$$

Now, adjusting this normalized value to the annual MWh consumption of this customer of 11,353 MWh,

Single interruption cost (SIC) = NIC × annual MWh consumption
= 4.31 × 11,353 = 48,931 USD/interruption of 3.63 hours.

The number of interruptions experienced on the average per year at the load bus is given by the value of SAIFI (2.52). The annual cost of sustained interruptions (ASIC) is therefore:

ASIC = 48,931 × 2.52 = 123,307 USD.

The annual cost of momentary interruptions (AMIC) is the adjusted NIC for the industrial class at zero interruption duration (3.02) from Table 8.4 times the average momentary interruption frequency (MAIFI) at the load:

AMIC = 0.72 × 9.28 × 11,353 = 75,856 USD.

The total annual interruption costs for the customers:

Total average annual interruption costs = ASIC + AMIC = 199,163 USD.

The interruption cost figures for duration developed with the Allen and Kariuk method were supported by the customer's own assessment of a 2-hour outage cost.

Comparison of Economic Value. Evaluation of the different economic values of the average annual interruption values gives perspective to the situation when the utility is faced with questions about the quality of supply. The utility concludes that a significant improvement to the level of reliability could be made at the bus by the installation of some automation in the form of reclosers and remote-controlled switches. The cost of this improvement was estimated at 37.2kUSD, which, when amortized over 5 years, would represent an annual cost of 7,440 USD. In contrast, the customer has evaluated the possibility of installing and operating its own generation on-site at a capital investment of 460 kUSD or 92,000 USD over 5 years, with an annual operating cost of 0.056 USD/kWh. A final consideration is that the pending deregulation of the industry in the country is considering adopting the penalty model of Norway, which set the value for industrial customers at an equivalent of 4.75 USD per kWh interrupted. The value of the various options are presented in Table 8.17 as incurred costs.

Review of this situation highlights the following:

- The pure economic value of reduction in energy sales lost, in this example, is insufficient by itself to cover the costs of implementing the network improvement project (case 2).

TABLE 8.17
Comparison of Different Outage Values and Their Impact on the Decision Process for DA

	Case 1	Case 2	Case 3	Case 4
	Initial Situation	Network Improvement	Customers Install Own Generation	Regulated Penalty
Utility	USD/year	USD/year	USD/year	USD/year
Loss of margin on energy sale resulting from annual interruptions	59	31	0	31
Revenue at risk/lost	567,650		567,650	
Profit at risk/lost	56,765		56,765	
Capital outlay for improvement		7,440		7,440
Penalty not paid due to improvement				26,971*
Customer				
Cost of energy (1)	567,650	567,650	635,768	567,650
Cost of interruptions (2)	199,163	102,586		102,586
Investment cost of generation (3)			92,000	
Total customer cost (4)	766,813	670,236	727,768	670,236
Total cost (4) as % of interruption costs (2)	25	16	0	16

* Difference between penalties before and after network improvement ($56,313 at SAIFI 2.52, CAIDI 3.63; $29,342 at SAIFI 1.44, CAIDI 3.31).

- The initial condition case for power quality exposes the utility to risk of significant loss of revenue. Action by other customers could be duplicated at other sites, further increasing revenue loss.
- The initial condition (case 1) has an economic cost to the customer above what it would cost to improve the situation by installing and operating their own generation.
- Although the utility cannot justify implementing a network improvement project through strict utility costing of savings in energy sales lost, failure to make the improvement exposes the business to a substantial risk of lost revenue on an annual basis should the customer install their own generation. This loss of opportunity is well above the cost of the improvement project, thus an opportunity cost approach to justification would seem valid in this case.
- A penalty-based environment based on the Norwegian model provides significant incentive, allowing the utility to recoup its investment within 2 years.

This example has served to illustrate that economical value cannot be based solely on the cost accounting practices of the utility, but must consider some of the surrounding subjective (soft) issues particularly the customers' perceptions of their business requirements and their loyalty. Although only one customer has been considered, the analysis could consider other customers within the industrial park who, though not yet conscious of the cost of poor quality, will add to the revenues at risk or may accept higher energy costs for better supply quality.

8.14 PRESENTATION OF RESULTS AND CONCLUSIONS

This chapter began with the categorizing of benefits into four types: direct and indirect, soft and hard. Once calculated on an annual basis, the investment cost and the annual benefits may be used to guide the investment decision using accepted methods such as

- Cumulative present worth of the costs versus savings
- Break-even analysis to show the payback period
- Internal rate of return (IRR) and so on

Whichever method is used, the concept of categorizing the component benefits into four categories, where the utility decides which benefits are based on reliable data and are thus "hard," as opposed to benefits that are "guesstimates" or "soft" will provide decision makers with a clear vision of the uncertainty in the business case. Figure 8.25 shows a chart of the cumulative present worth of benefits as they accrue over a 10-year period and, in this case, the relatively small contribution of benefits in the soft category.

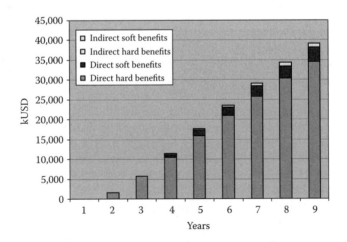

FIGURE 8.25 Cumulative present worth of a DA project benefit stream.

Creating the Business Case

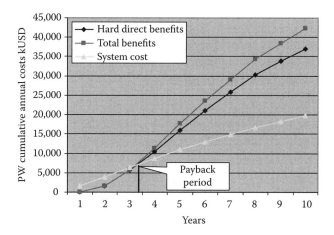

FIGURE 8.26 Illustration of payback period using cumulative present worth analysis of annual benefits and investment payments.

The same benefit information can be compared with the annualized investment cost amortized over a 10-year period as shown in Figure 8.26. In this implementation and financial assumptions, the system payback period is approximately 3 years, and the indirect and soft benefits make very little difference to the decision.

The internal rate of return is that interest rate where the cumulative present worth of the annual costs equals the cumulative present worth of the savings over the amortization period. This is shown in Figure 8.27 for the same example. The IRR is the test interest rate at which the ratio of the present value of costs to present value of benefits becomes unity, which in this case is 47.3%.

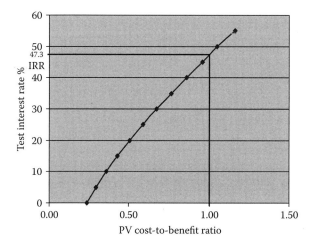

FIGURE 8.27 Relative costs versus test discount (interest) rate to determine internal rate of return.

This chapter has aimed at bringing together all the components that are needed to prepare a business case for distribution automation. Each case will be different, and assessment of all the factors such as available data, the operating environment and business priorities will be necessary. Certain central applications functions deliver only incremental improvement in terms of response times, and thus the only way of justification is to consider the benefits from these small improvements. The treatment of economic evaluation principals in terms of carrying charge development, depreciation, and taxation is very basic, and if more sophisticated analysis, which is beyond the scope of this book is required, readers are advised to review texts devoted entirely to profitability and economic choice.

Two case studies will be used in the final chapter of this book to illustrate developing the business cases covering two very different implementations.

REFERENCES

1. Allan, R.N. and Kariuki, K.K., Applications of customer outage costs in system planning, design and operation, *IEE Proceedings Generation, Transmission, Distribution,* 143, 4, July 1996.
2. Allan, R.N. and Kariuki, K.K., Factors affecting customer outage costs due to electric service interruptions, *IEE Proceedings Generation, Transmission, Distribution,* 143, 6, Nov. 1996.
3. Kennedy, M., *IEE Power engineering Journal,* London, Dec. 2000.
4. Billinton, R. and Pandey, M., Reliability worth assessment in a developing country — residential survey results, *IEEE Transactions on Power Systems,* 14, 4, Nov. 1999.
5. Tobias, J., Benefits of Full Integration in Distribution Automation Systems, Session Key Note Address, CIRED 2001, Amsterdam, June, 18–21, 2002.
6. Ackerman, W.J., Obtaining and Using Information from Substations to Reduce Utility Costs, ABB Utility Engineering Conference, Raleigh, NC, March 2001.
7. Bird, R., Substation Automation Options, Trends and Justifications, DA/DSM Europe, *Conference Proceedings,* Vienna, Oct. 8–10, 1996.
8. EPRI, Guidelines in Evaluating Distribution Automation, Final Report, EPRI EL-3728, Nov. 1984.
9. Delson, M., McDonald, J. and Uluski, R.W., Distribution Automation: Solutions for Success, Utility University, DistribuTECH 2001 Preconference Seminar, San Diego, Feb. 4, 2001.
10. Chowdhury, A.A. and Koval, D.O., Value-Based Power System Reliability Planning, *IEEE Transactions on Industry Applications,* 35, 2, March/April 1999.
11. Bird, R., Business Case Development for Utility Automation, DA/DSM Europe, *Conference Proceedings,* Vienna, October 8–10, 1996.
12. Kariuki, K.K. and Allan, R.N., Evaluation of reliability worth and value of lost load, *IEEE Proceedings, Generation, Transmission, Distribution,* 143, 2, March 1996.
13. Laine, T., Lehtonen, M., Antila, E. and Seppanen, M., Feasibility Study of DA in a Rural Distribution Company, *VTT Energy Transactions,* Finland.
14. Burke, J., Cost/Benefit Analysis of Distribution Automation, American Electric Power Conference.

15. Clinard, K., The Buck Stops Here — Justifying DA Costs, *DA&DSM Monitor,* Newton Evans Inc., May 1993.
16. Born, J., Can the Installation of GIS be Decided by Cost/Benefit Analysis, *Proceedings AM/FM/GIS European Conference VIII,* Oct. 7–9, 1992.
17. Walton, C.M. and Friel, R., Benefits of Large Scale Urban Distribution Network Automation and their Role in Meeting Enhanced Customer Expectation and Regulator Regimes, CIRED, 2000.
18. Cepedes, R., Mesa, L., and Schierenbeck, A., Distribution Management System at Epressas Publicas de Medellin (Colombia), CIRED 2000.
19. Jennings, M. and Burden, A.B., The Benefits of Distribution Automation, DA/DSM Europe, *Conference Proceedings,* Vienna, Oct. 8–10, 1996.
20. Staszesky, D. and Pagel, B., International Drive Distribution Automation Project, DistribuTech, San Diego, Feb. 2001.
21. Isgar, P., Experience of Remote Control and Automation of MV (11 kV) substations, CIRED, 1998.
22. Ying, H., Wilson, R.G. and Northcote-Green, J.E., An Investigation into the Sensitivity of Input Parameters in Developing the Cost Benefits of Distribution Automation Strategies, DistribuTech, Berlin.
23. Burke, J.J., Cost/Benefit Analysis of Distribution Automation, *IEEE Power Engineering Proceedings.*
24. Phung, W. and Farges, J.L., Quality Criteria in Medium Voltage Network Studies.
25. Wainwright, I.J. and Edge, C.F., A Strategy for the Automation of Distribution Network Management Functions, DA/DSM Europe, Vienna, Oct. 8–10, 1996.

9 Case Studies

9.1 INTRODUCTION

Inevitably the application of the concepts developed in the previous chapter has to be adapted to the real world to account for the business priorities and pressures of a particular utility. The availability and quality of data needed to calculate the hard benefits is one area where compromises have to be made in developing a business case that will withstand management scrutiny. In this chapter, two case studies will be used to illustrate justification for distribution automation based on the Chapter 8 methods. In developing these business cases, an attempt will be made to prioritize the benefits according to their contribution and hardness.

9.2 CASE STUDY 1, LONG RURAL FEEDER

9.2.1 EVALUATION OF PERFORMANCE

The system in this case study is a real network in northern Europe. It comprises a single source substation with a single 16 MVA transformer supplying busbars at 20 kV. The network is loaded to 8159 kVA maximum demand, supplying 2,272 customers. The source substation is in the center of Figure 9.1, and we will select one circuit of the three (highlighted) that heads north from the substation. This feeder comprises the following:

- 3.9 km cable at the source end
- 39.2 km overhead line
- 725 customers
- 1508 kVA maximum demand
- 17 switches and switchfuses
- SCADA-controlled feeder source circuit breaker at the substation

There is also a 20/10 kV transformer approximately halfway along the circuit, resulting from a previous voltage upgrading from the original 10 kV system to 20 kV for the sections nearest the source. This circuit has no extended control (FA).

Although the performance of the network was considered acceptable except in times of severe weather, the utility, in preparation for deregulation, was considering introducing feeder automation on its networks and wanted to see whether the benefits were sufficient to justify the implementation.

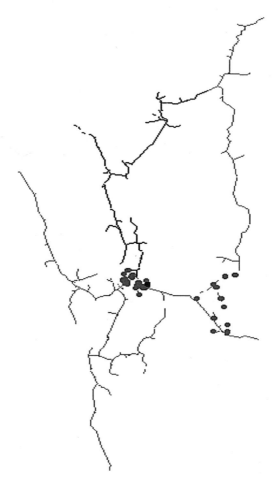

FIGURE 9.1 Geographical network diagram of the example long rural feeder used in case study 1.

The first step in any study is to formally tabulate the network performance and predict the improvement in performance with extended control. Present performance levels are usually determined from outage records, which are used to calibrate reliability-based planning models used for predicting performance under different AIL. The results of this preparatory analysis gave the following results shown in Table 9.1 for increasing levels of automation, where

- AIL Level 0 — No extended control
- AIL Level 1 — Addition of extended control and autoclose to the source circuit breaker
- AIL Level 2 — Supplement Level 1 with addition of extended control to normally open switch

Case Studies

TABLE 9.1
Feeder Performance Levels with Different Levels of AIL

Performance Level	Level of Extended Control				
	Level 0	Level 1	Level 2	Level 3	Level 4
MAIFI	0	31.39	31.39	31.39	31.39
SAIFI	27.44	7.29	7.29	7.29	7.29
SAIDI	96.28	25.42	20.24	18.33	12.44
CAIDI	3.51	3.49	2.78	2.51	1.71

- AIL Level 3 — Supplement Level 2 with addition of extended control to midpoint switch
- AIL Level 4 — Supplement Level 3 with addition of extended control to 10 in line switches

And the data used in this calculation were as follows:

- Temporary fault rate is 72 faults per 100 km pa
- Permanent fault rate is 24 faults per 100 km pa
- Switching time for manual switches is 1, 1.5 or 2 hours depending on the distance from the source
- Switching time for all remote-controlled switches is 10 minutes
- Repair time for all line faults is 5 hours

9.2.2 Crew Time Savings

Crew time savings (CTS) was considered to be one of the primary benefits to be obtained and was selected as the first benefit to estimate. This was calculated using the expressions derived in Chapter 8. The savings were calculated over a 10-year period at an interest rate of 6%.

Inspection of the circuit diagram shows that the selected feeder has three ends and two normally open points (NOPs). According to the formula, its NCF is 3.

The length of this circuit is 43.6 km, and the permanent fault rate is 24 faults per 100 km per year, and the temporary fault rate is 72 faults per 100 km per year. The utility crew costs 84 USD per hour and, taking into account the local terrain, an average speed of 20 km per hour is appropriate. We can now calculate $D(m)$ for each automation intensity as shown in Table 9.2 and apply this to the formula from Section 8.9.3.

- AIL 0 — No extended control (initial condition or base case):

$D(m)$ is 1.27; therefore, annual cost is $2 \times 1.27 \times 43.6 \times 43.6 \times 84 \times 72/100 \times 1/20 = 14{,}601$ USD

TABLE 9.2
Comparisons of Crew Time Savings for Different Levels of Automation AIL

	Level of Extended Control			
	AIL 1	AIL 2	AIL 3	AIL 4
Total savings per year, compared to AIL 0, kUSD	10.9	12.2	12.8	13.4
Present value of savings over 10 years at 6% interest rate in kUSD compared to AIL = 0	80.4	90.0	94.2	98.7

- AIL 1 — Autoclose and extended control of source breaker only:

    D(m) is 0.96; therefore, annual cost is $2 \times 0.96 \times 43.6 \times 43.6 \times 84 \times 24/100 \times 1/20 = 3679$ USD

- AIL 2 — Extended control of source breaker and NOPs:

    D(m) is 0.62; therefore, annual cost is $2 \times 0.62 \times 43.6 \times 43.6 \times 84 \times 24/100 \times 1/20 = 2376$ USD

- AIL 3 — Extended control of source breaker, one switch and NOPs:

    D(m) is 0.47; therefore, annual cost is $2 \times 0.47 \times 43.6 \times 43.6 \times 84 \times 24/100 \times 1/20 = 1801$ USD

- AIL 4 — Extended control of all switches and NOPs:

    D(m) is 0.31; therefore, annual cost is $2 \times 0.31 \times 43.6 \times 43.6 \times 84 \times 24/100 \times 1/20 = 1188$ USD,

which is summarized in Table 9.2, together with the present value of these savings at an interest rate of 6% over a 10-year project life.

The comparison of savings against cost of FA implementation for different AILs is plotted in Figure 9.2.

It is interesting to see that, at least for this long rural overhead feeder, the benefits over 10 years exceed the capital costs up to AIL 3, from which we can conclude that, in this example at least, the investment in extended control is clearly justifiable.

9.2.3 NETWORK PERFORMANCE AND PENALTIES

We have already seen that a regulatory authority might impose penalties on a utility for low performance of its distribution network, and it is helpful to consider

Case Studies

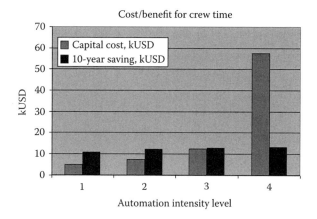

FIGURE 9.2 Comparison of savings against implementation costs for different AILs on the sample feeder.

how typical penalties might apply to this network. Three of the most likely penalty structures are based on

- Outage duration, where a penalty is paid by the utility if the annual outage duration exceeds the specified amount.
- Interruption frequency, where a penalty is paid by the utility if the number of interruptions during the year that exceed the specified amount. The interruptions might be momentary (for example, less than 3 minutes) or permanent or some combination of both.
- Maximum outage duration, where a penalty is paid by the utility for every customer who experiences any outage of a duration that exceeds the specified amount.

Figure 9.3 shows, for the sample network, the scatter plot of SAIDI and SAIFI for two levels of extended control, AIL 0 and AIL 3, clearly indicating some clustering. Grouping A to D has been added to the plot for clarity; for example, Group A shows loads that are off supply 35.2 times a year and where the outage duration ranges from 100.3 hours to 132.1 hours per year. The average interruption time, CAIDI, therefore ranges from 2.85 hours to 3.75 hours.

If we consider an interruption penalty at 10 interruptions per year, then we can see that, for AIL 0, a penalty would be payable in respect of all the customers (Group A and Group B) on this network. But by adding AIL 3, we move to Group C and Group D, both of which are below the penalty level, thereby creating a financial benefit of adding extended control to AIL 3.

If we now consider an annual outage duration penalty at 50 hours per year, then we can see that, for AIL 0, all the Group A load points and one in Group B would be the subject of penalty payments. By adding AIL 3, we move to Group

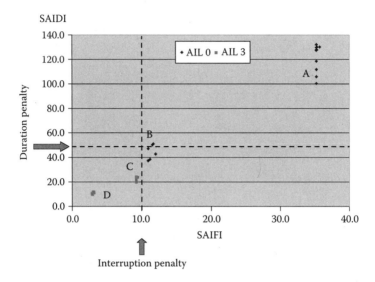

FIGURE 9.3 Scatter diagram for annual outage duration and interruption frequency for load points on the feeder.

C and Group D, both of which are below the penalty level, thereby creating an additional financial benefit of adding extended control to AIL 3.

The penalty for maximum outage duration might occur if any single outage over 24 hours occurred, and this duration is usually associated with extreme weather conditions. For the network under consideration, there have been no outages of this severity.

Suppose that the penalties payable were made up of 40 USD to each customer who experiences more than 10 outages per year and 35 USD to each customer who experiences more than 50 hours off supply per year.

Now, because there are 45 load points contained within Group A and Group B, who would qualify for the interruption frequency penalty, and there are, on average, 12 customers at each load point, then the interruption frequency penalty can be calculated as follows:

Frequency penalty = 45 × 12 × 40 USD per year = 21.6 kUSD.

And because there are 35 load points contained within Group A and (part of) Group B who would qualify for the outage duration penalty, then this penalty can be calculated as

Outage duration penalty = 36 × 12 × 35 USD per year = 15.12 kUSD,

which can be summarized in Table 9.3, which shows that the investment of 12.5 kUSD is clearly justified.

Case Studies

TABLE 9.3
Comparison of Costs of Implementing an AIL 3 over Payment of Penalties for Existing System

	Level of Extended Control	
	AIL 0	AIL 3
Frequency penalty, kUSD	21.60	0
Outage duration penalty, kUSD	15.12	0
Total penalty, kUSD	26.72	0
NPV of total penalty for 10 years at 6%, kUSD	197.00	0
Estimated scheme capital cost	NA	12.5

9.3 CASE STUDY 2, LARGE URBAN NETWORK

The second case study is based on a very large urban system of a developing country in Asia. The entire network (subtransmission and distribution) owned by the utility is manually operated. Recent privatization has put pressure on the management to improve system performance and modernize operating practices. The network consists of 66 kV and 33 kv sub transmission with 11 kV medium-voltage distribution. The 11 kV is predominantly cable serving small switching stations and ring main units. There are approximately 100 grid substations and 8000 distribution substations. The utility has a peak load of 2500 MW* and supplies approximately 2 million consumers. Control is performed out of five dispersed control centers. The control of the main incoming supply points (liaison with the supplying grid company) and subtransmission is coordinated out of a main central control room for all the subtransmission and grid substations, where switching is initiated by telephone to the grid substations, all of which are fully operated. Coordination of the 11 kV system is from four control centers in each of the four operating districts using telephones to the operated grid substations and radio to the line switching and repair crews.

Benefits derived from the introduction of a DMS will be examined, using derived or available data, in the areas of reduction in manpower costs, reduction in energy not supplied and asset management.

9.3.1 PREPARATION ANALYSIS — CREW TIME SAVINGS

The business case for a DMS study had to cover the entire network; thus, a practical screening approach had to be used. A section of the network considered to be typical of the configuration was selected, on which a basic analysis was undertaken to develop outage statistics, network complexity factors and thus crew time savings for increasing automation intensity levels. The area selected,

---

* All utility statistics are for example only to disguise the actual location.

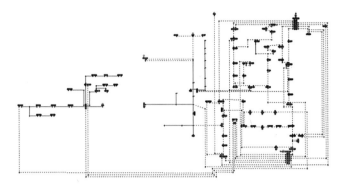

FIGURE 9.4 Model of 1/50th sample of entire 11 kV network.

comprised two grid substations and 29 11 kV feeders representing 1/50th of the network, was modeled from the single-line operating diagram (Figure 9.4).

Although feeder segment lengths were not available experience has shown that the assumption that distances in an operational schematic are in the majority comparative to be sufficiently accurate. The NCF was calculated for all 29 feeders. Five AIL cases were considered and represented as follows:

- Case 1, existing system (no remote control)
- Case 2, existing system with SCADA control of the source circuit breaker*
- Case 3, existing system with SCADA control of the source circuit breaker and the circuit breakers at the first switching substation
- Case 4, existing system with SCADA control of the source circuit breaker, the circuit breakers at the first switching station and all normally open points
- Case 5, existing system with SCADA control of the source circuit breaker, the circuit breakers at the first switching station and all normally open points and one additional switch on the main feeder between the first switching station and the NOP

Data were limited, and thus, the analysis was one of comparison of crew time savings with increased automation for the different NCFs. The comparative savings curves developed in Chapter 3 were calculated for this example including an automation level that represented remote control of the first switching station in the network, because this type of feeder layout had not previously been examined. The results followed the same form as previously described and are shown in Figure 9.5.

The percentage of crew time saved for each of the 29 feeders was derived from the relationships in Figure 9.5 and the results for feeders 11–29 summarized as an example in Table 9.4.

---

* This represented SCADA only being implemented at all grid substations and controlling all circuit breakers down to the 11 kV feeder source breakers.

Case Studies 439

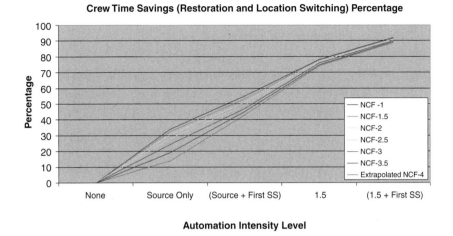

FIGURE 9.5 Percentage of crew time savings as a function of AIL and NCF.

The results of this analysis yielded an average crew time savings for the sample 1/50th network that will be applied to develop estimates for the entire system. In summary, the introduction of SCADA at the grid substations produces an immediate 31% reduction in crew travel costs for fault location, isolation, and restoration. The automation of the first switching stations adds another 20% improvement, and the NOPs yet another 26%. The marginal improvement in CTS from feeder automation begins to diminish for very high AIL — Figure 9.6.

9.3.2 Preparation Analysis — Network Performance

The availability of outage data was limited, and statistics had only been collected in a consistent systemwide manner for a limited time. The statistics were reviewed, interpolated, and averaged to develop a norm on which to base the analysis. This showed that the sample network should account for SAIDI of 16 hours per year, SAIFI of 4 per year, leading to CAIDI of 4 hours. The analysis model was accordingly calibrated using a switching time of 90 minutes and repair times (which affect customer restoration in few situations only) of 8 hours to give satisfactory results. For a dense city network, the switching time of 90 minutes is relatively long but did reflect the practice of local control of switching operations and the slow travel in large Asian cities, even though the distance between switching locations is relatively short. This estimation process is only necessary in the absence of a well-established fault and outage reporting system that would provide such statistics accurately. An example of results of the analysis is given in Table 9.5 for the same feeders as previously given (11 to 29) plus the total.

As in every power system, there is a variation of performance about the average, and this is shown in Figure 9.7 for the two-grid substations modeled.

TABLE 9.4
Comparative CTS for Different AIL against the Present System Operating Procedures for Feeders 11–29

Feeder Name	Calculated NCF	Asian Power System Crew Time Saved Percentage for Each Level of Automation				
		Source CB Only	Source CB + First SS	Source + 1.5	Source + First SS + 1.5	All
11	1.0	34	55	78	92	100
12	3.5	19	45	75	89.5	100
13	1.5	32	53	78	91.5	100
14	**4.0**	**14**	**43**	**70**	**89**	**100**
15	1.5	32	53	78	91.5	100
16	3.0	24	47	76	90	100
17	1.5	32	53	78	91.5	100
18	1.0	34	55	78	92	100
19	**4.0**	**14**	**43**	**70**	**89**	**100**
21	2.0	31	51	77	91	100
22	3.0	24	47	76	90	100
23	1.5	32	53	78	91.5	100
24	1.5	32	53	78	91.5	100
25	3.0	24	47	76	90	100
26	1.0	34	55	78	92	100
27	1.0	34	55	78	92	100
28	1.5	32	53	78	91.5	100
29	1.0	34	55	78	92	100
	2.02	31	51	77	91	100

Note: Numbers in **bold** for NCFs of 4 are extrapolated.

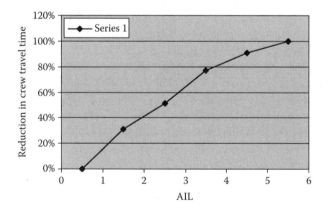

FIGURE 9.6 Percent improvement in CTS with increase in AIL.

Case Studies 441

TABLE 9.5
Example Results from the Sample Network

Circuite Name	Case 1 No SCADA			Case 2 Source DB only				Case 3 Source CB and First SS				Case 4 Source CB + 1.5				Case 5 Source CB, First SS + 1.5			
	SAIFI	SAIDI	CAIDI	SAIFI	SAIDI	SAIDI % Imp	CAIDI	CAIDI % Imp	SAIFI	SAIDI	SAIDI % Imp	CAIDI	% Imp	SAIFI	SAIDI	SAIDI % Imp	CAIDI	CAIDI %Imp	SAIFI	SAIDI	SAIDI % Imp	CAIDI	CAIDI % Imp

Circuite Name	SAIFI	SAIDI	CAIDI	SAIFI	SAIDI	SAIDI % Imp	CAIDI	CAIDI % Imp	SAIFI	SAIDI	SAIDI % Imp	CAIDI	% Imp	SAIFI	SAIDI	SAIDI % Imp	CAIDI	CAIDI %Imp	SAIFI	SAIDI	SAIDI % Imp	CAIDI	CAIDI % Imp
11	1.8	10.6	3.78	2.8	10.39	1.80	3.71	1.85	2.8	10.36	2.08	3.7	2.12	2.8	5.55	47.54	1.98	47.62	2.8	5.55	47.54	1.98	47.62
12	6.31	36.2	5.73	6.31	35.94	0.69	5.7	0.52	6.31	35.91	0.77	5.69	0.70	6.31	31.65	12.54	5.02	12.39	6.31	31.65	12.54	5.02	12.39
13	5.51	19.5	3.53	5.51	19.26	0.98	3.49	1.13	5.51	19.23	1.13	3.49	1.13	5.51	14.07	27.66	2.55	27.76	5.51	14.07	27.66	2.55	27.76
14	2.08	8.1	3.89	2.08	7.85	3.09	3.77	3.08	2.08	5.55	31.48	2.67	31.36	2.08	5.69	29.75	2.73	29.82	2.08	4.95	38.89	2.38	38.82
15	3.22	12.5	3.87	3.21	12.2	2.01	3.79	2.07	3.21	12.11	2.73	3.77	2.58	3.21	7.02	43.61	2.18	43.67	3.21	7.02	43.61	2.18	43.67
16	7.09	19.6	2.77	7.09	19.39	1.17	2.73	1.44	7.09	19.25	1.89	2.71	2.17	7.09	10.9	44.44	1.54	40.40	7.09	109.9	44.44	1.54	44.40
17	1.84	6.72	3.65	1.84	6.49	3.42	3.52	3.56	1.84	3.56	47.02	1.93	47.12	1.64	4.22	37.20	2.29	37.26	1.84	3.03	54.91	1.64	55.07
18	—	—	—	—	—	—	—	—	—	—	—	—	—	—	—	—	—	—	—	—	—	—	—
19	6.04	27.7	4.58	6.04	27.46	0.83	4.54	0.87	6.04	27.43	0.94	4.54	0.87	6.04	24.61	11.12	4.07	11.14	6.04	24.61	11.12	4.07	11.14
21	2.29	7.98	8.34	2.29	7.75	2.88	3.38	2.87	2.29	5.18	35.09	2.26	35.06	2.29	4.59	42.48	2	42.53	2.29	2.57	67.79	1.12	67.82
22	1.71	5.49	3.21	1.71	5.25	4.37	3.07	4.36	1.71	2.79	49.18	1.63	49.22	1.71	3.25	40.80	1.9	40.81	1.71	2.42	55.92	1.41	56.07
23	1.16	3.86	3.32	1.16	3.62	6.22	3.12	6.02	1.16	3.01	22.02	2.6	21.69	1.16	2.36	38.86	2.03	38.86	1.16	2.14	44.56	1.84	44.58
24	0.75	2.98	3.96	0.75	2.79	6.38	3.71	6.31	0.75	1.63	45.30	2.17	45.20	0.75	2.02	32.21	2.68	32.32	0.75	1.72	42.28	2.29	42.17
25	2.33	15.1	6.49	2.33	14.88	1.59	6.38	1.69	2.33	13.42	11.24	5.76	11.25	2.33	13.51	10.65	5.8	10.63	2.33	13.49	10.78	5.79	10.79
26	—	—	—	—	—	—	—	—	—	—	—	—	—	—	—	—	—	—	—	—	—	—	—
27	0.64	3.17	4.93	0.64	2.94	7.26	4.57	7.30	0.64	2.91	8.20	4.52	8.32	0.64	2.5	21.14	3.88	21.30	0.64	2.5	21.14	3.88	21.30
28	3.59	11.4	3.16	3.59	11.1	2.20	3.09	2.22	3.59	7.9	30.40	2.2	30.38	3.59	7.05	37.89	1.96	37.97	3.59	7.05	37.89	1.96	37.97
29	2.21	5.61	2.54	2.21	5.41	3.57	2.45	3.54	2.21	5.39	3.92	2.44	3.94	2.21	5.23	6.77	2.37	6.69	2.21	5.23	6.77	2.37	6.69
Total	3.58	14.2	3.96	3.58	13.92	1.63	3.89	1.77	3.58	12.74	9.96	3.56	10.10	3.58	10.44	26.22	2.92	26.26	3.58	10.15	28.27	2.84	28.28

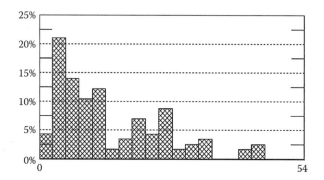

FIGURE 9.7 Distribution of SAIDI over the two-grid substation model.

The estimate for the savings in kWh is given in Table 9.6. This has been evaluated for each load point, taking into account the value for SAIDI for each load point together with a uniform load factor of 0.55. However, it must be noted that, because the load point values for actual load, and the load factor for each load point are not known, then the estimates given can only be regarded as indicative. Further work is, of course, possible to get more reliable data.

This preparatory analysis provides the foundation to proceed building the business case of the prospective benefits obtained from automation.

Manpower Savings. Manpower savings are obtained from the following areas:

- Reduced worker levels in grid substations as a result of implementation of SCADA (automation case 1)
- Crew time savings for fault location, isolation and restoration of the 11 kV feeder system for different AIL (cases 2–5)
- Reduced control room staff from consolidation of control from five centers to one central control room
- Increased efficiency in logging and fault reporting now done by operators in grid substations and preparation of central consolidated system wide reports

Reduced Manning Levels in Grid Substations. Benefits are derived from the costs of manning before and after automation and are developed by considering the mix of staffing used for different shifts in the grid stations and any change in philosophy of operator categories that follow automation implementation.

$$\text{Annual cost savings} = (\text{man-hours saved per grid substation}) \times$$
$$(\text{number of grid substations}) \times (\text{man-hour rate per hour}) +$$
$$(\text{man-hours to operate and maintain the SCADA system}) \times$$
$$(\text{SCADA operator man-hour rate per hour}).$$

Case Studies

TABLE 9.6
Savings in Energy Not Served for Increased Level of Automation

Constant Load Each Substation
Load Factor Each Substation

Location		kWh Lost per Year							kWh Saved per Year Ref Case 1				
		Case 1	Case 2	Case 3	Case 4	Case 5	Case 2	Percentage	Case 3	Percentage	Case 4	Percentage	Case 5
Circuit total	11	6983	6855	6836	3664	3664	128	1.8	147	2.1	3320	47.5	3320
Circuit total	12	41,805	41,511	41,478	36,553	36,553	294	0.7	327	0.8	5251	12.6	5251
Circuit total	13	16,049	15,886	15,863	11,608	11,608	163	1.0	186	1.2	4440	27.7	4440
Circuit total	14	12,022	11,658	8247	8449	7344	364	3.0	3775	31.4	3573	29.7	4678
Circuit total	15	8214	8051	7995	4636	4636	163	2.0	219	2.7	3578	43.6	3578
Circuit total	16	51,809	51,186	50,813	28,771	28,771	623	1.2	997	1.9	23,038	44.5	23,038
Circuit total	17	4438	4285	2351	2786	1997	154	3.5	2087	47.0	1653	37.2	2441
Circuit total	19	45,695	45,301	44,244	40,610	40,610	394	0.9	441	1.0	5085	11.1	5085
Circuit total	21	6587	6390	4277	3783	2117	197	3.0	2310	35.1	2804	42.6	4470
Circuit total	22	8151	7795	4149	4827	3592	356	4.4	4002	49.1	3324	40.8	4559
Circuit total	23	5727	5375	4476	3506	3177	353	6.2	1251	21.8	2221	38.8	2550
Circuit total	24	2952	2763	1618	2000	1703	190	6.4	1334	45.2	953	32.3	1249
Circuit total	25	37,424	36,827	33,214	33,429	33,377	597	1.6	4210	11.2	3995	10.7	4047
Circuit total	27	524	484	480	412	412	39	7.5	44	8.4	111	21.3	111
Circuit total	28	16,848	16,490	11,731	10,472	10,472	358	2.1	5118	30.4	6377	37.8	6377
Circuit total	29	925	893	889	864	864	32	3.4	36	3.9	61	6.6	61
Circuit total	Sample network	266,155	261,750	239,672	196,371	190,898	4405	1.7	26,483	10.0	69,785	26.2	75,257

TABLE 9.7
Grid Station Manpower Costs

Classification of Manpower	Position/Level/Category	Number of Staff	Monthly Salary $	Monthly Total $
Before Automation	Operator	600	220	132,000
	Asst. Engineer	25	550	13,750
	Junior Engineer	100	330	33,000
			Total	178,750
After Automation	Software Engineer (SCADA)	10	550	5,500
	Hardware Engineer (RTU)	10	550	5,500
	Asst. Engineer	8	550	4,400
	Junior Engineer	24	330	7,920
			Total	23,320
Automation Benefits (monthly)			Difference	155,430

Data from this example is as follows:

In each of the 100 grid substations there are two operators per shift and three shifts are worked per 24 hours. In addition, there are 100 junior engineers assigned to each substation and 25 assistant engineers with responsibilities for four substations each.

Management has determined that reassignment of personnel to allow introduction of unmanned substations will be possible with the introduction of SCADA. The calculation of the cost difference on a monthly basis in staffing levels before and after implementing automation is given in Table 9.7. The new staff will replace the existing grid substation three-shift router and be responsible for substation maintenance management, troubleshooting and providing data for the SCADA system (data engineering) as well as maintaining the RTUs and SCADA interfaces.

The manpower savings benefit from reduced grid substation manning per year is

$$\text{Annual benefit} = (\text{manpower costs before automation}) - (\text{manpower costs after automation})$$
$$= (178{,}750 - 23{,}320) \times 12 = \$1{,}865{,}160$$

Savings in Crew Time for Fault Location, Isolation and Restoration. Savings in crew travel time are derived from information from remote locations about the status of the network, which allows faster fault location and isolation. The improvements vary according to the particular voltage network. The subtransmission network now with fully manned grid substation operates like a slow SCADA system with manual reporting of any circuit breaker operation, whereas the 11 kV system is entirely dependent on field crew reporting. Consequently the improvement in crew savings will be achieved predominantly from automation of this network.

TABLE 9.8
Manpower Rates

		Quantity	Rate in USD per Month	Total Cost, USD Per Month	Total Cost, USD Per Annum
Crew van	Fixed costs	1	556	556	
	Operating costs	1	2,667	2,667	
Crew costs per van	Worker	6	220	15,840	
	Assistant worker	6	160	11,520	
	Helpers	10	110	13,200	
Total cost per van		75		43,782	3,283,667

TABLE 9.9
Incremental Crew Time Savings for Each AIL over the Base Nonautomated System

Crew Savings	Case 1	Case 2	Case 3	Case 4	Case 5
% saving from study	0	31	51	77	91
Delta between cases		31	20	26	14
Incremental USD	3,283,667	1,017,937	656,733	853,753	142,511

The present annual crew costs were estimated for the entire network by calculating the present annual crew costs (Table 9.8). The percentage savings estimated in the preparation study Table 9.2 were used to generate the actual savings given in Table 9.9.

Reduction in Control Room Staff. Implementation of a central DMS will allow control of the entire network from one point, thus the activities of the four district control centers for medium-voltage operations can be consolidated and moved to the main central location presently occupied by the subtransmission control team. The present activities of the main control room can also be conducted with reduced staffing. The manning levels before and after automation consisted of four districts with three shifts and three operators per shift. The main control room was staffed by four persons per shift. Consolidation into one control room, possible with a multioperator DMS, will allow staffing to be reduced and consolidated to deliver savings as calculated below.

Before Automation

Total number of operators   48
Average monthly burdened manpower costs $270/worker/month
Total annual control room manpower costs $48 \times 270 \times 12 = \$155,520$

After Automation

        Day Shifts (2)   Operators, 8   Salary $360/month
                              Supervisors, 2 Salary $400/month
        Night Shift   Operators, 2
                              Supervisors, 1

Total manpower costs of new control center =
$[(8 + 2) \times 360 + (2 + 1) \times 400] \times 12 = \$57,600$

Annual benefits = (present control room manpower costs) − (manpower costs for new control room) = $\$155,520 \times \$57,600 = \$97,920$

Reduction in Report Preparation and Operating Diagram Maintenance. Presently, the grid station staff makes all daily logs and incidents associated with every substation. A dedicated team then compiles these logs centrally. This team also maintains the single-line operating diagrams. This function will be facilitated by the implementation of the DMS and maintained by the new control room support staff.

Savings due to elimination of central report coordinating team composed of one assistant engineer and three technician/clerks.

Annual savings = $(1 \times 650 + 3 \times 330) \times 12 = \$19,680/\text{year}$

Savings in kWh Lost Due to Faster Restoration. Savings in kWh lost is achieved by faster restoration time resulting from automation. The improvements vary according to the particular voltage network.

Analysis of the outage statistics and considering experience from typical improvements from SCADA implementations, a 6% potential improvement was estimated for the subtransmission network. The savings for the 11 kV network were estimated for different AILs in the preparation study. Utility statistics were available for 1 year only and were used to quantify lost energy due to outages. The average margin on sales of energy was determined from billing records as 0.017 $/kWh (unit). Subtransmisson network annual energy losses due to outages were logged as 230 M units (M kWh) and for the 11 kV network 535 M units (M kWh). The energy saved from the implementation of SCADA and the four cases of feeder automation is given in Table 9.10.

9.3.5 SUMMARY OF COST SAVINGS

The manpower cost and energy not supplied savings are summarized in Table 9.11, which also shows the DA function that will deliver the savings.

TABLE 9.10
Tabulation of Saving in Energy Not Supplied by AIL

	Units Lost Million kWh	Estimated Savings %	Units Saved Million kWh	Value on Margin $
Subtransmission	233	6.00	14	237,660
Dist. 11 kV network Automation AIL	535	Incremental		
Case 2		1.7	9	154,615
Case 3		8.3	44	754,885
Case 4		16.2	87	1,473,390
Case 5		1.80	10	163,710

TABLE 9.11
Summary of Cost Savings Benefits by Category and AIL

Savings Category	Amount, $/Year	DA Function
Reduced manning levels in grid SS	1,865,160	SCADA in Grid SS (SA)
Reduction in control room staff	97,920	
Reduction in report preparation and operating diagram maintenance	19,680	
Crew times savings — 11 kV network	1,017,937	Case 2
	656,733	Case 3
	853,753	Case 4
	142,511	Case 5
Reduction in energy not supplied		
Subtransmission network	237,660	SCADA
	154,615	Case 2
	754,885	Case 3
	1,473,390	Case 4
	163,710	Case 5

9.3.6 COST OF SCADA/DMS SYSTEM

The cost for implementation of any DMS can be developed from the bottom-up and can be divided into a five main categories:

- Primary system adaptation/preparedness
- Central SCADA (hardware platforms, software, data engineering, installation)
- DMS applications (basic and advanced) — including data engineering of distribution network model

TABLE 9.12
Automation System Implementation Costs

			USD
Phase 1			
Control room hardware			220,000
Software (SCADA, DMS)			1,300,000
Power supplies			450,000
Communication interfaces			600,000
Large screen mimic			450,000
Miscellaneous			220,000
RTUs for grid station (case 2)			3,350,000
			6,590,000
Grid substation adaptations (transducers, etc.)			1,750,000
			8,340,000
Phase 2			
Feeder automation			
FSWS	800	0.15	2,6000,000
RMUs	2000	0.10	4,444,000
			7,044,000

- Remote data collection (RTUs, DTUs, etc.) — including installation and testing
- Communication system

The cost for the SCADA DMS system for this case was estimated as shown in Table 9.12.

9.3.7 Cost Benefits and Payback Period

The payback period was calculated using the interest rate of 7.5%, and the system was amortized over a 10-year period with a levelized annual carrying charge of 19%. The DA system was assumed to be implemented in phases shown in Table 9.13 with the benefits occurring at a later time, also in stages.

The payback period, as shown in Figure 9.8, given the amortization period used, was 3 to 4 years.

The decision process could also be made on the basis of what should the amortization period be considering the rapid obsolescence of IT systems, and in this case, it is estimated to be 6 years to achieve full coverage of the investment.

9.3.8 Conclusions

These two case studies have illustrated how the ideas in Chapter 8 can be applied to different projects, each requiring very different perspectives. The challenge is

Case Studies

TABLE 9.13
Project Investment Payment Dates and Resulting Timing of Benefits

Automation Implementation Phase	Investment Year	Benefit Year Partial/Full Benefit
SCADA (grid substations and 11 kV feeder breakers (AIL Case 2)	1 & 2	2/3 onwards
FA Case 3 first switching stations	2 & 3	3, 4/5 onwards
FA Case 4 normally open points	2 & 3	3, 4/5 onwards
FA Case 5 additional intermediate selected feeder switches	3 & 4	5, 6/7 onwards

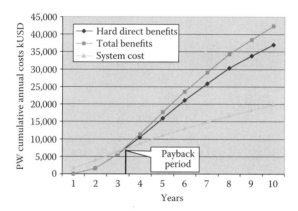

FIGURE 9.8 Curves of cumulative present worth of system costs and benefits considering and interest rate of 7.5% and an amortization period of 10 years.

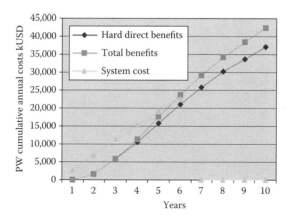

FIGURE 9.9 Curves of cumulative present worth of system costs and benefits considering and interest rate of 7.5% and an amortization period of 6 years.

to assemble the appropriate technical benefits that correspond to the major functions to be implemented and then convert them into monetary returns. First, there is the ranking of major benefits to see if it is necessary to search for more, and harder to determine savings, particularly because each additional benefit above the core savings will only be incremental and, in all probability, will be a shared benefit. The categorization of hard and soft benefits allows the decision makers to exercise judgment over the final evaluation.

Glossary

A of O Area of outperformance
AAD Automation applied device
ABUS Automatic bus sectionalizing
ACCP Average contingency capacity provision
ACD Actual crew distance
Acost Average linear cost per MVA for substation expansion
ACRF Average contingency rating factor
ACSC$_0$ Average contingency substation capacity
ADS Automated distribution system
ADVAPPS Advanced applications (for DMS)
AENS Annual loss of energy supplied
AGC Automatic generation control
AID Automation infeasible device
AIL Automation intensity level
AMDD Average maximum diversified demand
AMIC Annual cost of momentary interruptions
AMR Automatic (automated) meter reading
APD Automation prepared device
API Application program interface
APMD$_0$ Area peak maximum demand in base year
ARD Automation ready device
ARL Annual revenue lost
ASCII American Standard Code for Information Interchange
ASIC Annual cost of sustained interruptions
ASRC$_0$ Average substation released capacity
AVC Automatic voltage control. Applies the control to a load tap changer to vary system voltage to match changing load currents
BCC Backup control center
Bit The smallest unit of digital data, computational quantity that can take on one of two values, such as false and true or 0 and 1
bps Bits per second. The unit in which data transfer rate is measured across a communication channel in serial transmissions. 9600 bps indicates that 9600 bits are transmitted in 1 second.
BVAR Bus VAR control
BVOC Bus Voltage control
Byte Often 8 bits and the smallest addressable unit of storage
capital cost The initial payment for a particular asset at the time of purchase
CC Annual carrying charge

Channel Bandwidth In addition to the direction of transmission, a channel is characterized by its bandwidth. In general, the greater the bandwidth of the assigned channels, the higher the possible speed of transmission.

CIM Common information model

CIS Customer information system now sometimes referred to as customer relationship management (CRM)

Closed Ring A design of feeder where there are more than one sources of supply and all are connected to the load at the same time

CLPU Cold load pick up

CML Customer minutes lost

CMMS Computerized maintenance management system

Combined Neutral and Earth (CNE) A low-voltage distribution system where the neutral conductor and earth conductor are combined with each other.

CPM Capacity planning margin

CR Crew hourly rate (including vehicle cost)

CRGS Control room graphics system

CRM Customer relationship management — see CIS

CROM Control room operations management

CSF Comma separated format

CTS Crew time savings

Customer Outage Term to describe when the electricity supply to a customer fails

Customer Outage Costs The financial cost to a customer of an interruption of electricity supply

D(m) Crew distance traveled, as a multiple of the circuit length

DA Distribution automation

DAC Distribution automation cost. Associated with the load transfer benefit

Data Telemetry Transmission of the values of measured variables using telecommunication techniques

dBm (decibels below 1 milliwatt) A measurement of power loss in decibels using 1 milliwatt as the reference point.

DCC Distribution control center

DCE Data communication equipment. The devices and connections of a communications network that connect the communication circuit between the data source and destination (the data terminal equipment, or DTE). A modem is the most common kind of DCE.

Decibel (dB) A unit of measurement of the strength of a signal.

DFC Load class diversity factor

Disturbances Any deviation from the nominal; for example, a voltage disturbance is a deviation from the continuous voltage that has been declared by the supply authority.

DLC Distribution line carrier

DMLO Data monitor and logging

DMS Distribution management system

Glossary

DNC Dynamic network coloring
DOS Digital operating system
Downstream When measured from a particular point on the network, downstream means away from the source of supply.
DSM Demand side management
DTE Data terminal equipment. A device that acts as the source or destination of data and that controls the communication channel. DTE includes terminals, computers, protocol converters, and multiplexers.
EIT Enterprise information technology
EMD Electro magnetic device
EMS Energy management system
ENS Energy not supplied
EPRI Electric Power Research Institute, California
FA Feeder automation
Fade Margin The greatest tolerable reduction in average received signal strength that will be anticipated under most conditions. This measurement provides an allowance for reduced signal strength due to multipath, slight antenna movement or changing atmospheric losses. A fade margin of 10...20 dB is usually sufficient in most systems.
FAST Feeder automation switching time
FAT Factory acceptance test
FLIR (FLISR) Fault location isolation and supply restoration
Flow Control The collection of techniques used in serial communications to stop the sender from sending data until the receiver can accept it. This may be either software flow control or hardware flow control. The receiver typically has a fixed buffer size into which received data is written as soon as it is received. When the amount of buffered data exceeds a "high water mark," the receiver will signal to the transmitter to stop transmitting until the process reading the data has read sufficient data from the buffer that it has reached its "low water mark," at which point the receiver signals to the transmitter to resume transmission.
FPI Fault passage indicator
FSK Frequency shift keying. The use of frequency modulation to transmit digital data, i.e., two different modulation frequencies are used to represent 0 and 1. More than two frequencies can be used to increase transmission rates.
FVAR Feeder VAR control
FVOC Feeder voltage control
GIS Geographic information system (not to be confused with GIS (gas-insulated substation). Used for primary equipment
Grounding (Earthing) Method of connecting power system to earth, grounding is U.S. term, earthing is U.K. term
GSM Global system mobile (type of mobile phone system)
Harmonic Distortion A measure of distortion of the 50 Hz (U.K.) or 60 Hz (U.S.) supply caused by industrial loads.

HMI Human-machine interface (formerly man-machine interface)
HV High-voltage network (230–33 kV and above), also known as sub-transmission
HVDC High-voltage direct current transmission system
I/O Input/output (counts for SCADA)
ICCP Inter control center protocol
IEC International electrotechnical commission
IED Integrated electronic device. Covers simple remote terminal units (RTUs), traditional protection relays without control or communication features, and relays with full communications interfaces.
IIP Information incentive program
Interruptions Any loss of supply of electricity, see also SAIDI, MAIFI, SAIFI, CAIDI
IR/MHR Inspectors manhour rate
IRR Internal rate of return
IS&R Information storage and retrieval system
IT Information technology
L Circuit length
LAN Local area network
LC$_F$ Load class factor
LFC Load flow calculation
LM Load model
Load Break Elbow A device for terminating a cable onto a bushing that can be disconnected live and under the specified load current to achieve a form of switching
LODS Load shedding
Low Voltage Voltage for most domestic and commercial supplies, up to about 1000 volts
LV Low-voltage network for most domestic and commercial supplies, up to about 1000 volts, usually (415/220 volts). Also known as secondary distribution.
Master The unique application entity within the distributed application that directly or indirectly controls the entire activity for automatic action.
MCC Main or master control center
MHR/site The duration of the visit at each site in man (worker)/hours
MHR$_a$ Manworker hours per shift after automation
MHR$_b$ Manworker hours per shift before automation
MNST Manual switching time, which includes travel time to switch location.
Multiple RTU Addressing Several RTUs share the same radio unit; each RTU has its own address in the radio unit network list.
MV (Medium Voltage) Medium-voltage network up to 50 kV (typically 33, 22, 11, 6 kV). Also known as primary distribution at the lower end of the voltage range.
N Year

Glossary

NCF Network complexity factor. A measure of the complexity of a distribution feeder or network.

NOP Normally open point. Switch that is normally in the open position but which can be closed to supply a load from an alternative path or source of supply.

NORC Network optimal reconfiguration

NS Number of staff per rate classification

NSH$_a$ Number of shifts after automation

NSH$_b$ Number of shifts before automation

O&M Operations and maintenance

OASIS Organization for the Advancement of Structured Information Standards

OD Outage detection

OLF Operator load flow (DLF dispatcher load flow)

OLTC On load tap changer. A device fixed to a transformer that can vary the turns ratio by a small amount only while the transformer is on load. These devices can be limited to a capability for off load tap changing only to save expense.

OM Outage management

Open Loop Network An electricity feeder that is supplied from one source of supply in the normal circumstance but which has an alternative supply that can be used when required.

OR Operator hourly rate (including control room overhead costs)

OR$_a$ Operatorpower rates/classification used after automation

OR$_b$ Operatorpower rates/classification used before automation

OTS Operator training simulator (DTS dispatcher/distribution TS)

Packet Switched A system whereby messages are broken down into smaller units called packets, which are then individually addressed and routed through the network.

Pantograph Disconnector A disconnector where the mechanical operation to achieve disconnection is achieved by the movement of a pantograph device.

Parity A one-bit quantity indicating whether the number of 1's in a word is even or odd.

PBR Performance/penalty-based rates

PC Personal computer

PDS Program development system

Peer-to-Peer A communications model in which each party has the same capabilities and either party can initiate a communication session. Other models with which it might be compared to include the client/server model and the master/slave model. In some cases, peer-to-peer communications is implemented by giving each communication node both server and client capabilities.

Permanent Fault A network fault that is sustained or present for a long time period.

PLC Programmable logic controller (sometimes acronym used for power line carrier)

PMR Post-mortem review

Point-to-Multipoint One-way or two-way communications from a central point to a number of subsidiary points, and vice versa.

Poll Method to check the status of an input line, sensor, or memory location to see if a particular external event has been registered. The communications control procedure by which a master station or computer systematically invites tributary stations on a multipoint circuit to transmit data.

PQ Power quality

Primary Substation The source of electricity supply for a distribution network, sometimes known as a zone substation.

PVF Present value factor

PVRR Present value of revenue requirements (annual)

Radial Feeder An electricity feeder that is supplied from one source of supply only.

RBM Risk-based maintenance

Repeater Device that will repeat serial communications on to the predetermined destination.

Revenue Cost The payment of operating costs of an asset after the time of purchase.

RFSW Remote feeder switching

RISC Reduced instruction set code

RMU Ring main unit. A form of switchgear for distribution substations, typically comprising switches for the incoming circuit, the outgoing circuit and the connected load. In United States sometimes similar configuration is achieve with pad-mounted switchgear.

Rotating Center Post Disconnector A disconnector where the mechanical operation to achieve disconnection is achieved by the rotation of the disconnector assembly.

Routing of Messages The selection of a path or channel for sending a message.

RSSI Received signal strength indication. A parameter returned from a transceiver that gives a measure of the RF signal strength between the mobile station and base transceiver station, either as an uplink or downlink measurement.

RTU Remote terminal unit. An electronic device that is physically remote from a main station or computer but that can gain access through communication channels.

SA Substation automation

SAT Site acceptance test

SCADA Supervisory control and data acquisition

Glossary

Separate Neutral and Earth (SNE) A low-voltage distribution system where the neutral conductor and earth conductor are separate from one another.

Slave A unit that is under the control of another unit (master).

SOE Sequence of events

SRN Survey number

STLB Substation transformer load balancing

Switched Alternative Supply A second source of supply that is made available by closing a normally open point.

Switching Points Positions on a distribution network where switches are located to control the power flow on that network.

TANC Total area normal capacity

TCM (TCMS) Trouble call management (system)

TCP/IP LAN protocol

Temporary Fault A network fault that is transient, or only present for a short time period.

Transceiver A terminal unit that can both transmit and receive information from a data transmission circuit.

TTD Time tagged data

UML Unified modeling language

Unsolicited System In an unsolicited response system, the RTUs generate all reporting messages required, without being polled by the master. Typically, such messages report a change of state or a fault, or simply pass data to the SCADA central host without being polled for the data.

Upstream When measured from a particular point on the network, upstream means towards the source of supply.

VCLC Voltage conservation load control

VDU Video display unit

WAN Wide area network

WG Working group (IEC, CIGRE)

WMS Work order management system

$\times$ Exponential load growth rate

XLPE Cross-linked polyethelene, a common form of insulation on underground cables.

XML Extensible markup language

λ Feeder annual outage rate/unit circuit length

Index

A

Actuator, 14, 150, 168, 171–173, 234, 237, 281
    magnetic, 169, 173, 174
Advanced Applications (ADVAPPS), 36, 57, 60–70, 371, 373
    Constrained vs. Unconstrained, 70
    Data models, 86
Alarms, 73
Alternative supply, 106, 130
    Continuous, 115, 116, 275
    Switched, 114, 260, 261, 264, 273
Antenna
    Gain, 294, 296, 299, 302
    Height Isotropic, 294
    Omni-directional, 296
Automated
    Distribution System, 23
Automatic Generation Control (AGC), 7
Automatic sectionalizer, 196, 249
Automation
    Applied device, 23
    Centralized, 14
    Decision making, 15
    Decision Tree, 14
    Infeasible Device, 22
    Intensity Level (AIL), 17, 50, 60, 390, 395, 432, 435, 436, 438, 447
    Local, 14, 20
    Prepared Device, 23
    Ready Device, 21, 23
    Stages of, 16
Autoreclose, 115
Autoreclosing
    Benefits, 397

B

Benefits
    Autoreclosing, 399
    Crew time savings, 401, 404
    CRM improvement, 410
    Customer related, 380
    Data monitoring and logging, 408
    Dependencies, 367
    Displacement of conventional equipment, 386
    Flow Chart, 367, 370, 374–377, 409
    General, 2, 4
    Generic Types, 364–366
    Hard and Soft, 5, 359, 428
    from NORC and load balancing, 407
    Opportunity Matrix, 367–369
    Reduction in Inspection visits, 391
    Reduction in substation manning, 391
    Repair and maintenance, 408
    Results presentation, 426
    Savings in manpower, 389–403
    Shared, 371
    Unshared, 372
    Volt/VAR control, 406–407
Breakeven Analysis, 426
Building blocks
    battery charger, 232, 238
    IED, 171, 226
    Interfaces, 238
    power supply, 181, 229
    Remote terminal unit, 226, 281
    types, 234, 236–238, 244
Bulk Supply, 8
Busbar voltage, 122
Business case, 357

C

CAIDI (customer outage duration), 135, 260, 420, 425, 435, 439
Capacitor, 126, 137
Capacity Deferral, 3
Capacity release, 383
    from cold load pickup function, 386
    NORC and STLB function, 388
    upstream network, 387
    from Volt/VAR Control, 383
CAPEX, 359, 421
Capital Deferral, 366, 378, 379
Case Study, 433, 439
Circuit Breaker
    Rating Limit Check, 65
Closed Loop, 21

Cold Load Pickup
  Capacity avoidance, 386
Common Information Model (CIM), 93–100
  Classes and Relationships, 95
  Interface standards, 100
  Model Structure, 94
  Specification, 96
Communication protocol
  DNP 3.0, 290, 336
  IEC 870-5-101, 342
  Modbus, 333
Communication system, 289, 350–351
  Response time, 353
Communications architecture, 346
Compensated earthing, 132, 198, 204, 254
Condition Monitoring, 2
Control
  boundaries of responsibility, 11
  Delegated, 9
  depth of, 11
  Distribution Systems, 9, 30
  Extended, 9
  Hierarchy, 9, 29
  Layer, 3, 10
  Requirement, 8, 27
Control Center
  manning level reduction, 389
Control logic, 244
Control Room Operations Management (CROM), 35
Cost Benefit, 428, 448
Cost per kVA, 111
Crew distance traveled
  estimation, 992, 394
  Savings, 396
Crew time, 131, 142
Crew Time Savings (CTS), 392–405, 435, 437, 440
  Interruption related benefits, 390
  Investment and Operation related savings, 402, 433, 444
  related to Cold Load Pickup, 402
  Savings estimation tool, 399
Critical Function, 79
Critical Length, 399
Cumulative preset worth (value), 426
CT (Current Transformer)
  burden, 177, 181
  errors (in CT), 180
  measuring, 177, 179
  protective, 177, 179
  Zero sequence, 175, 176, 198, 210
Current limiting, 195
Current loading, 128

Customer
  Cost (outage value), 421, 423
  Loyalty/retention, 410
Customer Relationship Management (CRM)
  Benefit calculations, 378, 412
  Improvements, 412

D

Data
  interfaces, 89, 100
  model standards, 93
  storage and archiving, 44
  Time tagged, 44, 73
  Types, attributes, 40, 43
Data acquisition *see also* SCADA, 39, 45, 48, 49
Data Interface Standards, 100
Data Monitoring and Logging Function, 375, 377
  Benefits, 410
  improved engineering, 411
  improved operating decisions, 411
Database
  structures, 86
Deferral period, 131
Dependencies
  Data, 68
  Loads Voltage/Current, 62
Deregulation, 1, 9
Distribution Automation (DA)
  Acceptance, 1
  Architecture, 17, 20
  Benefits (see benefits)
  Benefits summary, 411, 412, 437, 448
  Concept, 11
  Definition, 13
  Industry Acceptance, 5
  Solution, 19
  Surveys, 5, 6
  System, 12, 19
Distribution (power) line carrier
  Broadband PLC, 308
  Characteristic impedance, 311
  Classical PLC, 306
  Coupling equipment for classical PLC, 306
  Equivalent circuit, 318
  Lines and cables, 310
  Modulation and coding, 319
  Narrowband PLC, 307
  Noise, 305, 328
  Ripple control system, 306
  Signal attenuation, 305
  Signal coupling for lines and cables, 314

Index

Signal interference, 305
Site measurement tool, 323
Site survey for lines and cables, 320
    Acceptance, 1
    Architecture, 17, 20
    Benefits (*see benefits*)
    Benefits summary, 411, 412, 437, 448
    Concept, 11
    Definition, 13
    Industry Acceptance, 5
    Solution, 19
    Surveys, 5, 6
    System, 12, 19
Distribution management system (DMS), 12, 30
    Evolution, 31
    functions, 35, 38
    Interfaces, 89
Distribution Network
    Operations environment, 29
    Structure, 21
Distribution system, 105
Downstream, 114, 261

E

Economic Value, 412, 422
    comparison of, 424
Economics
    capital costs, 141, 142
    Installation costs, 141, 142
    Iron losses, 135
    life cycle cost, 109, 142
    penalties, 142
    revenue costs, 141, 142
Energy Not Supplied, 405, 414
Energy related savings, 378, 446
    due to controlled load reduction, 404
    due to faster restoration, 403, 446
    due to loss reduction, 405, 407
Engineering and Planning
    improved data, 409
Enterprise IT systems
    data sources, 90, 357
Event Burst, 80
Extended control, 131, 136–137, 142, 172, 180–181, 266, 281

F

Factor "X", 6
Factory Acceptance Test (FAT), 22
Fault Calculation, 63

Fault location, 168
    Switch and test method, 209
    Switching method, 208
Fault Location, Isolation, Supply Restoration (FLIR), 57
Fault passage indication
    autoclosing, 222
    connections, 214–216
    directional, 213
    grading, 223
    grounding, 218
    Local indication, 147
    proximity type, 219
    remote indication, 147
    selection, 225
Fault Passage Indicator (FPI), 58, 181, 187, 207, 209, 211, 238, 261
Faults
    earth, 155
    ground (earth), 197, 251
    high resistance, 251
    open circuit, 251
    phase to phase fault, 251
    self clearing, 252
    self extinguishing, 252
    Transient, 221, 252, 254
Feeder Automation, 14, 19, 77, 368
    Benefits (*see benefits*), 363
Firm capacity, 129, 131
Free market, 7
Fuse types, 7

G

Gateway, 17
GIS
    Centric TCM, 60, 86, 89
    DMS interface, 91
    DMS interface levels, 92
Grounding (earthing), 105, 109, 131
    Compensated earthing, 132, 198, 204, 254
    Multigrounded, 137
    Petersen coil (*see compensated*), 200
    resistance, 132
    solid, 132
    unearthed, 132, 203
    unigrounded, 137

H

Hardware Configuration Matrix, 80
Hospital Supply, 276

I

ICCP, 88
Implementation
   incremental, 4
Indications, 39
Inferred System Responsibility, 23
Information
   for planning, 3
Inspection visit
   reduction, 389
Installation costs, 141, 142
Instrument transformer, *also see CT*, 176–177, 181
Integration
   device, 359
   horizontal and vertical, 34
Intelligent Electronic device (IED), 9
Internal rate of Return, 426
ISO Seven layer model, 290
   communication protocol, 290
   physical layer, 290

L

Legislation, 106, 116–117
   regulation, 117
Life cycle cost, 109, 142
Load factor, 135
Load Flow, 61
   Load Models, 62
Load growth, 129
load transfer, 130
Loss Minimization, 66, 407
Losses
   Copper, 135
   Electrical, 128
   Iron losses, 135
   minimization, 136
   reduction, 126
LV cable
   combined neutral and earth, 142
   separate neutral and earth, 142
LV system
   automation, 146
   distribution board, 138, 146
   distribution network, 105, 142

M

MAIFI, 260, 264, 420

Management
   Crew, 30
   Functional Organization, 31
   Outage, 37
Marginal
   Improvements, 17
Methods to improve performance, 271
Monitoring and Event
   Processing, 41
   Trending, 41

N

Network capacity release, 383
Network complexity, 109
Network Complexity Factor (NCF), 390, 392, 400, 438
Network Performance, 109, 116, 259, 268, 280, 436, 441
   customer effects, 255, 257
   energy not supplied, 133
   switching time, 274
Network structure, 105
Network type
   closed ring, 115
   open loop, 114, 116, 160, 273
New Customer Services, 3
Noncritical function, 79
Normally Open Point (NOP), 21

O

OECD, 421
Operating Conditions, 30
Operating decision improvement, 409
Operating flexibility, 139
Operation and Maintenance
   Cost Reduction, 2
OPEX, 359, 421
Out performance bonus, 420
Outage Duration
   Reduction, 3
Outage Management, 37, 50, 54, 57
Overhead network, 109–110, 115
Overload factor, 128

P

Payback Period, 429, 448
Performance
   Evaluation, 431

Index

Measures and response times, 79–86
Scenario Definitions, 79
Performance-based penalties (PBR), 5, 413, 415, 434
   England, 418
   Examples, 418
   North America, 418
   Scandinavia, 418
   Types, 415, 416, 417
Planning
   Operations, 28
Polling and Report by Exception, 348
Post Mortem Analysis, 44
Power delivery system, 7
Power Quality (PQ), 1, 3
Power system
   operation, 28
Preparedness, 22
Privatization, 1, 27
Procurement practices, 22
Protection
   directional, 155, 205, 206
   discrimination, 188
   instantaneous relay, 192
   overcurrent, 155
   protection relay, 155
   relay coordination, 188
   sensitive earth fault, 155, 190, 222
   setting study, 188

Q

Quick estimation tool
   crew time savings, 399

R

Radio link calculations
   Antenna height, 298
   Egli model, 298
   Fade margin, 296, 299, 302
   Free space loss, 300
   Noise, 296
   Propogation loss, 297
Recloser, 149, 168–170, 253
Reliability, 110, 259–260, 270
   Improvement, 3
Remote Switch Control, 374
Remote Terminal Unit (RTU) *see* SCADA
Report by exception *see* SCADA, 48
Reporting
   Operations, 29

Response Times
   calculation of, 81–86
   Data acquisition and processing, 85
   indications, 82
   metered values, 83
   output command, check back, 84
Retrofitting, 21
Revenue at risk, 419
Revenue costs, 141–142
Ring Main Unit (RMU), 115, 138–139, 151, 158, 160, 164, 273
Rogowski coil, 182, 227

S

SAIDI (sustained interruption duration), 134, 259, 261, 264, 266, 280, 417, 420, 421, 441
SAIFI (sustained interruption frequency), 133, 259, 263–264, 417, 420–421, 425, 435, 437
Savings
   AIL related, 449
   in Manpower, 378, 390, 389
   Repair and Maintenance, 408
SCADA, 9, 14, 33, 36, 39, 47
   Centric TCM, 60
   Control Functions, 44
   Data Models, 87
   Event burst, 80
   Event processing, 43
   GIS interface, 92
   Hardware Configuration, 45
   Hardware Configuration Matrix, 80
   Performance *see also* Response times
   Polling Principles, 48
   Point data, 88
   SCADA, 156
   Sequence of Events (SOE), 41, 44
SCADA Ready, 21
Scatter Diagram, 436
Scenario Definitions
   Performance, 79, 81
   States, 80
Screening Method, 379
Selection of communication option, 350
Sensor, 175, 182, 184
   combisensor, 184, 227
   current, 170, 182
SF6, 151, 164, 168–169
Substation
   Capital deferral, 379–380
   Capital deferral-screening, 381

Ground mounted, 143, 158
Pole mounted, 144
Primary, 157
Reduction in manning levels, 389
Tee off, 115, 139
Unit type, 138
Substation Automation (SA), 20, 70–77, 368
  Annunciator units, 75
  basic designs, 70
  benefits, 358–363
  communication, 77
  Control Units, 75
  Disturbance recorders, 76
  feature comparison, 72
  Local Automation, 73
  Supervisory Control, 74, 75
Summary of communication options, 331
switchgear
  air break disconnector, 170
  air insulated switches, 167
  automation ready, 234
  autosectionalizer, 150
  circuit breaker, 149, 154–155, 160
  disconnector, 149, 158, 160
  elbow, 162, 163
  ground-mounted fuse, 193
  load break elbow, 139, 162
  MV network, 105
  MV switchgear, 157
  normally open switch, 130, 135,
  Pad mount, 163
  pole-mounted fuse, 193, 195
  pole-mounted switch, 151, 167
  primary, 154
  recloser, 149, 168, 169, 170, 253
  remotely controlled network switches, 124
  ring main unit, 115, 138, 139, 151, 158, 160, 164, 273
  sectionalizing, 272
  switch, 149
  switch disconnector, 150
  switchfuse, 149, 160
  vacuum interuppter, 151, 169, 175
switchgear operating mechanism
  dependent manual, 171
  independent manual, 172
  magnetic actuator, 174
  motor wound spring, 172
  solenoid, 172

T

transformer windings, 129
travel time, 282
Trouble Call, 30, 33, 37, 52
  Outage Inference Engine, 53, 54

U

Underground network, 109, 110
Upstream, 114, 261
Utility
  Cost (outage value), 413, 423

V

Values
  economic, 422
  Energy, 39, 40
  Measured, 39, 40
VAR control, 66, 374, 407, 408
Volt control, 67, 374, 407
Voltage, 109
voltage control
  automatic voltage regulation, 122
  in line voltage regulator, 124
  line drop compensation, 123
  off load tap changer, 121, 122
  power factor correction, 121, 124
  tap changer, 121, 124, 136, 137
  Transformer taps, 121
  Voltage control, 109, 121, 124, 126
  voltage control relay, 122
  voltage regulator, 121
VT (Voltage Transformer)
  voltage divider, 170
  voltage sensor, 183
  voltage transformer, 175, 177, 180, 168

W

Wire communication
  Distribution line carrier, 304
  Fiber optics, 304
  Telephone line, 304
Wireless communication
  Cellular, 303
  Narrowband VHF/UHF (Data radio), 290
  Satellite, 303
  Trunked system, 302
  Unlicensed spread spectrum, 290